Take a Journey of Disco
and the Deeper Realms of H

As the recovery research continues to accum
understanding of schizophrenia and psychosis — credibility:

- After over 100 years and billions of dollars spent on research looking for schizophrenia and other related psychotic disorders in the brain, we still have not found any substantial evidence that these disorders are actually caused by a brain disease.
- We have learned that full recovery from schizophrenia and other related psychotic disorders is not only possible but is surprisingly common.
- We've discovered that those diagnosed in the United States and other "developed" nations are much *less* likely to recover than those in the poorest countries of the world; furthermore, those diagnosed with a psychotic disorder in the West today may fare even worse than those so diagnosed over 100 years ago.
- We've seen that the long-term use of antipsychotics and the mainstream psychiatric paradigm of care is likely to be causing significantly more harm than benefit, greatly increasing the likelihood that a transient psychotic episode will harden into a chronic psychotic condition.
- And we've learned that many people who recover from these psychotic disorders do not merely return to their pre-psychotic condition, but often undergo a profound positive transformation with far more lasting benefits than harms.

In *Rethinking Madness*, Dr. Paris Williams takes the reader step by step on a highly engaging journey of discovery, exploring how the mainstream understanding of schizophrenia has become so profoundly misguided. He reveals the findings of his own pioneering research of people who have fully recovered from schizophrenia and other psychotic disorders, weaving the stories of these participants into the existing literature and crafting a surprisingly clear and coherent vision of the entire psychotic process, from onset to full recovery.

As this vision unfolds, we discover…

…a deeper sense of appreciation for the profound wisdom and resilience that lie within all of our beings, even those we may think of as being deeply disturbed.

…ways to support those struggling with psychotic experiences while also coming to appreciate the important ways that these individuals can contribute to society.

…that by gaining a deeper understanding of madness, we gain a deeper understanding of the core existential dilemmas with which we all must struggle, arriving at the unsettling realization of just how thin the boundary really is between madness and sanity.

More Acclaim for *Rethinking Madness*

"In this eye-opening book, Paris Williams effectively challenges the prevailing myths about the origins and treatment of psychosis, suggesting that it is a natural, although precarious, process of self-restoration that should be protected, rather than a hopeless lifelong degenerative brain disease to be managed and medicated. The mounting evidence for the abject failure of the medical model to treat psychosis is presented alongside six case studies of people who fully recovered despite psychiatric treatment and who felt more deeply in touch with hope, meaning, a sense of aliveness and the interconnectedness of life as a result of their difficult journeys. Williams also offers an innovative and profound model that synthesizes current Existential theory, attachment theory and Buddhist mindfulness perspectives. *Rethinking Madness* is an important and hopeful book."

John J. Prendergast, Ph.D., adjunct professor of psychology, senior editor of *The Sacred Mirror* and *Listening From the Heart of Silence*, and editor-in-chief of *Undivided: The Online Journal of Nonduality and Psychology*

"With his groundbreaking new book, *Rethinking Madness*, Paris Williams takes us into a world in which he joins psychology with Buddhism and Western philosophy to give us a panoramic view of how madness is born, matures, and may be resolved. Backed by an extensive and engaging survey of historical and contemporary views of psychosis and its etiology, Williams presents an integrative, deep and ultimately humane body of theory and practice that will be of great use to anyone working in this intriguing and difficult area."

Joe Goodbread, Ph.D., author of *Living on the Edge* and *Befriending Conflict*

"Every page of this book was exciting to me, offering clear, profound insights not only into the processes of psychosis/alternative realities, but also into philosophical views about human experience, including the spiritual elements of the psychotic process. . . . While Dr. Williams never trivializes the anguish and psychic and sometimes physical pain mentally ill people endure, he is never without hope for their relief. His help/harm equation in the recovery process had me enthralled with its truth.

This book should be a part of the training of every physician, psychiatrist, and pastoral counselor, and owned by the family and friends of every mentally ill person as well as the sufferers themselves."

Joanne Greenberg, author of the international bestseller
I Never Promised You a Rose Garden

"This book contains a brave, well-researched, and invaluable new approach to the vexing subject of psychosis. The case studies and the conclusions are novel and unique in their formulations. The insights and theoretical postulates derived from this research are important and likely to move the field forward in unexpected ways. For people who have experienced psychosis or altered states, it is a ray of hope in their struggle to thrive."

Peter Stastny, M.D., lecturer of Epidemiology and Co-Author of
The Lives they Left Behind—Suitcases from a State Hospital Attic

---

"Paris Williams has written a much needed and extremely thoughtful critique of the major approaches to psychosis. Current psychiatric treatment, while helpful for some, has proven inadequate for most psychotic patients. This book helps us to understand why. But beyond offering a critical appraisal of current methods, Williams also offers a powerfully hopeful vision of new possibilities for the treatment and transformation of this puzzling disorder."

Brant Cortright, Ph.D., professor of psychology, and author of
Psychotherapy and Spirit* and *Integral Psychology

"Dr. Paris Williams presents a clearly written comprehensive treatise on madness. Deceptively easy to understand, yet thought provoking and challenging, his work offers plausible reasons to overcome the too simple historical medical approaches that ignore the richness of the human experience and the positive potential inherent in one's journey through madness. Dr. Williams' book will expand the reader's view of this quintessential and ubiquitous human experience that we have come to call madness."

Ronald Bassman, Ph.D., author of *A Fight to Be: A Psychologist's Experience from Both Sides of the Locked Door*

"This is an important book. It states boldly what many of us working in the field and following research based on lived experience have come to suspect: 'the mainstream vision of psychosis currently held in the West is somehow seriously missing the mark.' This book provides compelling research evidence to support this conclusion as well as gathering and developing more hopeful alternatives that offer real healing."

Isabel Clarke, author of *Madness, Mystery and the Survival of God*, and editor of *Psychosis and Spirituality*

"Rollo May once said, 'One does not become fully human painlessly.' Dr. Paris Williams' search for what it really means to be human and how to fully reinvent oneself after being diagnosed with a so-called mental illness echoes May's assertion. Moreover, Williams' book teaches us that there is a person behind the label of madness and that madness can just be one's motion towards healing. This book brings hope to many people who suffer from so-called mental illness and who struggle with the concept of illness, and opens up the dialogue in psychology beyond just the medical model."

Doris Bersing, Ph.D., professor of psychology and director of clinical training at Saybrook University

---

"*Rethinking Madness* is clear, thorough, fascinating and bold. It is written in easy to understand English, not postmodern jargon; it thoroughly examines and debunks the psychiatric model; and Williams is not afraid to make controversial affirmations. Anyone who is skeptical about the mental health system will be convinced after reading Williams' book: There is an alternative."

Seth Farber, Ph.D., author of *Madness, Heresy, and the Rumor of Angels,* and *The Spiritual Gift of Madness: The Failure of Psychiatry and The Rise of the Mad Pride Movement*

Rethinking Madness is one of those rare works that successfully challenges psychiatry's widely accepted but objectively incorrect beliefs about the extremes of human experience, and does so in a manner that is both scholarly yet accessible. Williams is a wonderfully clear writer dealing expertly with a complex range of theory and deep human experience, and *Rethinking Madness* is an important book that cannot help but leave the reader with new insights and a sense of wonder at the self-healing power within each of us.

Darby Penney, co-author of *The Lives They Left Behind: Suitcases from a State Hospital Attic*

Rethinking Madness

Towards a Paradigm Shift in our Understanding and Treatment of Psychosis

Paris Williams, Ph.D.

SKY'S EDGE
SAN FRANCISCO
2012

Sky's Edge Publishing
101 Vendola Drive
San Rafael, CA 94903
www.skysedgepublishing.com

ISBN: 978-0-9849867-0-5

Library of Congress Control Number: 2012901521
Library of Congress subject headings:
Schizophrenia -- Treatment.
Psychoses -- Treatment
Schizophrenia -- Case studies
Schizophrenia -- Etiology
Ex-mental patients -- Case studies

Contents (brief)

Contents (detailed)

List of Tables

List of Figures

Acknowledgments

Now that the writing is finally finished, I have the great pleasure of expressing my appreciation to all of those who have supported me in this long process. In many ways, the journey that culminated in my series of research studies and finally in this book began long before I ever set foot in a university. There are so many individuals who inspired me, challenged me, and/or touched me in some important way along this journey that I couldn't possibly name all of them here. Indeed, one of the greatest lessons I have taken from this journey is a greater appreciation for the very intricate web of interdependence that shapes my every thought and action. I express my deepest gratitude to all who have contributed a strand to this web.

I want to express special appreciation to the members of my doctoral research and dissertation committees–Drs. Linda Riebel, Kirk Schneider, Tom Greening, and my wonderful chair, Doris Bersing.

I also want to express my deepest gratitude to all of the participants of the three research studies that culminated in the writing of this book. I hold the highest admiration for the courage that every one of you exhibited in your willingness to plunge so deeply into what were often very painful memories in order to share the details of your journeys and the tremendous wisdom you've gained with the rest of us.

I also want to thank the many people who provided important feedback related to the structure of this book. I am especially grateful to Melissa Grabanski for her excellent guidance in this regard.

Finally and most of all, I want to express my deepest gratitude to my wife, Toni. There is no doubt that her combination of rare insight into the human condition and (nearly) inexhaustible patience have been my most valuable resources.

And the day came when the risk to remain tight in a bud was more painful than the risk it took to blossom.

–Anais Nin

The goal of the recovery process is not to become normal. The goal is to embrace our human vocation of becoming more deeply, more fully human . . . to become the unique, awesome, never to be repeated human being that we are called to be.

–Patricia Deegan[1]

Preface

During the past several years, I conducted a series of research studies that inquired deeply into the experiences of people who had made full and lasting recoveries from schizophrenia and other long term psychotic disorders. As the participants revealed their stories one by one, I became increasingly astonished by what I was learning. The deep meaning within these participants' experiences and the profound positive transformation that each of them had gone through flew completely in the face of virtually everything I had ever learned in the mainstream texts about psychosis. I dove deeper and deeper into the existing research on schizophrenia, psychosis, and recovery, trying to make sense of what I was learning. The common beliefs about psychosis and schizophrenia that are held so strongly in the West quickly began to slip away like so much sand through my fingers. I realized that what I was learning from these participants was taking me so much further than I had ever imagined possible, sending me on a journey not unlike Alice's descent into the rabbit hole.

As I attempted to disentangle the complex and vast web of research, I found myself descending ever further into a world where the truth appeared to be much stranger than fiction, a world riddled with contradictions, paradox and hair-pulling conundrums. The journey began with the complete dismemberment of the brain disease theory of psychosis, and continued beyond the point where even the construct of "schizophrenia" itself blew away like mere dust in the wind. Ever deeper this journey took me until eventually I had no other choice but to arrive at the conclusion that the condition we generally think of as psychosis is not the result of a diseased brain after all. Rather, it is probably much more accurate to see psychosis as a desperate survival strategy brought on *intentionally* by one's very own being.

It seems that all of us, and indeed all living organisms, are imbued with an unfathomable intelligence and force that strives constantly for our survival and our growth; and it appears that it is this very same organismic intelligence that intentionally initiates psychosis in a desperate attempt to survive what would otherwise be intolerable conditions. The evidence in this regard is surprisingly robust. Every participant in all three of my studies (and within many accounts of others who have gone through similar journeys) experienced the onset of psychosis after finding themselves overwhelmed by such intolerable conditions; and every participant also underwent a profoundly healing and positive transformation as a result of the full resolution of their psychotic process. It is clear that all of these individuals now experience a sense of wellbeing and resilience that far surpasses that which existed prior to their psychosis. Furthermore, every participant found that the psychiatric treatment they had received caused significantly more harm than benefit in their recovery, a finding that may come as a surprise to many readers but is actually in very close alignment with the other recovery research.

This book, then, is a documentation of the journey I took and the many surprising gems I found along the way as I chased the answer to the question, "What is it that occurs at the deepest level of one's being as one works through the entire psychotic process, from onset to full recovery?" One of the most significant and least expected gems I stumbled upon is the realization that this exploration not only brings us to a much deeper understanding of the nature of psychosis and important implications for supporting those who struggle with it, but it has also opened the door to a much deeper understanding of the core dilemmas with which we all must struggle.

So I invite you to suspend everything you think you know about psychosis and the deeper levels of human experience and join me in this plunge down the rabbit hole.

Paris Williams

Introduction: The Case of Sam

Sam began his eighteenth year a passionate young man, driven by a strong desire to contribute to a more peaceful world and eager to move into a life of his own. Shortly after reaching this milestone, however, his experience and understanding of the world began to shift and grow increasingly unsteady. The solid "ground" of his connection to consensus reality—something he had always taken completely for granted—was now crumbling away beneath him. He tried desperately to retain his footing, but to no avail. After just several short months, his final foothold gave way and he fell headlong into the chaotic seas of madness.

The year was 1971, the Vietnam War was in full swing, and Sam had just received his military lottery number. "I received a 31 in the 1971 draft at age 18 guaranteeing that I would be inducted into the military at 19. I was an anti-war activist and had a lot of stress from this situation." A few months later, the dreaded draft notice arrived. His stress shot to an overwhelming level and he soon found himself struggling to maintain his sanity. "I got drafted and I was kind of going crazy. I was just having a lot of difficulty around that. . . . I was trying to work on a valid conscientious objector claim, and I remember when I was writing up some of that stuff, getting it together, I was writing in spirals [laughs], so that was kind of an indication, I guess, that I wasn't doing so well."

He began to lose sleep, his mental condition deteriorated dramatically, and he soon found himself living in a very different world from that experienced by those around him, a world that was at times filled with tremendous meaning and exhilaration and at times with overwhelming confusion and terror. Sam had fallen into a deep and powerful state of psychosis. He would continue to struggle with psychosis for the next twenty years, going on to receive a diagnosis of paranoid schizophrenia and descending into profound and sometimes prolonged states of psychosis numerous times.

In the initial stages of Sam's first period of psychosis, he had a number of unusual experiences involving a sense of impending disaster, first with regard to his own life: "I remember I was at a friend's house and I felt for my pulse, and I felt like I didn't have a pulse for a long period of time. That was kind of distressing. I was still walking around conscious [laughs]. [At other times,] I thought that people had put spells on me..um..witches" *. Then, these feelings expanded to include a sense of impending disaster for the entire world, and along with this came a powerful compulsion to act in some way to save it: "I had some sort of a..um..I don't know, vision or something like that, that unless I did certain things, the world was gonna end. And I focused on the song, *Bye, Bye, American Pie* [laughs]."

As Sam moved further away from consensus reality, more elaborate belief systems began to emerge. Perhaps the most prevalent types of these involved his interacting in different ways with what he refers to as the "initials agencies": "Either I was being hunted by the initials agencies, or else I was part of the initials agencies, or else I was doing independent operations for initials agencies. And when I say initials agencies, I mean things like the FBI and CIA."

One such mission involved his playing an important role in *Desert Shield* in Iraq:

> Before Desert Storm started . . . I think it was Desert Shield . . . I was part of a group . . . Iraqis were kidnapping people and using them as human shields, and I had this feeling like I did a mission over there and helped get those people released. I used hand signal communications with the geosynchronous satellites. At any given time, the military geosynchronous satellites have real-time views of the world, and I was using hand signals to bring in air strikes.

Another mission involved capturing the infamous D. B. Cooper:

> I was an independent operative that had contracted with the FBI, and . . . just before the statute of limitations ran out, I captured D. B. Cooper [the infamous D. B. Cooper who hijacked a Boeing 737 airplane in 1971, received

* The direct quotes from the participants come from several sources—both oral and written. When ellipsis are used within these quotes, ".." refers to a significant pause within speech without any omission, ". . ." refers to an omission of irrelevant words within the sentence, ". . . ." refers to an omission of irrelevant material that spans beyond the present sentence, and ellipsis in forms other than these were directly copied from the participant's written material.

> a ransom of $200,000, and then parachuted from the tail of the plane never to be seen again]. . . . And so I met this guy who had some experience as an aerospace engineer, and he fit the description of D. B. Cooper, and he also had a lot of receipts on him for spending a lot of money, and so I was yelling, questioning him in the middle of a Greyhound bus depot in [laughs] downtown, and there was a security guard, and he didn't like what I was doing, so I kind of jumped all around and evaded him, and then I was walking in the downtown area, and that's how I got caught by the police. He'd given the police a description. I was wearing Korean paratrooper jump boots that have a lot of resilience on the bottom, so I was able to do a lot of evasive stuff by bouncing around [laughs].

Unfortunately, this encounter ended with Sam getting severely beaten by the police and then placed in the state psychiatric hospital for several months.

Playing major roles within movies was another prevalent theme within Sam's anomalous belief systems. He had taken a particular interest in filmmaking prior to these experiences, something he believes may have contributed to his having these particular types of experiences. He describes one such experience in which he believed he was a demolitions expert: "The Earth was going to be used as a [movie] set, and . . . all the humans were replaced by cyborg types, and . . . I set up a lot of real time demolitions to do a lot of, let's say, special effects using the Earth as a movie set [laughs]." In another such experience, Sam experienced himself as a director of a movie and was "using hand signals to signal production staff so that they would get the right feeling of the movie." In yet another episode, he experienced himself as an actor involved in a high-speed car chase. This episode unfortunately resulted in severe consequences: "I got in a car crash doing 120. . . . I thought there was a camera crew on top of us."

Sam went through a significant period of time in which he believed he was caught in the middle of a war: "There was a time when I was with my girlfriend and we were going grocery shopping, and I had the belief that there is war going on on planet Earth all over the world. . . . For some reason, I thought it had kind of come over to the United States, and so I did a lot of taking cover. . . . Wherever I went, I'd always be in some sort of cover from live rounds."

One particularly interesting aspect of Sam's experience with these various belief systems is that he would quite often experience several of them taking place at once:

> I felt like I was jumping between universes, or, as somebody in the modern physics world would say, parallel worlds. . . . Well, it wasn't always jumping from one to another. They could be multi-layered. . . . Like the thing with

> D. B. Cooper and all that stuff seemed like it was, you know, I was doing both. I was doing both the independent work for the FBI, but it also seemed like I was in a movie that Sam Peckinpah was directing."

A Path to Recovery. Beginning with his first period of psychosis at age 19, Sam was placed on antipsychotic drugs and he remained on them for approximately ten years. It was not until he came off the drugs, however, that he began to make significant progress towards a full and lasting recovery. While he does believe that the drugs played a supportive role at times, he feels that, overall, they were more of a hindrance than a benefit in his recovery.

One major hindrance of these drugs was their impact on his cognitive abilities which resulted in a major interruption of his academic studies: "I had trouble thinking when I was on the medicine. I was on and off of it for about ten years. Then, once I got off of it, I was able to get my thinking faculties back and go back to college and complete my bachelor's degree." He believes that another feature of the antipsychotic drug use that may have been particularly detrimental to his recovery was their tendency to bring on psychosis after sudden withdrawal. Sam recalls having discontinued psychiatric drugs "cold turkey" on several occasions, and then each time becoming lost in psychosis again shortly thereafter.

Beginning in his early thirties, Sam was able to taper off the antipsychotics and begin the long road to recovery that eventually culminated in a full and lasting recovery. He describes the essence of his recovery as having recovered the "spark" that he had prior to the terror of being drafted into Vietnam and the subsequent psychosis:

> I was a growing person, a young adult, had been a teenager and was in the end of my teenage years, and I had explored a lot of different things having to do with direction in my life and that kind of thing. But, you know, back then, there was kind of this spark in me that made me the personality of who I was, and . . . when I started having the experiences early on . . . . partly it was the experiences, quite a bit of it was the medication—they put me on Thorazine—it took a lot of that spark of who I was and kind of dampened it a little bit. So, the person Sam was not really there. It was more like a . . . sort of zombie-like [laughs] guy who didn't really have the spontaneity or sense of humor, that kind of thing, that I had before I had these experiences. . . . [So] I had like a ten to twelve year period of what somebody told me was like an interrupted life. . . . I had to go through some of the development things that most people go through at 20 to 25 or whatever, I had to do those at 32 to 38, you know. I think the key is that I was able to re-find the

> person I used to be, but also developing other parts of myself at the same time once I was able to get off the medicine.

Sam believes that a major factor in reconnecting with his spark was in making the shift from receiving harmful treatment to receiving helpful treatment. Initially, he was placed on a heavy prescription of antipsychotic drugs, with the instructions that he was to remain on them for the rest of his life: "As the episodes occurred, I'd be medicated, but I was medicated between the episodes, too, and you know, it was just the incorrect choice the doctors made to continue me on the medication." After about ten years of being on the antipsychotics, at the age of 29, he was very fortunate in that he found a doctor who was willing to help him come off of them, an event he describes as a major turning point in his recovery, the point when he began to feel like he was getting his "spark" back and to develop other parts of himself.

About ten years after coming off the antipsychotics, in his late 30's, Sam had two minor relapses over a period of about two years. This time, however, he managed to find doctors with alternative treatment philosophies. Rather than insist that he maintain a lifelong regimen of antipsychotic drug use, they suggested he use antipsychotics at minimal doses and for minimal duration, and he found the results of such judicious use much more beneficial: "My experience was that the use of antipsychotics is okay for acute symptoms, but as long-term prophylactics, they have debilitating effects."

Sam also found other resources in his life that served him in his recovery, including self-help methods for developing his sense of self worth, coming to appreciate the importance of prioritizing self care (which included stopping the use of recreational drugs and becoming much more diligent about sleep hygiene), and developing self awareness, especially with regard to personal behavior that indicated he might be losing touch with consensus reality and stressors that might trigger this. He also discovered the great benefit of developing a variety of creative outlets, including writing, music, and art.

Sam feels that a final crucial factor in his recovery was the presence of people in his life who were willing to stand by his side throughout his process: "When I was going through these times, I was having some difficulty in relating with people that were close to me, but having gone through them, it helps me appreciate how some of the people that are close to me have kind of stood by me and accepted me both when I was having the problems and now that I'm healed from the problems, so I think that's the key."

A Profound Transformation. As a result of having gone through such extraordinary experiences, Sam found that he has undergone a profound transformation with regard to his experience and his understanding of the world and of himself, what I will refer to as one's "personal paradigm."

One significant change in this regard is the sense that he has developed a stronger connection with others and the world in general, and closely related to this is an increased sense of the interconnectedness of all things: "I think, spiritually, I had always thought that all experiences and manifestations of the universe are interconnected. These experiences helped me to solidify these ideas in my mind."

Although Sam has not lost contact with consensus reality for many years now, he has come to believe that there is some validity to his perception of parallel worlds and the possibility of the paradoxical coexistence of different realities:

> You know, I have never ruled out the parallel worlds theory. . . . I don't know if a lot of people understand the idea that these experiences were real for me. . . . I could say, well, at one level, these things actually did happen, and then on the other level, the one within the time-space continuum which you and I share, these things are things that happened due to the way my neurons were firing. . . . I was experiencing all the stuff that I described, but I can look back on it and say, you know, there was..uh..a little leakage between the vibrational process of this universe and the vibrational processes of other universes. So I can say, you know, that happened, and I can also look back and kind of laugh and be part of this universe and say, boy, that's pretty wild crazy stuff there, Sammy [laughs]. . . . So I can look at it in both ways.

Sam has also come to appreciate just how limited the average person's understanding of the world really is: "Each person's view of reality is so narrow that they have no idea of what's going on in the universe. You know, our view of things that go on in the world is filtered. We think we know what's going on in the world, [but] we don't."

An important aspect of Sam's transformation is his development of a strong desire and capacity to serve others who are going through similar processes. He now works in a state hospital supporting patients in their reintegration back into the community, and he finds that his own experiences have provided him with a rare gift in providing such service: "What I hope is that others can benefit from maybe some of the things that happened with me. That is my goal in the work that I do. I try to help people using some of my experiences." Not only have Sam's experiences increased his ability to have more empathy and understanding for what many of his clients are going through, but they have also helped him to connect with his clients on a more personal level: "I have a way to work with people that have had these things that is, you know, I've been there type thing, and just connect on a human level rather than I gotta try to fix you type level."

Closely related to this is a strong ability to see the humanity in others, even in those for whom this is particularly difficult to do, something that Sam also attributes to his recovery process. He explains that one important aspect of this capacity is the ability to see someone's harmful behavior as an ignorant response to having been wounded:

> I work with people who have murdered people, I work with all kinds of people, and one of the things that I know is that if somebody goes through a process like that and they've done something very terrible in their life, that they need as much healing of their soul as they can get. So my goal is to help people heal as much as possible, at not only the level of becoming part of shared consensus reality but being able to make sure that their souls are being sewn back together again.

Having been on both sides of the locked door, Sam has significant insight into the serious problems of the current mental health care system and he strives to make a real difference: "I would like to do what I can, not only to help people, but change the way we do our business in the mental health world. It's not quite recovery focused yet. We still have a ton of work to do." One particular change Sam would like to see is with regard to the standard practice in the field of prescribing (often forcefully) psychiatric drugs heavily and over very long periods of time. As mentioned earlier, Sam felt that antipsychotics helped him to some extent when used judiciously and during the times of severe disconnection from consensus reality, but that the use of them outside of these times and particularly the long-term use of them was primarily harmful. After working extensively in the mental health care field, he has come to believe that the harms and benefits that he personally experienced from antipsychotic drug use also apply more generally to others.

Sam also finds that he experiences a greatly enhanced sense of wellbeing in his life now as a direct result of having gone through his psychosis, something that is likely to come as a major surprise to many people:

> I think that [these experiences have] given me probably an appreciation of where I am, so that I'm almost in a state of..uh..I don't know, constant bliss? I don't know how you'd describe it, but, you know, it's very rare that things in the world bug me much [laughs]. I think things do bug me. I don't like

> people getting beat up, and I hate injustice, but at least for my own personal wellbeing and how I am, I have a pretty good attitude about things in the world, and how I'm able to make sure that I'm doing all right.

Sam feels strongly that his experiences played an important role in his development of self acceptance and a fruitful sense of meaning and purpose in his life: "I'm glad I'm who I am, glad I went through the experiences I went through. They helped build the character of who I am now. I have a meaning and a purpose in my life, and I wouldn't trade it with anybody." Closely related to this renewed sense of meaning is a greatly expanded sense of possibility: "There are infinite possibilities and I am eager to learn all I can about all possibilities."

As of the writing of this book, Sam is 58 years old and has been completely free from all psychiatric drugs for 19 years without having had any relapses since. He is involved in a committed, loving relationship, and he continues to serve patients in a state hospital, bringing rare passion and insight into his work.

◇◇

It is likely that many readers are quite surprised by and perhaps even skeptical of Sam's story. There are many aspects of it that simply do not conform to our cultural understanding of madness and schizophrenia. When confronted with such a story, many of us may find ourselves asking questions such as: "How could he have fully recovered? Isn't schizophrenia a degenerative brain disease?"; "How could the psychiatric drugs have been more of a hindrance than a benefit in his recovery? Don't they correct a biochemical imbalance in the brain?"; and "So, if he really did experience genuine recovery, how on earth could his sense of wellbeing and ability to meet his needs have actually *improved* as a result of his psychosis?"

Yet, Sam's story is far from unique. Longitudinal studies[*] and numerous first-person accounts[†] have demonstrated that many, many individuals have experienced full and lasting recovery from long-term psychosis (the condition most of us generally think of as "schizophrenia"). In fact, the World Health Organization has concluded that in some countries, particularly those we often refer to as "third world" countries, full and lasting recovery is actually the most common outcome for people diagnosed with schizophrenia[1]. And when we look more closely at the recovery stories of many of these people, such as that of Sam, we discover that very often, recovery does not simply entail the return to the person's

---

* See Table 4.1

† For example, Bassman, 2007; Beers, 1981; Dorman, 2003; Greenberg, 1964; Modrow, 2003.

pre-psychotic way of being, but that a profoundly healing transformation has taken place resulting in a much healthier state of mind than that which existed prior to the psychosis.

When we recognize the very real possibility of full recovery and profound healing and transformation, it becomes quite clear that something truly extraordinary is taking place within the psychotic process. While we must not forget the potential for tremendous pain and tragedy that is all too often associated with psychosis, in order to move towards a more complete vision of what psychosis really is, we must also account for the possibility of great healing and renewal.

In order to do this, however, we must come to terms with the fact that the mainstream vision of psychosis currently held in the West is somehow seriously missing the mark. We must question everything we think we know about psychosis and schizophrenia and be willing to approach this topic with a beginner's mind, finding the courage to venture into the terrifying vortex of madness while not losing sight of the very real possibility of genuine healing and life-renewing transformation. This book is an attempt to embark on just such a journey.

As we move through this book, we will look at five more stories of individuals who have journeyed far into the realms of madness* and back, all of which are accounts taken from my own recent research†. These stories will be used as the main reference points as we look at the other research and develop a deeper understanding of the psychotic process.

In Part One, we will explore some of the largest and most entrenched myths‡ about madness that have developed within Western society, taking a particularly close look at the research that is purported to support the mainstream psychiatric understanding of psychosis.

In Part Two, we turn to look at a number of alternative models of psychosis, especially those that appear to most accurately reflect the evidence that has been accumulating with regard to the nature of psychosis and recovery.

---

* *Madness* and *psychosis* are used interchangeably throughout this book, representing the same phenomenon.

† I conducted a total of three case studies of people who have fully recovered from long-term psychosis—one single-case study and two multiple-case studies (see "Williams, P." in the Bibliography section for a link to the full text version of my dissertation which outlines the details of these studies). The third and final study was the largest, involving six participants. This book includes the stories of all six of these participants—Sam's story is presented above, and the remaining five stories are presented in Part Two and Part Four.

‡ By use of the term *myth* in this context, I don't mean to imply a belief that is necessarily false—it may or may not have validity; rather, my use of the term *myth* here refers to a belief system that is so entrenched in a society that it is generally assumed to be valid without ever being seriously questioned.

In Part Three, I present a novel and highly integrative model—the *duality/unity integrative model*—which resulted from a thorough analysis of the stories of the six participants presented in this book. This model integrates both Western and Eastern understandings of human experience taking place at the most fundamental levels of our beings. My aim in introducing such a model is to provide a framework that can help us more fully understand what is happening at these most fundamental levels of experience during psychosis, as the evidence suggests that psychosis entails a profound reorganization of the self at these deepest levels.

In Part Four, we use the duality/unity integrative model as a framework to more closely examine the six stories of madness and recovery presented in this book. We look closely at each stage of the psychotic process, seeking a fuller and more comprehensive understanding of the entire psychotic process, from onset to full recovery, and exploring what factors were most influential in each stage.

Finally, in the Conclusion section, we stand back and take in the big picture, reflecting upon how this exploration may serve us in moving towards a more accurate and helpful paradigm in our understanding of psychosis, one that is much more supportive for those suffering from psychosis and one that is much more beneficial for our society as a whole.

Part One

Deconstructing the Myths of Madness

Chapter 1

First, Some Terminology

Before embarking on a serious exploration of psychosis and recovery, it is important that we first clarify the terms that we will be using. Unfortunately, just as this topic is so often filled with ambiguity, confusion, mystery, and even fiery controversy, the terms most commonly used for this topic are also often ambiguous and/or controversial. So, in an attempt to avoid becoming entangled in philosophical debate about the various terms, I offer a set of basic definitions that will facilitate our discussion while minimizing controversy:

Consensus reality: The set of beliefs and experiences considered to be valid according to an individual's society or group (more on this shortly).

Anomalous experience: A subjective experience (typically either a belief or a sensory experience such as a sound, a vision, a taste, a smell, or a tactile sensation) that is considered invalid within the framework of consensus reality (according to the particular individual's society or group).

Psychotic experience: An anomalous experience that causes significant distress and/or limitation.

Psychosis: An ongoing condition in which psychotic experiences predominate. (This condition also typically entails significant instability with regards to one's psychotic experiences.)

Long-term psychosis: A psychotic episode that lasts for one month or longer; or a series of psychotic episodes, the total duration of which is at least one month.

(While the duration of time may seem somewhat arbitrary, it is based on the criteria used for diagnosing schizophrenia—more on this shortly.)

Recovery: The condition of experiencing a general diminishment of the distressing and/or limiting aspect(s) of one's anomalous experiences.

Full recovery: The condition of having achieved a homeodynamic* balance in which the overall distress (and not necessarily the anomaly) of one's subjective experiences is the same or less than that which preceded the psychosis.

Below, I explain these terms in more detail, and I also include the definitions of several other closely related terms.

THE MEDICAL MODEL

In the late 1800's, Emil Kraepelin was the first to clearly articulate the system of assumptions that underlies the field of biological psychiatry and to formally assert that schizophrenia and other mental disorders are products of a diseased brain. In 1886, the year of his first appointment as a professor of psychiatry, Kraepelin began to shape his ideas on psychiatric classification. Based on his clinical work and theoretical speculation, he concluded that psychiatric disorders, like physical illnesses, are discrete entities with distinct physiological causes. He is perhaps best known for labeling one particular cluster of symptoms *dementia praecox* (meaning literally "senility of the young"), which eventually evolved into today's diagnostic label, *schizophrenia* (meaning literally "split mind"). Though Kraepelin failed to identify any form of physical pathology associated with dementia praecox or any of the other diagnostic labels he coined, he remained confident that such pathology would one day be found.

Through Kraepelin's life and afterwards, the field of psychiatry continued to gain political power, and for many years now, it has been generally considered to be the highest authority in the field of mental health in the West, especially regarding the psychotic disorders such as schizophrenia. As psychiatry's power grew, so did its allegiance to Kraepelin's model. In 1978, Klerman, considering himself to be a member of the *neoKraepelinian movement*, clearly articulated this model in the following manifesto:

* The term *homeodynamic* will be defined at the end of this chapter in the section defining *recovery*.

Tenets of the Medical Model
(as put forth by the NeoKraepelinians)

- Psychiatry is a branch of medicine.
- Psychiatry should use modern scientific methodologies and base its practice on scientific knowledge.
- Psychiatry treats people who are sick and who require treatment for mental illness.
- There is a boundary between the normal and the sick.
- There are discrete mental illnesses. Mental illnesses are not myths. There is not one, but many mental illnesses. It is the task of scientific psychiatry, as of other medical specialties, to investigate the causes, diagnosis and treatment of mental illnesses.
- The focus of psychiatric physicians should be particularly on the biological aspects of mental illness.
- There should be an explicit and intentional concern with diagnosis and classification.
- Diagnostic criteria should be codified, and a legitimate and valued area of research should be to validate such criteria by various techniques. Further, departments of psychiatry in medical schools should teach these criteria and not depreciate them, as has been the case for many years.
- In research efforts directed at improving the reliability and validity of diagnosis and classification, statistical techniques should be utilized.

Source: Bentall, 2009, p. 53.

Although these principles are rarely articulated as clearly as they are here, it is evident that they continue to lie at the foundation of the dominant paradigm in the mental health field today, a paradigm that is often referred to simply as the *medical model*. When researching long-term psychotic disorders, it's important to acknowledge that most of the assumptions we have in the West regarding the extreme states of consciousness to which we have given labels such as *psychosis*, *schizophrenia*, *schizoaffective disorder*, *bipolar disorder*, etc., come directly from this model[1].

SCHIZOPHRENIA

Since the time of Kraepelin, schizophrenia has remained the hallmark diagnosis for long-term psychosis. Clinicians use a set of criteria based on the reported and observed perceptions, beliefs, and behavior of an individual (the same protocol used for diagnosing any of the other so-called "mental illnesses"). This set of criteria is outlined within the Diagnostic and Statistical Manual (DSM), which is produced by the American Psychiatric Association (APA). Following is a summary of the criteria for schizophrenia listed in the most recent edition of the DSM—the DSM-IV-TR*:

> *Characteristic Symptoms:* Two (or more) of the following, each present for a significant portion of time during a 1-month period (or less if effectively treated):
>
> (1) delusions
>
> (2) hallucinations
>
> (3) disorganized speech (e.g., frequent derailment or incoherence)
>
> (4) grossly disorganized or catatonic behavior
>
> (5) negative symptoms, i.e., affective flattening, alogia, avolition
>
> *Note:* Only one [of these] symptoms is required if delusions are bizarre or hallucinations consist of a voice keeping up a running commentary on the person's behavior or thoughts, or two or more voices conversing with each other.[2]

Several qualifications are given in addition to these characteristic symptoms, including the existence of significant social and/or occupational dysfunction, and the existence of "signs of the disturbance" persisting for at least six months, even though the actual characteristic symptom(s) only need to last for a minimum of one month (or less if treated)[3]. In addition, a person is usually assigned a particular subtype of schizophrenia, depending upon the most predominant experiences and/or behavior (*paranoid type, disorganized type, catatonic type, undifferentiated type*, and *residual type*). For psychotic episodes of relatively brief duration, a diagnosis of *schizophreniform disorder* is used when "signs of the disturbance" last for less than six months, and a diagnosis of *brief psychotic disorder* is used when these last for less than a month. Those suffering from long-term

* A new edition of the DSM is scheduled to be released shortly after the publication of this book; however, the criteria for schizophrenia are likely to remain very similar, if not exactly the same.

psychotic conditions are sometimes assigned an altogether different diagnosis than schizophrenia if emotional disturbances are markedly present or if the criteria for a schizophrenia diagnosis are not well met in other ways. The most common of these are *schizoaffective disorder*, severe cases of *bipolar disorder*, *major depressive disorder with psychotic features*, and the catch-all diagnosis often used when nothing else seems to fit—*psychotic disorder NOS* ("not otherwise specified").

There's no doubt that these criteria require highly subjective judgment on the part of the clinician, and in fact, the diagnosis of schizophrenia is highly controversial. Despite over a century of intensive research, no biological markers or physiological tests that can be used to diagnose schizophrenia have been found[4], its etiology continues to be uncertain[5], and we don't even have clear evidence that the concept of schizophrenia is a valid construct[6]. However, diagnosis and treatment based upon the diagnosis continues unhindered by these serious problems.

Psychosis

The concept of psychosis, like the concept of schizophrenia, does not have a clear-cut, mutually agreed upon definition, and the definitions that have been put forward are highly controversial. Highlighting the ambiguity of the term psychosis, even the DSM-IV-TR, often referred to as the "bible" of the mental health care field, admits that "the term *psychosis* has historically received a number of definitions, none of which has achieved universal acceptance" [7]. The DSM then goes on to give its own extremely vague definition, one that it never further clarifies: "the term psychosis refers to the presence of certain symptoms. The specific constellation of symptoms to which the term refers varies to some extent across the diagnostic categories" [8]. For the purpose of discussing recovery from psychosis, the DSM's definition is not particularly helpful. Unless we are able to arrive at a working definition for psychosis, however, any discussion of recovering from it is essentially meaningless. I propose that a helpful way to view psychosis involves first making the distinction between anomalous experiences that are distressing and those that are not.

Anomalous experiences. As mentioned above, the content and form of the subjective experiences of psychosis are generally described as being delusions and/or hallucinations[9]. There is the assumption that these experiences are not valid because they do not conform to consensus reality, and so according to this reasoning, the ultimate goal should be to bring the individual's experiences back into alignment with consensus reality as quickly as possible. However, consensus reality does not necessarily correspond to some objective truth, and indeed may be vastly different from one culture to another. Therefore, I find it more useful to define subjective experiences that are not in alignment with consensus reality as simply *anomalous experiences*, rather than assuming that they are psychotic.

Psychotic experiences. Just as we can cite numerous examples of mental experiences that are considered normal and yet are clearly distressing and limiting (e.g., bereavement, worry, stage fright), there are also numerous examples of mental experiences that do not conform to consensus reality (in other words, anomalous beliefs and perceptions) and yet clearly do not result in distress and/or limitation. For example, there are many people who hear voices but are not at all distressed by them, and some even find them to be helpful[10]. Ordinarily, based on the criteria found within the DSM-IV-TR[11], such experiences would be considered psychotic based purely upon the degree of their dissonance with consensus reality, but if such experiences do not cause harm to oneself or others, then what is the benefit of labeling them as such? Doing so leads to the difficult and sometimes impossible task of trying to determine whose beliefs and perceptions are closer to so-called objective reality. Therefore, I believe it's much more useful to determine the degree to which an individual's anomalous experiences cause distress and/or limitation*: how much they interfere with one's ability to experience satisfying relationships, meet one's basic needs, maintain a sense of wellbeing, and support the wellbeing of those with whom one interacts.

Considering, then, that both anomalous and so-called normal experiences may or may not cause distress, I suggest that a useful definition for *psychotic experiences* is "anomalous experiences that cause significant distress." There are two aspects of the way I'm using the term distress here that are important to emphasize: (1) ultimately, distress is subjective—no one except for the individual experiencing distress is capable of determining what is and what is not distressing for that individual; and (2) this distress may either arise directly from the anomalous experiences themselves (for example, fear occurring as a direct result of auditory hallucinations) or arise from behavior that was a direct result of the anomalous experiences. A hypothetical example of this is the case of someone believing that he or she can fly and then happily jumping out of a ten-story window. The anomalous experience itself (believing one can fly) may not cause distress, but the resulting behavior almost certainly will (if not to the individual, then certainly to the observers and/or any loved ones left behind).

---

* Distress and limitation typically show up hand in hand. That is, distress is generally associated with the inability to meet one's needs, whether it be the need for ease, joy, connection with others, autonomy, self worth, etc. Therefore, I will hereafter refer to both distress and limitation as simply distress.

Long-term psychosis. As will be discussed in more detail in Chapter Three, there is significant debate within the mental health care field regarding the validity of the concept of schizophrenia. However, it's clear that many do people suffer from psychotic experiences for relatively long periods of time. Therefore, I will often use the somewhat less controversial term *long-term psychosis* to refer to any episode of psychosis that lasts for a minimum of one month (drawing from the DSM-IV's criteria for schizophrenia) or a series of psychotic episodes the total of which is at least a month. This will therefore most often include cases that have been diagnosed as schizophrenia, schizoaffective disorder, bipolar disorder, major depressive disorder with psychotic features, and psychotic disorder NOS.

Recovery

Having made a distinction between distressing and non-distressing anomalous experiences, I propose that *recovery* must refer to the abatement of the distressing aspects of anomalous experiences. In other words, it is distress that one is recovering from, not necessarily anomaly. Therefore, someone is recovering when the distress associated with these experiences is diminishing, regardless of whether or not the anomalous aspects of these experiences are diminishing. For example, there are many cases of people who had experienced severe distress as a result of hearing voices, but over time, their relationship with the voices changed to the point that the voices were no longer distressing or limiting, even if they continued to exist to some degree[12]. According to the definition used here, those who have experienced some abatement of the distress and limitation caused by the voices would be *recovering*, and those for whom the voices no longer cause significant distress or limitation at all would be considered *fully recovered*, at least assuming there are no other psychotic experiences with which they continue to struggle.

Another important aspect of this definition of recovery entails the concept of *homeodynamics*. In the life sciences, it is common to think of living organisms as existing in a state of homeostasis, which is an organism's resistance to change and its ability to maintain a stable internal environment. Mainstream psychology and psychiatry evidently draw from this model when attempting to return a psychotic individual to a state that is as close to his or her pre-psychotic state as possible. It has been suggested, however, that it is actually more accurate to consider organisms to be living in a homeo*dynamic* state rather than a homeo*static* one. The term *homeodynamic* suggests that "once a new stressor is encountered, the organism never returns to its previous dynamic state, but establishes a new dynamic balance appropriate to this newly integrated experience"[13].

In the context of psychosis, then, this concept suggests that it may be terribly problematic to attempt to return someone to their pre-psychotic condition rather than to support them in integrating their anomalous experiences as they move into an altogether new way of being in the world, an idea that is in very close accord with the recovery research. This is a distinction that will be particularly important as we continue to explore psychosis and recovery throughout this book, and to highlight the importance of this, I've incorporated it into my working definition for *full recovery*: "The condition of having achieved a *homeodynamic* balance in which the overall distress (and not necessarily the anomaly) of one's subjective experiences is the same or less than that which preceded the psychosis."

CHAPTER 2

Myth #1: Schizophrenia is a Brain Disease

In modern Western society, as discussed above, the large majority of those diagnosed with schizophrenia receive their primary care from psychiatrists who subscribe to the theory that schizophrenia and other psychotic disorders are caused by diseases of the brain (i.e., the medical model)[1]. Unfortunately for these patients, the prognosis for their condition when viewed through this theoretical lens is extremely poor. According to this theory, schizophrenia is a brain disease not unlike Alzheimer's or Parkinson's, and just as there has never been a documented case of someone making a full recovery from either of these disorders[2], so it is generally assumed that no one can ever fully recover from schizophrenia, at least not without some biological cure still awaiting discovery. Therefore, when discussing the topic of recovery from schizophrenia, it's important to look closely at the validity of this theory.

The belief in the brain disease theory is predominant within the Western mental health field, with the most influential leaders of the field widely promoting it and citing research which purports to support it[3] in spite of their admission that the cause of schizophrenia is still undetermined[4]. The most common sub-theories of the brain disease theory are (1) that schizophrenia is caused by a biochemical imbalance within the brain, (2) that schizophrenia is caused by or at least closely related to abnormal brain structures, and (3) that schizophrenia is a genetic disorder. If we use the term *myth* to describe beliefs that are generally held within a society without generally being seriously questioned, then there is no question that these theories have grown to myth status within Western society. So, in the spirit of bringing an attitude of the beginner's mind into an exploration of the mysteries of madness, it's essential that we take a closer look at the brain disease theory of schizophrenia and its major sub-theories.

THE BIOCHEMICAL IMBALANCE THEORY

The first hypothesis suggesting that schizophrenia may be caused by biochemical means was put forward over 50 years ago. In 1938, the Swiss chemist Albert Hoffman accidentally ingested a very small amount of a substance he had synthesized from the fungus ergot[5]. The substance was lysergic acid diethylamide (LSD), and the result was a hallucinogenic state apparently quite similar to that of many psychotic states. This discovery led some researchers to speculate that perhaps schizophrenia is caused by a self-created (endogenous) substance[6]. In 1962, it was thought that a breakthrough had occurred in substantiating this theory when a pink spot was discovered on the chromatography paper used to test schizophrenia patients' urine. The initial belief was that this pink spot was an indication of the speculated endogenous hallucinogenic substance; however, it was soon discovered that this spot was correlated with the ingestion of a combination of foods commonly consumed by psychiatric patients, and support for the "endogenous hallucinogen" theory soon faded away[7].

Soon after the endogenous substance theory had begun to fade away, the first dopamine imbalance hypothesis was posited. In 1951, the French naval surgeon Henry Laborit accidentally stumbled upon a drug which had such a powerful numbing effect on his surgical patients that they needed almost no anesthetic. Laborit also noticed that the drug seemed to put his patients into a strange daze, in which they seemed completely indifferent to anything going on around them, and yet they maintained enough cognizance to answer questions. One of his colleagues noticed that this effect was very similar to that seen in lobotomized patients, and he suggested it may be useful in psychiatry. Much research was soon conducted with this drug, soon to be named chlorpromazine, and two important features were soon discovered about it: (1) it reduces a number of psychotic symptoms, at least in the short term, and (2) it blocks dopamine receptors (as did all of the other antipsychotic drugs that became available over the next several decades)[8]. These findings eventually led to the hypothesis that schizophrenia is caused by an excess of dopamine within the individual's brain[9].

This promising new hypothesis led to a vast amount of research, most of which seemed to support it very well. First, trials demonstrated a correlation between the amount of chlorpromazine given and the reduction of psychotic symptoms in the short term[10]. Second, trials conducted in the early 1970s showed that drugs such as amphetamines and L-DOPA (used to treat Parkinson's disease), which are known to *increase* dopamine in the brain, sometimes led to psychotic states in otherwise non-psychotic people[11]. And third, patients diagnosed with schizophrenia have been found to be generally more vulnerable to psychotic effects as a result of these dopamine-increasing drugs[12]. The result of these and other similar studies has been a continuation of the dopamine hypothesis to the present day, with ongoing controversies over the details, such as which dopaminergic neural pathways are most affected and whether or not other imbalances exist in entirely different pathways (especially in neural pathways using serotonin, GABA, and glutamate neurotransmitters[13]).

As research has continued into this area, however, serious doubts about the dopamine hypothesis have arisen. For example, although it is known that an individual's D2 dopamine receptors (the type of receptors most affected by typical antipsychotic drugs) are completely blocked within hours of consuming a sufficient dose of an antipsychotic drug, the actual antipsychotic effects often do not become apparent for up to several weeks[14]. If psychotic symptoms are the direct result of too much dopamine, it is argued, then why don't we see a more immediate abatement of these symptoms as soon as the dopamine levels have been effectively reduced?

Another finding that challenged the validity of the dopamine hypothesis came about during the search for increased D2 dopamine receptor levels in the brains of schizophrenic patients, using postmortem studies and PET scans of live patients. Initially, the studies seemed to indicate support for the theory by demonstrating that a higher level of dopamine receptors existed in these patients. Lee and Seeman[15] were among the first researchers to validate this finding. However, Lee and Seeman and others suggested that the dopamine receptors of these patients may have increased as an adaptation response to the antipsychotic drug treatment itself. In other words, these neurons may have attempted to increase their number of dopamine receptors in an attempt to compensate for the dopamine-inhibiting action of the antipsychotics. Subsequently, several different research teams, using animal and postmortem studies, all confirmed that when the effects of antipsychotics on dopamine receptors were taken into account, there was no significant difference in dopamine receptor levels between the patients and healthy controls[16].

These multiple challenges to the validity of the dopamine hypothesis have led a number of researchers to lose heart in it altogether. John Kane, a well-known researcher at the Long Island Jewish Medical Center, was one of these, confessing that "a simple dopaminergic excess model of schizophrenia is no longer credible. . . . Even Carlsson, who first advanced the hypothesis, [has] concluded that there is 'no good evidence for any perturbation of the dopamine function in schizophrenia'" [17].

Hyman, neuroscientist, provost of Harvard University, and ex-director of the National Institute of Mental Health (NIMH), summarized over 40 years of research on the mechanism underlying the effects of antipsychotic drugs. One of the main conclusions he arrived at was that the use of antipsychotics actually *creates*, rather than corrects, a biochemical imbalance within the brain. Prior to treatment, those diagnosed with schizophrenia have no known biochemical imbalances within the functioning of their neurons, but once they are placed on antipsychotics, the brain goes through a dramatic modification that results in abnormal neurotransmission[18].

As support for the dopamine hypothesis has waned, many researchers have begun to look for other possible biochemical imbalances (for example, with the neurotransmitters GABA, glutamate, and serotonin), but numerous peer reviews have found that the research has so far not supported these hypotheses either[19]. In short, then, after more than 50 years of intensive research, the biochemical imbalance theory remains unsubstantiated.

ABNORMALITIES IN BRAIN STRUCTURE

Another popular generator of hypotheses has been the idea that abnormalities in brain structure may be the cause of schizophrenia, or at least indicative that schizophrenia is the direct result of brain disease. The first attempts to study the postmortem brains of psychiatric patients were conducted by Kraepelin, the original founder of the concept of schizophrenia, around the turn of the twentieth century. He asserted that "partial damage to, or destruction of, cells of the cerebral cortex must probably occur . . . which mostly brings in its wake a singular, permanent impairment of the inner life" [20]. In the end, however, his search did not result in any significant findings.

It was not until about seventy years later, in the early 1970s, with the invention of CT scanning, that an apparent breakthrough in this quest had finally arrived. CT scans allow views of cross-sectional slices of the brains of living patients, and in 1976, Johnstone and her colleagues published the first report showing significantly enlarged lateral ventricles in the brains of patients who had been diagnosed with schizophrenia, something that strongly implied atrophy of the cerebral cortex and perhaps other regions of the brain[21]. Numerous other studies have since come out (using both CT scanners and, arriving in the 1980s, the more technologically advanced MRI scanners), nearly all of which have replicated these findings. Wright et al. conducted a meta-analysis of 58 studies with 1,588 schizophrenia patients and found that the mean total ventricular volume of these patients was 126% that of the control group, a clear difference suggesting significant brain atrophy[22].

Similar studies continue to accumulate today, with the most recent studies being able to distinguish with very precise detail exactly which areas of the brain are experiencing atrophy. For example, one of the most recent of these studies was able to isolate shrinkage of the thalamus, "especially of medial nuclei and the adjacent striatum and insular cortex, [and therefore concluded that these] appear to be important contributors to ventricular enlargement in schizophrenia" [23]. The impressive collection of studies corroborating the correlation between ventricular enlargement and the diagnosis of schizophrenia makes this one of the most consistent findings of a neurobiological correlate with schizophrenia. On the surface, all of this evidence appears to provide substantial support for the hypothesis that schizophrenia is the product of a brain disease; but upon closer inspection, this support quickly unravels.

The first finding that challenged the validity of the brain structure abnormality hypothesis came with the realization that numerous factors have been found to lead to ventricular enlargement. Among these are depression, alcoholism, early childhood trauma[24], water retention, pregnancy[25], advancing age, educational achievement, social class, ethnicity, and head size[26]. It was also discovered that ventricle size can actually fluctuate quite rapidly within even healthy individuals, leading to varying results even within the same individual[27]. Perhaps the most relevant factor that has been demonstrated to cause ventricular enlargement (and other significant brain damage) is the use of antipsychotic medication itself, and virtually all of the research that discovered increased ventricular volume in those diagnosed with schizophrenia did not account for this obviously important factor[28]. It is almost certain that the majority of the participants of these studies had been using a significant amount of antipsychotics and thus likely had some degree of ventricular enlargement as a direct result of antipsychotic use when their brains were evaluated.

A second serious challenge to the validity of the brain structure abnormality hypothesis came when it was recognized that the majority of those diagnosed with schizophrenia do not show any obvious brain abnormality at all. Lewine found that "there is no brain abnormality in schizophrenia that characterizes more than 20-33% of any given sample. The brains of the majority of individuals with schizophrenia are normal as far as researchers can tell at present"[29]; and this in spite of the fact that most of these participants were likely exposed to other brain atrophying factors including especially antipsychotic medications. Conversely, it is common to find healthy individuals who have no schizophrenic symptoms and yet have brain abnormalities similar to those sometimes found in schizophrenics[30].

There has been another highly significant challenge to the theory that any link between schizophrenia and brain structure abnormality must imply that schizophrenia is a brain disease. It has been well recognized that unusual modifications of the brain are not always indicative of disease. Numerous studies have shown that various experiences modify our brains all of the time. For example, the volume of the posterior hippocampus has been shown to increase in London taxi drivers as they memorize the streets of the city[31], while the hippocampus and corpus callosum have been shown to atrophy as a result of post-traumatic stress following warfare[32] or sexual abuse[33].

When we look at the sum total of all of this research relating to brain abnormality, we find we are left with yet another unsubstantiated hypothesis. First, we realize that there are many factors that can lead to ventricular enlargement, and that most people who are diagnosed with schizophrenia have also been exposed to one or more of these other factors (different types of trauma, depression, alcoholism, and especially the use of antipsychotic medication). Then, when we add to this the fact that the majority of patients diagnosed with schizophrenia do not show brain abnormalities at all (even when exposed to these other factors) while a number of healthy subjects do, the support for this hypothesis becomes highly questionable indeed.

HEREDITY

So, the two major hypotheses purporting to support a link between schizophrenia and a disease of the brain (the biochemical imbalance hypothesis and the abnormal brain structure hypothesis) have so far failed to achieve any significant validation; but what about the issue of heredity in schizophrenia? Many researchers have suggested that schizophrenia is a biological disease with a strong genetic component. This is a fairly complex topic, and to look at it critically requires that we break it down into several parts. We must look at the research that suggests a genetic basis of schizophrenia, the research that suggests an environmental basis for schizophrenia, and we must also look critically at the assumption that finding a genetic link for schizophrenia confirms that schizophrenia is indeed a biological disease. Let's now look at each of these in turn.

The genetic research. Most mainstream psychiatric and mental health authorities refer to the heredity of schizophrenia as a well established fact. The NIMH, for example, said:

> Scientists have long known that schizophrenia runs in families. It occurs in 1 percent of the general population but is seen in 10 percent of people with a first-degree relative (a parent, brother, or sister) with the disorder. People who have second-degree relatives (aunts, uncles, grandparents, or cousins) with the disease also develop schizophrenia more often than the general population. The identical twin of a person with schizophrenia is most at risk, with a 40 to 65 percent chance of developing the disorder.[34]

While these are some highly significant figures, and they are said from a very high authority in the field, some scholars in the field have challenged their validity[35]. Jay Joseph is one of the most active researchers in the field regarding the topic of genetics and schizophrenia, having published over thirty articles in peer-reviewed journals and two books on this and closely related topics. He pointed out that the basis for all of the current genetic research that attempts to isolate genes for schizophrenia (more on this later) is contingent upon the validity of only a small handful of twin and adoption studies, all of which he concluded have significant validity problems.

Twin studies. The logic behind the twin studies is quite straightforward: if reared-together identical (monozygotic) twins have a significantly higher rate of concordance for schizophrenia than reared-together fraternal (dizygotic) twins, then, according to the logic behind these studies, the genetic basis for schizophrenia should be confirmed. This is because the identical twins share 100% of their genes, while the fraternal twins share an average of only about 50% of their genes (the same as any two siblings). All such twin studies (published after 1963, when the research methods were considered much more sound), when pooled together, showed that the identical twins have a concordance rate of 22.4%, while the fraternal twins have a concordance rate of 4.6%[36]. These are much lower rates than those cited above by the NIMH, but they are still significant. The identical twins have a concordance rate that is four to five times higher than that of fraternal twins, and few would question the significance of this. However, these figures are significant only if the validity of the studies is significant, and Joseph has made a very strong case that they are not. He listed numerous methodological problems with these studies, including:

> (1) lack of an adequate and consistent definition of schizophrenia; (2) non-blinded diagnoses, often made by investigators strongly devoted to the genetic position; (3) diagnoses made on the basis of sketchy information; (4) inadequate or biased methods of zygosity determination (that is, whether twins are [identical or fraternal]); (5) unnecessary age-correction formulas; (6) non-representative sample populations; and (7) lack of adequate descriptions of methods.[37]

However, even considering all of these methodological problems, Joseph conceded that the concordance rate is clearly higher for the identical twins, just as it is for most psychological and biological traits. Where the fatal problem to this type of study lies, he claims, is in the assumption that the two different types of twins share similar environments. He argued that this assumption is not at all true—identical twins are treated much more similarly, encounter more "identity confusion," and encounter significantly more similar environments than fraternal twins. Because of these significant differences, Joseph argued that the difference in concordance rates between the two types of twins can be explained *solely* by environmental differences: "There is no reason to accept that [identical-fraternal] comparisons measure anything more than the *environmental* differences distinguishing the two types of twins [author's emphasis]" [38].

A second method of twin studies consists of single-case reports of identical twins that have been raised apart. However, because of small sample sizes, there have not been any systematic reared-apart twin studies for schizophrenia or any other psychiatric condition[39].

Adoption studies. The second research methodology that has played a major role in establishing the theory of schizophrenia as a genetic disorder is the adoption study. To date, there have been only seven major adoption studies on schizophrenia[40]. In three of the studies, the main goal of the study was to research the outcomes of adoptees whose biological parent(s) had been diagnosed with schizophrenia, comparing them with a control group (adoptees whose biological parent(s) had *not* been diagnosed with schizophrenia[41]. In the fourth study, adoptees with schizophrenic biological parents who had been adopted to normal parents were compared with adoptees with normal biological parents who had been adopted by parents with schizophrenia[42]. The three remaining studies (which are separate parts of one expanded study) began with adoptees who had been diagnosed with schizophrenia, and then involved the search for their biological parents to find the percentage of them who had also been so diagnosed (again, a control group was used for comparison)[43]. In all of these studies, the researchers concluded that they found evidence that supported a significant genetic factor for schizophrenia[44]. In spite of these apparently robust findings, Joseph suggested that there are serious problems with regard to both the methods used and the conclusions drawn for all of these studies[45].

One particularly potent criticism is that "most of these studies would not have found statistically significant differences without greatly expanding the definition of schizophrenia to include non-psychotic 'schizophrenia spectrum disorders'"[46]. For example, Kety et al.'s first study[47] found *zero* cases of chronic schizophrenia within the first-degree relatives of those diagnosed with schizophrenia, and Rosenthal's study[48] found just one of the 76 offspring adopted away was later diagnosed with schizophrenia, which is not any more than what would be expected in any random sampling of the population. When Joseph narrowed the definition to the DSM diagnosis for schizophrenia and recalculated the results, he found that only two of the seven studies show significant differences between the experimental groups and the control groups, and he argued that even these two have significant validity issues from one or more of the other methodological problems mentioned earlier[49].

A second significant problem with the validity of these studies arises from the issue of selective placement of these adoptees. Joseph pointed out that in all three regions and time periods in which these studies were conducted, compulsory eugenic sterilization of people diagnosed with schizophrenia and other "mental disorders" was in effect, meaning that most of the offspring of such parents would have been labeled as being offspring of schizophrenics, and this would almost certainly have influenced the placement of these infants: "One can conclude that the most qualified potential adoptive parents, who were usually informed of 'deviance' in the adoptee's family background, would not have selected children with a biological family history of mental disorders"[50]. Conversely, it's likely that infants with such a history would have been given to the least "qualified" parents. The researchers of these studies generally did not consider this very important factor and attempt to control for it, an obvious oversight that significantly compromises the validity of their findings.

The search for genetic linkage. Unfortunately, Joseph is one of only a small handful of researchers who have taken the courageous position of seriously challenging the research that purports to support the heredity of schizophrenia. As solid as his critiques are, and as scant and faulty as the genetic research is, there are no doubt many who would understandably be hesitant to place significant weight on one researcher's work in the face of such a staunchly held belief within the field. If Joseph's work does not give pause to the skeptic, however, there is an interesting development that very well may—many researchers who are clearly in *favor* of the genetic position have *also* expressed doubts and concerns.

Contemporary research into the heredity of schizophrenia has moved away from twin and adoption studies and towards efforts to locate the specific genes themselves within DNA, but so far no such genes have been found. Tim Crow, one of the most widely published researchers in this area, said in one of the most recent reviews of genetic research that was published in the *American Journal of Psychiatry*, that "recent meta-analyses have not identified consistent sites of genetic linkage. The three largest studies of schizophrenia fail to agree on a single locus . . . and there is no replicable support for any of the current candidate genes"[51]. Williams and his large team of genetic researchers admitted, "Our results suggest that common genes of major effect . . . are unlikely to exist for schizophrenia"[52]. Most leaders in the field are losing hope that they will ever find a single gene that is linked to schizophrenia and are instead hoping that they will have more success by looking for many associated genes of small effect, each of which increases susceptibility for schizophrenia[53]. Delisi, another schizophrenia genetic researcher, concluded that "psychiatric genetics appears to be at a crossroads or crisis"[54]. Joseph named the elephant in the room to which Delisi's comment was probably pointing: "The 'crisis' facing psychiatric genetics is that investigators are looking for genes that probably do not exist"[55].

The environmental research. While the research purporting to show a strong genetic influence on the development of schizophrenia has clearly been overstated, the research showing environmental influences* on the development of schizophrenia has generally been greatly *under*stated, in spite of the fact that this research has consistently shown a strong correlation between childhood abuse/trauma and the later development of schizophrenia and related psychotic disorders.

One of the most important findings correlating childhood trauma and the development of schizophrenia emerged, ironically, within one of the adoption studies mentioned above—a study that is ordinarily used to prop up the genetic position. In the adoption study in which adoptees with schizophrenic biological parents who had been adopted to normal parents were compared with adoptees with normal biological parents who had been adopted by schizophrenic parents, it turns out that only 4% of the children raised by the normal parents were later diagnosed as "severe + psychotic," even though their biological parents had been diagnosed with schizophrenia. On the other hand, 34% of those raised by the "disturbed" adoptive families were later diagnosed[56], even though their biological parents had been "normal." Statistical analysis revealed that "the dysfunction of the family, and the maltreatment of the child implied thereby, had *7 times* more explanatory power than genetic predisposition [emphasis added]"[57].

There have been a number of other studies that have demonstrated similar results, of which the following is only a relatively small selection: One study looked at 524 child guidance clinic attendees over 30 years and discovered that 35% of those later diagnosed with schizophrenia had been removed from their

---

* The term *environmental influences* in this context refers to any factors other than those that are genetically inherited that affect the development of an individual. Research and common sense suggest that genetic and environmental factors work together synergistically to influence a person's development (a recent theoretical approach known as *epigenetics* suggests that these are even more closely intertwined than has been previously recognized); however, for the purpose of our exploration here, it's helpful to treat genetic and environmental influences as somewhat distinct.

homes due to neglect, a percentage twice as high as that for any other diagnostic category[58]; another study found that 46% of women hospitalized for psychosis had been victims of incest[59]; another study of child inpatients found that 77% of those who had been sexually abused were diagnosed with psychosis compared to only 10% of those who had not been so abused[60]; and yet another study found that 83% of men and women who were diagnosed with schizophrenia had suffered significant childhood sexual abuse, childhood physical abuse, and/or emotional neglect[61]. Bertram Karon, researcher and acclaimed psychosis psychotherapist, has found evidence of a significant correlation between the experience of intense feelings of loneliness and terror within childhood and the later onset of schizophrenia, a finding that is clearly closely related to the findings of these other studies[62].

After a thorough literature review of all of the research studies that have been conducted since 1872 that have looked at correlations between childhood abuse and psychosis and/or schizophrenia, John Read and his research team concluded that "child abuse is a causal factor for psychosis and 'schizophrenia,' and, more specifically, for hallucinations, particularly voices commenting and command hallucinations"[63]. This research clearly presents a potent challenge to the brain disease model. After all, if schizophrenia really is caused by a degenerative disease of the brain, then why would social factors such as childhood abuse and neglect so greatly increase one's likelihood of developing it?

A number of professionals have suggested that we can be open to the idea that schizophrenia has *both* a genetic and an environmental component if we consider that certain individuals are born with a genetic predisposition to schizophrenia, which may then require distressing environmental circumstances to actually bring the disease on. At first glance, this hypothesis does seem to hold some merit—after all, it's readily apparent that some individuals are simply more resilient in the face of distressing circumstances than others. And since we can often discern these differences in early infancy, it's quite plausible that differences in innate temperaments make some individuals more susceptible to being overwhelmed by distressing circumstances than others. An irony occurs when following this line of reasoning, however, in that it quickly precludes the need to incorporate a hypothetical brain disease at all. When taken through to its most logical conclusion, this line of reasoning suggests that psychosis is likely caused simply by overwhelming distress, and that different temperaments, different types and degrees of distress, and different developmental periods in which the distress occurs result in different likelihoods of developing a psychotic disorder. This is an idea we will take up further in Part Two and then frequently for the remainder of the book.

The effects of recreational drugs. Another line of research that has been presenting a serious challenge to the genetic theory of schizophrenia is that which has established a high correlation between the use of certain recreational drugs and the development of schizophrenia and closely related psychotic disorders. The correlation between marijuana use and the onset of psychosis/schizophrenia has been particularly robust[64], though there have also been some findings suggesting that the use of LSD, methamphetamines, and PCP[65] may also be correlated with the development of schizophrenia. Advocates of the hypothesis that schizophrenia is a genetically based brain disease have generally suggested that such drug use merely acts as a catalyst, bringing on the full force of the "disease" in those who are biologically predisposed to it. A common component of this line of reasoning is the suggestion that these individuals may merely be using the drug(s) in an attempt to "self medicate" the initial symptoms of the disease.

The latest research, however, has cast serious doubts on this hypothesis: First, a recent longitudinal study, which identified adolescents who regularly smoke marijuana and then tracked them for ten years, found that most of those who went on to develop "persistent psychotic symptoms" (in other words, those who essentially met the criteria for schizophrenia) did not report having any psychotic symptoms at all prior to the use of marijuana. The researchers concluded that these results essentially nullified the "self-medication" hypothesis[66]. Second, other studies found that childhood trauma/mistreatment[67] and/or urbanicity[68] work synergistically with marijuana use to increase the likelihood of the later development of schizophrenia.

This line of research presents us with several important implications: First, it reinforces the other environmental research in suggesting that environmental factors alone (e.g., child abuse/mistreatment, urbanicity, and/or recreational drug use) may be enough to bring on psychotic experiences that meet the criteria for schizophrenia. In other words, we see continuing evidence refuting the hypothesis that schizophrenia is a genetically based disease of the brain. A second implication we find in this research is that some individuals may indeed use marijuana (and/or other recreational drugs) to "self-medicate"; but rather than self-medicating any kind of biological disease, they may simply be trying to experience some relief from the distress caused by other factors. The tragic irony with this particular strategy, however, is that while drug use is often a very powerful way to experience temporary relief, it often causes greater distress in the long run, a result very similar to what we see with psychiatric drug use, as will be discussed in further detail in Chapter 5.

Does heredity imply a biological disease? Returning to look at the heredity research so commonly cited to support the brain disease theory, what we discover when we take a close look at this research is that it is surprisingly limited and potentially seriously flawed. And we find that the research supporting the environmental influence on the development of schizophrenia is significantly less limited and relatively robust in comparison. In spite of this, of the three major branches of the brain disease theory of schizophrenia—the biochemical imbalance theory, the abnormal brain structure theory, and the genetic theory—the genetic theory is arguably the one with the highest likelihood of receiving some degree of validation. As mentioned above, it seems somewhat plausible that we may discover some degree of genetic vulnerability to the development of schizophrenia/psychosis. And yet heredity is arguably the factor which is *least* likely to be indicative of biological disease. Those who support the theory that schizophrenia is the product of a brain disease often argue that if we do someday find significant evidence of a genetic link to schizophrenia, then this will confirm that it is in fact a biological disease, but is this really so?

There are many psychological characteristics that are genetically influenced but that are clearly not biological diseases. One obvious example is high intelligence, which is estimated to have genetic influence that is at least as strong as that reported in the schizophrenia studies[69], and yet it is doubtful that anyone would consider genius to be a biological disease. Another example is shyness, which has also been shown to have a genetic basis in some studies[70]. But again, would anyone really consider shyness to be a biological disease? Evidently, it is because the symptoms that we associate with the diagnosis of schizophrenia are generally considered so unusual and undesirable that we are so quick to make the assumption that any evidence of its heritability would automatically imply biological disease; yet, considering that psychological traits with heritable components such as shyness and intelligence are almost certainly not biological diseases, the assumption that schizophrenia can be proven to be a brain disease based merely on heritability is seriously flawed. In other words, if we ever do demonstrate with certainty that schizophrenia has a significant genetic component, this alone would not validate the claim that schizophrenia is the result of a brain disease.

Correlation is not Causation

A final important question to ask when examining the validity of the brain disease hypothesis is, how valid are the assumptions that gave rise to it? One of the most important assumptions generally held in this regard is that when and if it becomes established that there is some clear correlation between some anomalous feature of the brain and the characteristics we label as schizophrenia, then it will also be established that it is this brain anomaly that causes schizophrenic experiences. However, this assumption involves the confusion between correlation and causation.

As already mentioned, there is substantial evidence that environmental factors can greatly increase one's likelihood of developing schizophrenia and other psychotic conditions. And we have seen that environmental factors can also lead to both immediate and permanent changes to the brain (both biochemically and structurally)[71]. We also know that just as changes to the brain (such as injuries, strokes, and lesions) can affect mental functions, the reverse is also true: consciously directed mental functions can lead to direct changes in the brain. Daniel Siegel, a pioneering neuroscience researcher at UCLA, has conducted significant research on the issue of brain/mind causation, and he concluded:

> We can say that brain and mind *correlate* their functions, but we actually don't know the exact ways in which brain activity and mind function mutually *create* each other. It is too simplistic to say merely that the "brain creates the mind" as we now know that the mind can activate the brain . . . the mind can directly stimulate brain firing and ultimately change the structural connections in the brain [emphases added].[72]

The brain disease hypothesis has relied heavily upon this mistaking of correlation with causation, particularly maintaining the assumption that correlations between schizophrenia and brain anomalies (when and if they are found) imply that these brain anomalies must be the primary causative agent of such experiences. Yet, our understanding that the mind and the environment can affect the brain, combined with our understanding that unusual conditions with regard to any of these have shown some correlation with psychotic experiences, have seriously undermined the validity of this assumption.

PERHAPS IT'S TIME WE FINALLY LET GO OF THE BRAIN DISEASE THEORY

So, what exactly have we learned from all of this research? Of the thousands of research studies conducted over more than a hundred years, what can actually be said about this phenomenon we call schizophrenia? When we look to the highest authorities on the matter, we find confusing mixed messages. For example, the NIMH, on its Schizophrenia home page, proclaims confidently that "schizophrenia is a chronic, severe, and disabling brain disorder"[73], a statement you find on nearly every major page or publication they have put out on the topic; and yet if you spend a little more time looking through their literature, you will find that they admit that "the causes of schizophrenia are still unknown"[74]. Similarly, the APA also confidently proclaims that "schizophrenia is a chronic brain disorder"[75], but then they acknowledge on the very same page that "scientists do not yet know which factors produce the illness"[76], and that "the origin of schizophrenia has not been identified"[77]. The strong bias towards the brain disease theory is clearly evident in the literature of these and other similar organizations, and yet the message comes through loud and clear that we still do not know the cause of schizophrenia. Even the U.S. Surgeon General began his report on the etiology of schizophrenia with the words, "The cause of schizophrenia has not yet been determined"[73]. It would appear, then, that it is simply not appropriate to claim with such confidence that schizophrenia is a brain disorder.

Considering the vast amount of research on the subject, it's truly impressive how few solid conclusions we have been able to draw about schizophrenia, and perhaps it is this lack of conclusiveness that is the most revealing thing of all. Many researchers have concluded that the lack of significant evidence in all of this research ironically provides us with highly significant evidence that schizophrenia is most likely *not* a disease of the brain[74]. These researchers have pointed out the flaw in the assumption that so-called abnormal behavior and perceptions must imply brain disease. Psychiatrist Peter Breggin, longtime critic of the medical model, offered the following analogy in this regard:

> To claim that an irrational or emotionally distressed state, however extreme, in itself amounts to impaired brain function is simply false. An analogy to television sets and computers may illustrate why this is so. If a TV program or Internet site is offensive or irrational, it does not indicate that anything is wrong with the electronics of the television set or the hardware of the computer. It makes no sense to attribute the bad programming or the offending Internet site to bad wiring. Similarly, a person can be very disturbed psychologically, without any corresponding defect in the wiring of the brain.[75]

When we consider the enormity of the effort to substantiate the brain disease theory of schizophrenia combined with the continued lack of support for it, it's difficult not to draw the same conclusion as Breggin and the others who share his perspective: The brain disease hypothesis must be seriously flawed.

Chapter 3

Myth #2: "Schizophrenia" is a Valid Construct

As we have seen, there is actually very little evidence that schizophrenia is the result of a biological disease, in spite of the pervasiveness of this myth held within our society. There are a number of researchers who have even taken this argument one step further, suggesting that schizophrenia may not even be a valid construct at all. There are three major elements to this argument: (1) all of the various major psychotic disorders may simply be variations of one phenomenon; (2) there may be no clear boundaries between psychosis itself and what we think of as sanity; and (3) it may simply be impossible to ever arrive at universally agreed upon definitions of schizophrenia, psychosis, or even of "madness" itself.

Using a Continuum Instead of Categories

The British Psychological Society (the BPS, Great Britain's counterpart to the American Psychological Association), in its official report summarizing their understanding of "mental illness" and "psychotic experiences," concluded that the research suggests that there are not actually clear boundaries between the major psychotic disorders (schizophrenia, schizoaffective disorder, and bipolar disorder):

> [One] way of examining the validity of diagnostic categories involves using statistical techniques to investigate whether people's psychotic experiences actually do cluster together in the way predicted by the diagnostic approach.

> The results of this research have not generally supported the validity of distinct diagnostic categories. For example, the correlation amongst psychotic symptoms has been found to be no greater than if the symptoms are put together randomly. Similarly, cluster analysis—a statistical technique for assigning people to groups according to particular characteristics—has shown that the majority of psychiatric patients would not be assigned to any recognisable [sic] diagnostic group. Statistical techniques have also highlighted the extensive overlap between those diagnosed with schizophrenia and those diagnosed as having major affective disorder.[1]

To put this in layperson's terms, the BPS has suggested that the various psychotic disorders are most likely not discrete entities at all, but may more appropriately be classified as variations of one phenomenon, a phenomenon that many have suggested we refer to simply as *madness.*

The BPS has taken these conclusions one step further, suggesting that "mental health and 'mental illness' . . . shade into each other and are not separate categories"[2]. In other words, they suggest that not only are the various psychotic disorders best understood as merely representing different points on a continuum of a single phenomenon (see the top of Figure 3.1), but that sanity and madness themselves are also best understood as merely being different points along a single continuum (see the bottom of Figure 3.1). They cite evidence suggesting that psychotic experiences are merely extreme expressions of more ordinary traits found within the general population.

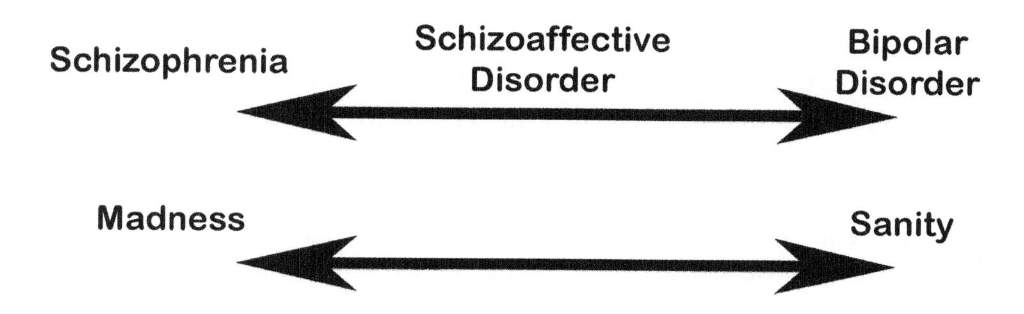

Figure 3.1. The psychotic disorders and sanity/madness as continuums.

SCHIZOPHRENIA AND *MADNESS* AS ESSENTIALLY CONTESTED TOPICS

After conducting a thorough grounded theory study using participants who have been diagnosed with schizophrenia, Jim Geekie and John Read (researchers who also have extensive clinical experience with this population) came to the conclusion that the most accurate way to define schizophrenia is as an *essentially contested topic*[3]. What they mean by this is that "madness is, quite simply, something about which we argue the meaning of, inevitably, and interminably"[4]. In other words, arriving at a universally agreed upon definition of schizophrenia in particular, or madness in general, is simply not possible:

> Even in mainstream psychiatric textbooks, the literature on schizophrenia is characterized by a confusing plurality of theories, each competing for dominance. . . . We see then that there is consensus about one thing: that schizophrenia has been, and continues to be, subject to a wide range of explanations.[5]

Geekie and Read suggested that a major reason that this debate is still as heated and unresolved as ever is that the terms schizophrenia, psychosis, and madness have been mistakenly perceived as pointing to some objectively verifiable entity, when they may merely be placeholders "in an ongoing debate that we have about who and what these terms actually refer to"[6]. In other words, Geekie and Read suggested that these terms do not refer to objective entities but rather to subjectively and socially constructed categories, and they are therefore just as prone to pluralistic interpretations and flux as are other subjective and social phenomena.

The BPS has recognized this problem to some degree and has attempted to mitigate it by suggesting that the term schizophrenia be dropped altogether and replaced with the more inclusive term psychosis[7]. However, Geekie and Read made the point that the term psychosis has been and continues to be subject to the same problems that have plagued the term schizophrenia in this regard, so they have argued that it would be more helpful to acknowledge that this is an essentially contested topic, to honor the role of pluralism when addressing it, and to devote most of our resources to determining what actually contributes to the wellbeing of those who are suffering from these kinds of experiences, regardless of what we choose to call them.

Chapter 4

Myth #3: People Cannot Fully Recover from Schizophrenia

Considering, then, that the etiology of schizophrenia is still unknown and that even the validity of the very concept of schizophrenia is questionable, how do we explore the topic of recovery from schizophrenia? Whether or not schizophrenia is a valid concept, it is clearly evident that many people do suffer from distressing anomalous experiences, and when such suffering becomes relatively chronic, these individuals will most likely be diagnosed with schizophrenia (or another major psychotic disorder, as discussed earlier). Therefore, when we look at the research on recovery from schizophrenia, while we cannot say with any certainty that there is any biological disease from which these participants are recovering, we can say with some degree of confidence that these participants have been suffering from long-term distressing anomalous experiences, and we can explore the issue of recovery from within this context.

While there continues to be the widespread belief in our society that people diagnosed with schizophrenia generally do not recover, the actual research tells a very different story. Table 4.1 provides a list of all of the major longitudinal recovery studies of at least 15 years duration that I was able to locate.

Going into the details of all of these studies would be quite lengthy and fall outside the scope of our discussion here; we will, however, look more closely at some of these studies in a later chapter. For now, there are several key points that are worth mentioning:

First, each study uses somewhat different criteria for determining what is meant by "significantly improved" and "fully recovered," and some have slightly different terminology to represent these classifications, yet they all essentially agree that fully recovered refers to participants being asymptomatic and self-sufficient

Table 4.1 15+ Year Longitudinal Schizophrenia Recovery Studies

STUDY	n*	AVERAGE FOLLOW-UP (YEARS)	RECOVERED OR IMPROVED	FULLY RECOVERED
The Burgholzli study (Bleuler, 1974)	208	23	53%	20%
The Iowa 500 study (Tsuang & Winokur, 1975)	186	35	46%	20%
The Bonn Study (Huber et al., 1975)	502	22.4	65%	22%
Lausanne study (Ciompi, 1980)	289	37	49%	27%
Chestnut Lodge study (McGlashan et al., 1984a, 1984b)	446	15	36%	not mentioned
The Japanese study (Ogawa et al., 1987)	105	21-27	77%	31%
The Vermont study (Harding et al., 1987)	269	32	68%	45%
The Cologne study (Marneros et al., 1989)	148	25	58%	7%
The Maine sample (DeSisto et al., 1995)	269	36	49%	not mentioned
The Dutch study (Wiersma et al., 1998)	82	15	77%	27%
WHO International Study —incidence cohort —prevalence cohort (Hopper et al., 2007)	 502 142	 15 25	 67% 63%	 48% 54%
The Chicago Study —off antipsychotics —on antipsychotics (Harrow & Jobe, 2007)	 25 39	15	 84% 51%	 44% 5%

* "n" is a common abbreviation in research studies for "number of participants."

in meeting their needs, both socially and financially, for some specified period of time.

Second, the finding that recovery rates are quite high is surprisingly robust. The authors of the largest such series of studies—the World Health Organization (WHO) studies—have concluded that the "overarching message [is that] schizophrenia is largely an episodic disorder with a rather favorable outcome for a significant proportion of those afflicted"[1]. Note also that while there is significant variation in the results of these studies, there is a general pattern that is somewhat consistent across these studies: Generally one half to two thirds of the participants in these studies have significantly improved over the long term, generally about a quarter of the participants are rated as having fully recovered, and generally less than a quarter remain permanently disabled. It is also interesting to note that many of the participants in these studies who have recovered were those who were considered to be the most profoundly disturbed[2]. Returning to the brain disease hypothesis for schizophrenia, it is illuminating to compare the high recovery rate for schizophrenia with the recovery rate for well-established diseases of the brain such as Parkinson's, Alzheimer's, or multiple sclerosis: There is no documented evidence of even a single individual making a full recovery from any of these well-established diseases of the brain[3]. Again, we find compelling evidence that schizophrenia is simply not a disease of the brain.

Finally, several of these studies have provided data that allow us to directly compare the outcomes of participants using the Western standard treatment for schizophrenia (typically the use of antipsychotics) with the outcomes for participants not using this treatment, and the findings have reliably been strongly in favor of those *not* using standard Western psychiatric treatment, something that is likely to come as quite a surprise to many.

Chapter 5

Myth #4: Mainstream Psychiatric Treatment Greatly Increases Beneficial Outcomes

Most of the research on schizophrenia has revolved around attempts to understand the etiology of schizophrenia and attempts to specifically define it. The result of over a century of these lines of inquiry have left us with little more than a greater appreciation of the mystery of schizophrenia and psychosis, and ongoing controversy over theory and semantics. But if our understanding of schizophrenia and the psychotic disorders has advanced remarkably little in all this time, then what can we say about our ability to support those suffering from these conditions? The myth that currently prevails in our society is that schizophrenia is a brain disease and that major ongoing advances have been and continue to be made in regards to the treatment of this disease. If, however, the research reveals that the first part of this myth (that schizophrenia is a brain disease) actually has very poor validity, then what does this say about the second part of this myth, that the treatment of schizophrenia has been steadily improving? The results of the research in this regard, it turns out, are in very close alignment with the research regarding the brain disease hypothesis—the evidence suggests that the effectiveness of the standard treatment of schizophrenia in the West is generally very poor.

International recovery studies conducted by the World Health Organization have revealed that those diagnosed with schizophrenia in the United States and other so called "developed" countries fare significantly worse than those in the poorest countries of the world[1]. Other reviews suggest that patients diagnosed with schizophrenia in the United States today may fare even worse today than did asylum patients in the early nineteenth century[2]. Why is this happening?

Western psychiatry claims to have made many advances in the treatment of the psychotic disorders, so why do we find these extremely poor results? In order to address this question, we need to look more closely at the research that assesses the harms and benefits of the standard Western treatment for schizophrenia, a treatment paradigm often referred to as biological psychiatry.

Biological psychiatry, the standard treatment for schizophrenia and the other psychotic disorders in practice today in most so-called developed countries, is based directly on the medical model theory of schizophrenia, and it currently consists primarily of hospitalization of the sufferer and manipulation of the brain, typically via the use of psychotropic drugs (especially antipsychotics, which are often combined with other classes of psychotropic drugs) and occasionally electroconvulsive shock therapy (ECT). Sometimes, patients seek this treatment voluntarily; at other times, the patients are coerced to "comply" with this treatment in various ways[3]; and quite often, this treatment is forced on patients against their will[4]. Typically, patients who receive such treatment are told that they have a brain disease (despite the lack of evidence) and that they will most likely have to remain on these debilitating drugs for the remainder of their lives[5], again despite significant evidence to the contrary[6]. The possibility of full recovery is rarely mentioned by advocates of this model, presumably because it is believed that these patients suffer from a degenerative brain disease in which genuine recovery is essentially impossible.

Because of the belief that genuine recovery is virtually impossible, the emphasis in this model is generally on symptom management via the use of highly debilitating antipsychotic drugs[7]; and the use of such debilitating drugs is justified since there is virtually no hope of genuine recovery anyway. This circular reasoning is likely to lead to a self-fulfilling prophecy of "no real recovery," which is something that we'll look at more closely later. The primary benefits typically espoused by the advocates of this treatment model are (1) that such treatment significantly reduces the symptoms and distress associated with this "disease"[8], and (2) that the patients so treated require less hospitalization[9].

But are these assertions really true? Does the research really support this widely held assumption that biological psychiatry improves the outcomes of those diagnosed with schizophrenia and other psychotic disorders? In order to assess the validity of this assumption, there are three major questions that we need to ask: (1) What are the actual harms and benefits of antipsychotic drug use—the primary treatment modality of biological psychiatry? (2) What are the effects of other aspects of this treatment paradigm? And (3) what does the research tell us about the outcomes of alternative treatment modalities? Let's turn now to look more closely at each of these questions in turn.

The Harms and Benefits of Antipsychotic Drug Use

Advocates of the medical model often claim that antipsychotics, by reducing symptoms and hospitalizations, have made community care possible, and they frequently refer to the reduction in the number of occupied psychiatric hospital beds that have apparently corresponded with the introduction of antipsychotics as evidence of this claim[10]. The actual research, however, tells a very different story. For many of the countries using psychiatric treatment at that time, the number of hospital residents actually increased significantly after the introduction of antipsychotics, and for those countries in which there was a decline, the decline generally began earlier and could be explained by other factors, usually economic—many of these societies simply could no longer afford to maintain so many hospital residents[11].

Because of the many different factors that affect the number of hospital residents at a given time, attempting to determine the effectiveness of treatment solely by looking at correlations with the number of hospital residents offers us little more than grounds for speculation. Fortunately, there has been a long line of research, extending back nearly 50 years, that addresses this issue more directly, and even though this research is relatively scarce (due primarily to ethical concerns), the findings are surprisingly reliable in showing that while the use of antipsychotics may provide some benefit in the short term, they are actually likely to increase the chronicity of psychosis over the long term[12].

The NIMH studies. In 1964, the NIMH conducted a study that followed 344 patients diagnosed with schizophrenia for six weeks[13]. The participants were divided into two groups—an experimental group that received antipsychotics and a control group that received a placebo. After six weeks, the members of the experimental group were faring significantly better, so it first appeared that the antipsychotics were really helping these participants. However, the one year follow-up assessment revealed that the members of the group who had been taking antipsychotics had a significantly higher relapse rate: "patients who received placebo treatment in the drug study were *less* likely [author's emphasis] to be rehospitalized than those who received any of the three active phenothiazines (thioridazine (Mellaril), fluphenazine (Prolixin), and chlorpromazine (Thorazine))"[14].

During the 1970s, the NIMH conducted three more similar studies, comparing antipsychotic treatment with different types of environmentally/socially oriented treatment that minimized the use of psychiatric drugs[15]. In all three studies, the members of the groups receiving antipsychotic treatment fared significantly worse. Carpenter and his research team concluded that "antipsychotic medication may make some schizophrenic patients more vulnerable to future relapse than would be the case in the natural course of the illness"[16].

The Agnews Hospital study. In 1978, a research group headed by Rappaport at the University of California in San Francisco conducted one of the most significant studies ever performed that directly compared antipsychotic use with a placebo. The study was a 3-year longitudinal study in which 80 male participants who had recently been diagnosed with schizophrenia were divided into 4 groups—those who remained on chlorpromazine (a typical antipsychotic) both in the hospital and afterwards (Antipsychotic/On); those who had been on chlorpromazine while in the hospital but were not "medication compliant" afterwards (Antipsychotic/Off); those who had been on a placebo while in the hospital and began taking chlorpromazine afterwards (Placebo/On); and those who began on placebo and remained off antipsychotics during the entire duration of the study (Placebo/Off)[17]. All participants remained in the hospital for an average of six weeks, and then were assessed at regular intervals over the following 3 years. They were assessed for *Severity of Illness* and *Change of Clinical Status* over time. Figure 5.1 shows the results obtained upon the final follow-up.

Rappaport found that after the first six weeks, those on the antipsychotics were faring significantly better than those on the placebo, a finding that matched similar findings in other studies. However, within a short time, those not taking the antipsychotics began to do significantly better, and after 3 years, the members of the group who had never been on antipsychotics (Placebo/Off) were faring by far the best; and the two groups still taking antipsychotics at the end of the study (Placebo/On and Antipsychotic/On) were clearly faring the worst. The group who had never taken antipsychotics had just *one-eighth* the percentage of rehospitalizations (only 2 of the 24 participants in this group during the entire 3 years) as the group who remained medication compliant for the full duration of the study (16 of the 22 participants in this group were rehospitalized), a truly striking finding. The Antipsychotic/Off group participants had begun antipsychotics in the hospital and were expected to remain on the drugs afterwards, but failed to be compliant to differing degrees, which could explain their relatively poor outcome*; however, they still demonstrated better outcomes on all measures than either group that had remained fully compliant with the use of antipsychotics.

---

* The sudden and/or erratic starting and stopping of antipsychotic drug use has been determined to be particularly problematic—more on this later.

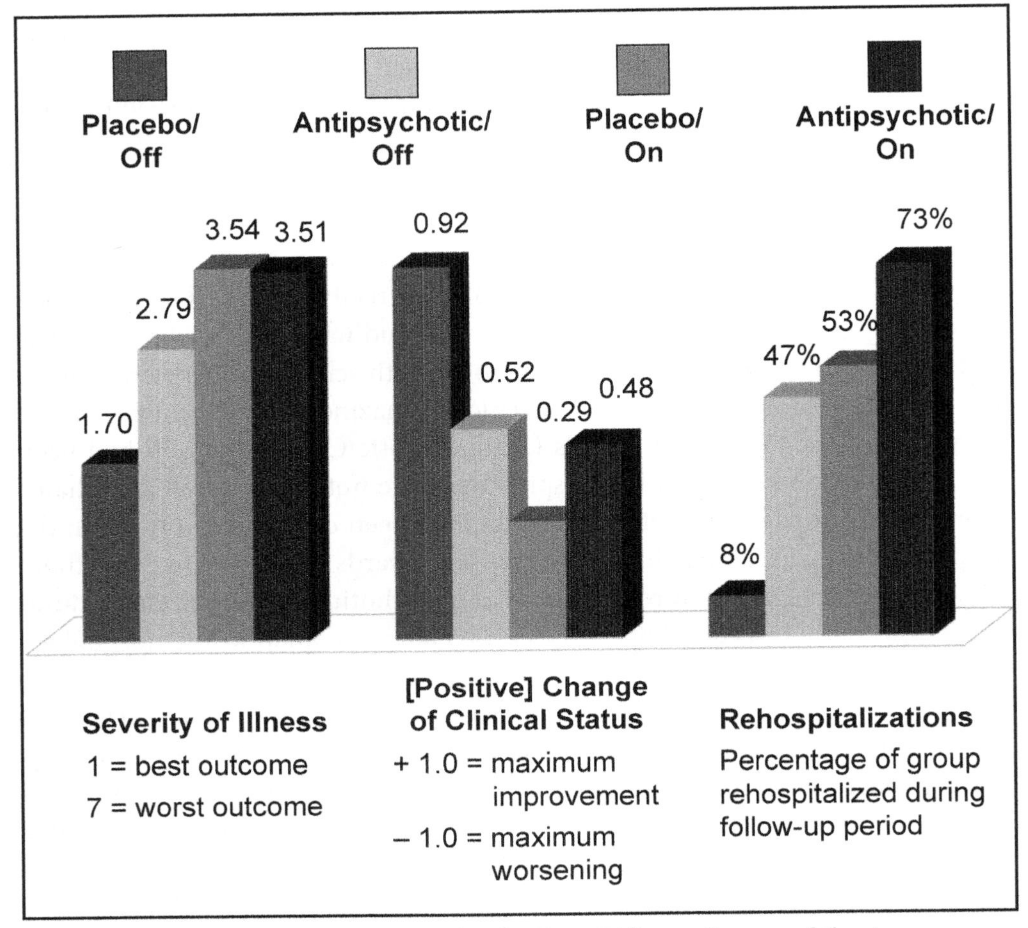

Figure 5.1. 3-Year Outcome Measures for the Four Different Groups of the Agnews Study.

Rappaport concluded that those who had never received antipsychotics "showed greater clinical improvement and less pathology at follow-up, fewer rehospitalizations and less overall functional disturbance in the community than the other groups of patients"[18]. This conclusion in itself is quite striking given its stark contrast to the assumptions generally held about the benefits of antipsychotic treatment; yet it is all the more striking when we consider that this study carries particularly high validity for such a study since it is one of the only such studies to use randomized *neuroleptic-naïve* participants (those who had never been exposed to antipsychotics), a method that has rarely been used since it has generally been considered unethical to do so[19,*].

* This is a truly ironic concern given that the findings of this study suggest it may be unethical *not* to consider avoiding the use of antipsychotics for many of these individuals.

The Vermont study. The Vermont Longitudinal Research Project, headed by Courtney Harding, was a schizophrenia recovery study that offers tremendous hope[20]. This study consisted of discharging 269 of the most chronically psychotic patients from the back wards of the Vermont State Hospital, placing them in the community with support in place, and then following them for an average of 32 years. The participants of this cohort had been suffering from schizophrenia for an average of 16 years, had been completely disabled for an average of 10 years, and were considered to be some of the most hopeless cases in the hospital. Upon the final follow up (an average of 32 years later), 68% of the 262 remaining alive were considered significantly improved, and nearly one half (45%) were completely asymptomatic and considered to be fully recovered. In an interview in the American Psychological Association journal, *The Monitor on Psychology*, Harding said that all participants who had recovered in this study had one thing in common: they all "had long since stopped taking medications"[21].

The Chicago study. The most recent schizophrenia recovery study conducted by the NIMH is an ongoing longitudinal study headed by Martin Harrow that began in 1992 and continues to the present day. At the 15-year point, the data was officially released in a peer-reviewed journal, and the results are quite striking. When comparing the outcomes of the group of participants who chose to remain on the antipsychotics with the group who had chosen to stop taking them, we find that the group who stopped taking them fared far better[22]. They had nearly *eight times* the recovery rate and only one-third the percentage of those considered to have a "uniformly poor" outcome (see Figure 5.2).

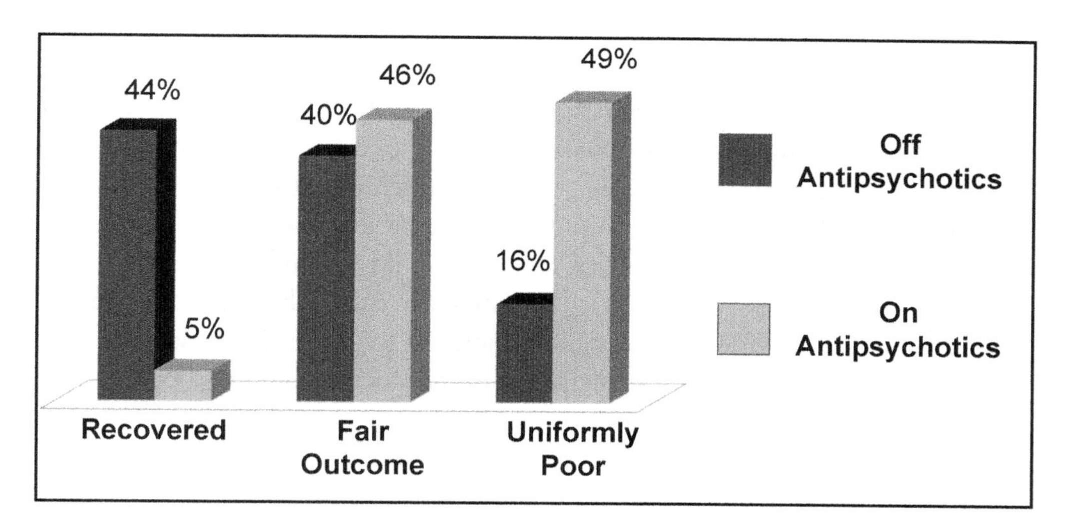

Figure 5.2. 15-Year Outcome Measures for the Two Schizophrenia Groups in the NIMH Chicago study.

It's important to note that this is a *naturalistic* study rather than a controlled experimental study. All of the participants were initially diagnosed with either schizophrenia or a milder psychotic disorder (the latter is not included in Figure 5.2, though the profile is very similar, with those coming off the drugs faring significantly better). They all had initially received conventional treatment and were then followed as they went about their lives in their own way with no direct intervention from the researchers except for infrequent follow-ups. Because of this, some people (including the researchers themselves) have expressed reluctance to believe that the use of the antipsychotics could really be as harmful as the data suggests. They argue that the reason those who stopped taking antipsychotics have fared so much better over time is because they were already doing much better initially, which is why they felt they could stop taking the antipsychotics. In other words, they argue that the group who remained on antipsychotics had much poorer recovery rates not because they stayed on the drugs, but because they were much more "sick" from the very beginning.

At first glance, this argument appears somewhat plausible. Upon closer inspection of the data, however, we find that it has one very serious problem: Harrow's participants had been divided into a number of different subgroups during statistical analysis, and in every subgroup division, regardless of whether the particular subgroup contained participants with a good initial prognosis or a poor one, those who got off the drugs consistently showed significantly better outcomes. This finding has been even further corroborated in Harrow's most recent 20-year follow-up in 2012, in which the group was divided statistically into still further permutations[23]. Again, in every case, regardless of how the group was divided, those who got off the antipsychotics showed significantly better outcomes.

There is one finding that has emerged from this research that is perhaps the most revealing of all. As mentioned above, this study included a second group of participants whose symptoms were not strong enough to qualify them for a diagnosis of schizophrenia. Because of their milder symptoms, their prognosis for recovery was considered to be substantially better than that of the schizophrenia group. It turns out, however, that those diagnosed with schizophrenia who got off the antipsychotics fared significantly *better* than those diagnosed with a milder disorder who remained on them. In other words, those with a significantly *worse* prognosis at the beginning of the study who then stopped taking antipsychotics demonstrated significantly higher recovery rates than those with a significantly *better* initial prognosis who remained on the drugs. The most likely conclusion, then, can hardly be more clear, and as would be expected, is in very close alignment with the conclusions of the other recovery studies: long-term antipsychotic use significantly reduces the likelihood of recovery.

Why the increased chronicity of psychosis from antipsychotics? Chouinard and his research team[24] were among the first to suggest that the reason these studies were showing a higher rate of relapse for those taking antipsychotics is that antipsychotics affect the brain in ways that make the individual more vulnerable to psychosis. Chouinard posited that the brain attempts to compensate for the effect of these drugs by increasing the number of dopamine receptors, which subsequently increases the individual's susceptibility to psychosis (a condition often referred to as *tardive psychosis* or *supersensitivity psychosis*—psychosis caused as a direct result of using the drugs), a hypothesis that still holds weight today[25]. A closely related hypothesis has been generated and validated with the invention of the MRI. It was discovered that antipsychotic drug use (both of the older *typicals* and the newer *atypicals*) causes atrophy of the cerebral cortex and enlargement of the basal ganglia[26], changes which have been demonstrated to increase the severity of both positive and negative psychotic symptoms[27]. So while antipsychotics may be useful to some individuals in reducing their symptoms during the initial weeks of acute psychosis, the evidence has been surprisingly robust in demonstrating that the long-term use of antipsychotics significantly increases the likelihood of developing a chronic psychotic disorder.

Other severe side effects of antipsychotics. Antipsychotic drug use has not only been shown to increase the likelihood of chronic psychosis, but it has also been shown to frequently cause a host of other physical, emotional, and cognitive problems that have the potential to be severely debilitating and may even lead to a significantly shortened life span.

Tardive dyskinesia. Tardive dyskinesia is a disorder of the voluntary nervous system that results from permanent damage to the basal ganglia, a region of the brain important for motor control[28]. The resulting symptoms are uncontrollable movements of the tongue and other parts of the body, which result in difficulties with speaking, eating, walking, and even sitting still. It is estimated that about 5% of patients taking antipsychotics develop this disorder within the first year, with an additional 5% of patients developing it with every subsequent year[29].

Akathisia. Akathisia is a term used to describe the condition in which one feels overwhelming agitation and restlessness on the inside while feeling trapped in a body that is heavily sedated and unresponsive, a common side effect of antipsychotic drug use. Those who have experienced akathisia often describe it as the most severe torment, the severity of which is virtually impossible for those who have not experienced it to fully grasp. Research has shown high correlations between akathisia and suicidality, homicidality, and other violent behavior[30].

Cognitive impairment. Research has found that antipsychotics typically cause a significant level of cognitive impairment, especially regarding the capacity to learn, retain information, and perform executive functions such as problem solving and planning[31].

Emotional impairment. It has long been recognized that antipsychotics cause emotional deadening, which on one hand is most likely the primary effect that reduces distressing emotions associated with psychosis, but on the other hand often results in a lack of joy and sense of meaning in life[32]. In a recent qualitative study in which 28 participants who were self-identified as recovered were interviewed, the authors concluded that "by far the most common and disturbing side effect was a chronic sense of 'numbness' and/or lack of emotion, associated particularly with the use of antipsychotics"[33]. These participants described the antipsychotics as resulting in "not feeling anything," "feeling isolated," "being completely numb," and feeling like "a complete, drooling zombie"[34].

Suicidality. Recent research suggests that schizophrenia patients who are given antipsychotic treatment today commit suicide at a rate *twenty times higher* than that of schizophrenia patients prior to the introduction of antipsychotic treatment[35]. This is an astonishing figure, and it is important to keep in mind that this represents a correlation, and not necessarily causation; nonetheless, given the magnitude of this correlation, it would be difficult to deny that antipsychotics almost certainly play a significant role in this greatly increased suicide rate.

Other physical health problems. A host of other serious physical problems has been associated with antipsychotic treatment, including diabetes, obesity, agranulocytosis (the potentially fatal loss of white blood cells), neuroleptic malignant syndrome (a life threatening neurological disorder), blindness, seizures, arrhythmia, fatal blood clots, heat stroke, swollen breasts, leaking breasts, sexual dysfunction, and debilitating skin rashes[36].

A significantly shortened life span. Recent research has revealed that "people with serious mental illness (SMI) [consisting primarily of those diagnosed with a psychotic disorder] die, on average, 25 years earlier than the general population"[37]. The authors of this research have concluded that a number of the risk factors leading to this extraordinary shortening of one's lifespan—particularly obesity and diabetes—may be directly attributed to the use of antipsychotic drugs. It is likely that antipsychotics also contribute to a shortened lifespan by increasing the likelihood of suicide, as mentioned above.

Aren't the newer atypical antipsychotics better? In the early '90s, the second generation of antipsychotics—the *atypical* antipsychotics—was introduced to the market. The atypicals target a somewhat different pattern of neuroreceptors than the typicals, and there was hope that these newer drugs would offer improved symptom management with fewer side effects. As the research has been accumulating, however, it appears that this hope has not panned out. While the side effect profile is somewhat different, it seems that the overall risk is relatively similar, and some argue that the overall risk of the atypicals may even be worse[38].

In 1994, risperidone (Risperdal) was the first atypical antipsychotic to be released in the U.S. as a first-line agent in the treatment of schizophrenia, and it is still one of the most widely used atypicals today[39]. Upon approval, the Food and Drug Administration (FDA) concluded that risperidone was neither more effective nor safer than any of the older "typical" antipsychotics:

> We would consider any advertisement or promotion labeling for RISPERDAL false, misleading, or lacking fair balance under section 501(a) or 501(n) of the Act if there is presentation of data that conveys the impression that risperidone is superior to haloperidol [a popular typical antipsychotic] or any other marketed antipsychotic drug product with regard to safety or effectiveness.[40]

Later, researchers concluded that risperidone had a higher incidence of Parkinsonian symptoms and akathisia than haloperidol[41]; and Mattes, director of the Psychopharmacology Research Association, concluded that risperidone may actually be less effective than the typical antipsychotics for managing positive* psychotic symptoms[42].

As other atypical antipsychotics entered the market, the research demonstrated that they performed similarly poorly. In 2000, a group of researchers at Oxford University conducted a meta-analysis of 52 studies, including a total of 12,649 patients and a variety of atypical antipsychotics, and the authors concluded: "There is no clear evidence that atypicals are more effective or are better tolerated than conventional antipsychotics"[43]. In 2005, the NIMH concluded that the atypical antipsychotics performed no better than the typicals regarding their efficacy or the likelihood that patients would tolerate them[44]; and in 2007, a British government study concluded that patients using atypical antipsychotics had

---

* "Positive" psychotic symptoms refer to anomalous sensory perceptions and beliefs (usually referred to in the field as "hallucinations" and "delusions." This is in contrast to "negative" symptoms, such as the significant lack of expressed emotion, speech, or motivation (referred to in the field as "affective flattening," "alogia," and "avolition," respectively).

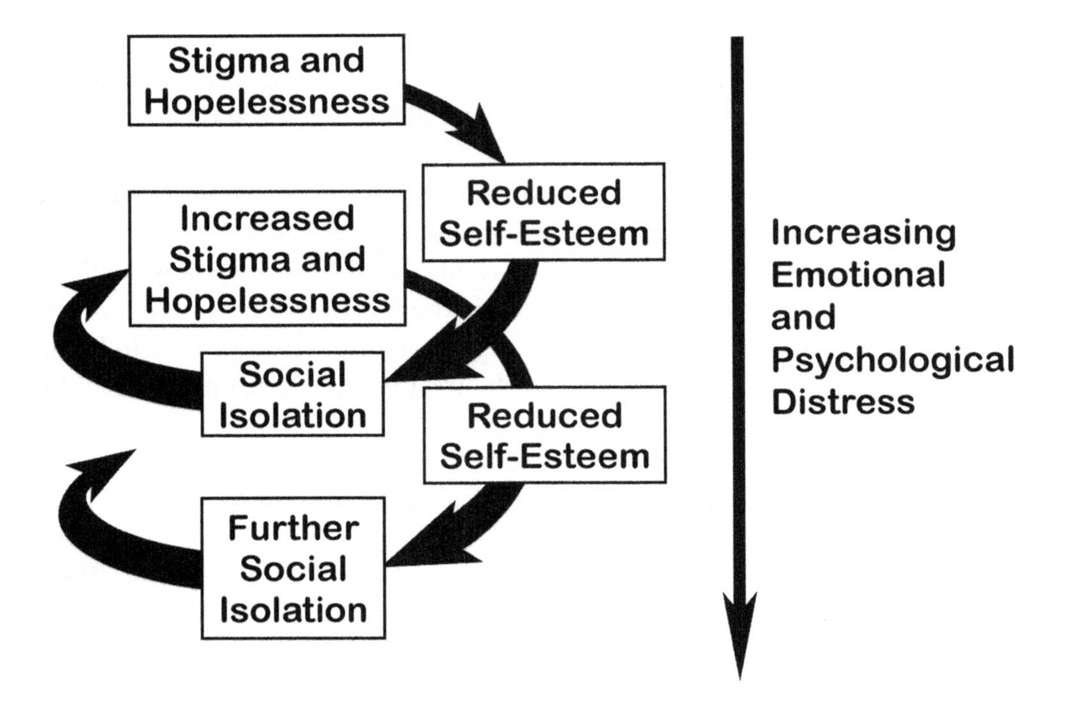

Figure 5.3. The downward spiral of stigma and hopelessness.

a worse quality of life than those who were using the older typical antipsychotics, and this in spite of the fact that this same group of researchers considered patients on the older antipsychotics to generally have a very poor quality of life[45].

Finally, the research suggests that the atypical antipsychotics may cause more adverse physical ailments than the typicals. While they do seem to have less impact on the dopaminergic pathways of the brain than the typicals, they affect other pathways (especially those that use serotonin and glutamate) in ways that the typicals do not, and this may lead to more serious physical health problems (especially agranulocytosis, metabolic dysfunction, obesity, and diabetes) with a correspondingly higher likelihood of early death[46].

Other Harmful Effects of the Medical Model Treatment Paradigm

Besides the use of debilitating drugs with severe side effects, it appears that there are other unfortunate consequences of applying the medical model paradigm of treatment to those suffering from extreme states of consciousness.

Generating an attitude of stigma and hopelessness. Perhaps one of the most harmful consequence of the medical model paradigm is the attitude of dehumanization, stigma (both from others and internalized), and hopelessness (also both from others and internalized) that arise when one is labeled with a "mental illness"[47]. It appears that there are several reasons that this occurs. First, in spite of claims to the contrary by mainstream psychiatry and the pharmaceutical industry, research has consistently shown that biological theories of psychosis and the very labeling of distressing or anomalous psychological conditions as "illness" are likely to increase the tendency for these individuals to be perceived by others as dangerous and unpredictable[48]. In other words, people generally feel significantly more threatened when they perceive someone's unusual behavior as arising from a diseased brain rather than from psychosocial factors.

Second, it has been demonstrated that such labeling based on medical model ideology can actually lead to a self-fulfilling prophecy. Markowitz has demonstrated in her research that certain negative stereotypes typically come with the label of "mental illness," and when one is so labeled, the individual is at risk of personally identifying with these stereotypes as well as trying to minimize being so stereotyped by others[49]. This often results in a self-fulfilling prophecy in which the individual begins to experience a reduction in self-esteem and confidence and then is likely to withdraw socially in order to avoid being rejected by others. This results in a reduction of friends and employment, which leads to further stress and reduced self-esteem, and a downward spiral towards ever increasing emotional and psychological distress (see Figure 5.3)[50].

Medical model treatment resulting in trauma. Another unfortunate consequence of the medical model treatment paradigm, one that almost certainly hinders recovery in the West, is that people diagnosed with schizophrenia and other psychotic disorders are often treated in a manner that is disempowering, frightening, and sometimes even downright violent. The following scenario may help to illustrate this point:

Imagine for a moment that you are struggling with a highly distressing extreme state of mind (in other words, psychosis). Upon seeking help (or perhaps having "help" forced upon you), it is possible that you may be physically restrained without first being given the opportunity to be really listened to, forced to take toxic and debilitating drugs, and have your freedoms and many of your rights taken away from you indefinitely[51]. This treatment clearly has the potential to result in feelings of terror, rage, and helplessness, the particular combination of experiences that is very likely to lead to trauma[52]. The challenges, unfortunately, do not end here. In addition to such traumatic treatment, it is very likely that you will be told that you have a brain disease, and that it will most likely last for your entire life with little hope of genuine recovery[53], even further exacerbating your feelings of helplessness and hopelessness. This kind of treatment may well lead directly to the development of posttraumatic stress symptoms[54], which will probably further compound the distress you were already experiencing prior to the "treatment." Now that you have been so labeled, because of the mental illness paradigm prevailing in Western society, it is likely that you will find yourself being stigmatized and seen by others as "crazy," and it is likely that you will internalize this stigma, seeing yourself as hopelessly damaged[55].

All of these factors so often exacerbated by the medical model paradigm—stigma, hopelessness, and trauma—are likely to collude with and reinforce the distress associated with psychosis, creating a downward spiral similar to that illustrated in Figure 5.3. In other words, when one is struggling with extreme states of mind, it is likely that the medical model paradigm within Western society actually exacerbates the distress, leading to an ever worsening spiral into distressing emotions, overwhelming experiences, social isolation, and intrapsychic conflict. Finally, in what is a tragic irony, the primary method that our society has ordained for dealing with this horrific dilemma is to actually *create* genuine brain disease (through brain-damaging drugs, electroshock therapy, and/or other similarly harmful means) in an attempt to numb these individuals from their unbearable suffering.

Comparing the Outcomes between Medical Model Treatment and Alternative Treatment Modalities

So, in spite of the fact that the medical model paradigm of psychosis remains unsubstantiated, and in spite of the fact that we have seen the potential for tremendous harm caused by medical model treatment, we must not forget that there is still the potential for some limited benefit from such treatment—specifically, if medications are offered in a way that does not cause further trauma or disempowerment, they have been shown to be effective in the short term for many individuals in reducing the distress caused by positive psychotic symptoms (anomalous beliefs and sensory perceptions). When looking at the sum total of this research, however, it is quite clear that this treatment frequently causes more harm than benefit.

But what else is there? Some of these people are clearly very lost and distressed. How can we ease their distress and also ensure their safety and the safety of those around them? As it turns out, we have significant research showing that there are a number of alternative treatment options that are likely to be much more supportive and much less harmful than medical model treatment. But before we look at this research, it will help if we first turn once again to an imaginary scenario, this time one that will help to illustrate the core principles that typically accompany the most successful of these alternative treatment modalities.

Imagine once again that you are struggling with highly distressing mental experiences—unusual perceptions and/or beliefs that others around you do not share. This time, however, you live in a society or a community that validates your experience. Your beliefs may be challenged, but not your underlying experience. You will not be locked up against your will or forced to ingest debilitating drugs. You will not be told that you have a diseased brain with no hope of real recovery, but rather, there is the assumption that you *will* recover, and there is even the assumption that your experiences may eventually allow you to contribute to your community in a unique and powerful way. Your needs for choice, dignity, and respect will be held—your mind, body, and spirit will not be invaded. You find that people listen to your suffering with empathy and compassion rather than fear and judgment.

It stands to reason that, given such an environment, distressing emotions that typically accompany psychosis (terror, rage, despair, and confusion) will have a difficult time becoming firmly established, and your distress will most likely not be further exacerbated. As Westerners, we may have a difficult time believing that such a response to someone suffering with such extreme states of mind could be possible, but this is actually much closer to the paradigm found in collectivistic societies (societies that are closer to their indigenous roots)[56], and it has been found that people diagnosed with schizophrenia and other psychotic disorders in such societies have much better rates of recovery than those diagnosed with similar disorders in the West[57].

The World Health Organization (WHO) studies. The WHO studies, including the 15-year and 25-year longitudinal studies listed earlier in Table 4.1, as well as the 2-year and 5-year studies which preceded them, have resulted in a very interesting conclusion with very high reliability (remaining highly significant across all four of these studies): Those suffering from schizophrenia in so-called developed countries have far worse outcomes than those who live in so-called developing countries[58]. When the 13 countries that participated in these studies were divided into two categories of developed countries (such as the U.S. and Russia) and developing countries (such as India, Nigeria, and Colombia), the residents of the developing countries fared significantly better, regardless of the criteria used to determine outcomes. For example, when looking at the participants who were assigned a Global Assessment Functioning (GAF) of greater than 60 at the end of the study, 70% of those in developing countries satisfied this criterion, as compared to only 43% of those in the developed countries[59].

It is also of great interest to note that the use of antipsychotic drugs (and presumably the medical model paradigm of treatment in general) is *inversely* correlated with good recovery outcomes: a far higher percentage of those in the developed countries were maintained on antipsychotics than were those in the developing countries[60]. In other words, the countries with the highest percentage of residents using antipsychotic drugs showed the *worst* recovery rates; and the countries with the lowest percentage of residents using antipsychotics showed the *best* recovery rates. It is particularly interesting to look at the statistics of the group who arguably showed the overall best results after all final follow-ups were taken into consideration—those of rural Chandigarh, India. They showed both the highest rate of complete remission, with 71% of them having shown no psychotic symptoms within at least 2 years prior to the final follow-up, and the highest overall assessment scores, with 79% of them being assessed as having *good* to *very good* overall functioning[61]. It turns out that they were also the group with the lowest percentage of participants maintained on antipsychotic drugs, with only 5% of them having been regularly maintained on antipsychotics[62].

While there are likely a number of factors that play a role in the large disparity in outcomes between developed and developing countries (for example, the strength and stability of family and community structures), there can be little doubt of the significance of such high correlations between medical model treatment and poor recovery rates revealed in the most thorough international studies ever conducted on recovery from schizophrenia.

Alternative residential communities in the West. Even in the West, there have been residential facilities designed to provide a similar kind of care for those recovering from extreme states of mind as that which is often naturally provided within many collectivistic societies such as those that are found within many developing countries. Generally speaking, residents of such homes are generally given maximum freedom and empathy, are not forced to take psychiatric drugs against their will (although they may be used judiciously during acute crisis), are not forced to remain in the home against their will, and are encouraged to be an active member of the community, both within the residence and in the community at large. There have not been many such homes, however, and so the research on them is somewhat limited. However, the research that does exist is quite promising.

In 1969, Loren Mosher, Chief of the Center for Studies of Schizophrenia at the NIMH, opened such a facility in San Francisco, which he named the *Soteria house*, with the intention of using it for research on recovery from schizophrenia[63]. In a two-year longitudinal study, patients who were hospitalized with their first psychotic episode and who volunteered to participate were randomized into two groups—with members of one group moving into the Soteria house and members of the other group receiving standard psychiatric treatment. After two years, the Soteria residents showed equal or better outcome measurements on all eight assessment points, having demonstrated the largest gains in the areas of psychopathology, social functioning, and employment. Also, a significantly higher percentage of the Soteria house residents were living independently, had fewer hospital readmissions, and far fewer of them were using psychiatric drugs[64].

Unfortunately, the Soteria house was closed down in 1983 due to a lack of funding. Since then, due to laws allowing forced antipsychotic drug treatment, which have come to be the norm in most states, there has been a legal barrier to opening another similar home. Fortunately, in 2006, Jim Gottstein, a lawyer in Alaska who founded the nonprofit organization, PsychRights, managed to persuade the Alaska State Supreme Court to overturn the forced antipsychotic drug treatment law, and in 2009, a Soteria house finally opened its doors for the first time in the US in 26 years (in Anchorage, Alaska)[65].

Other similar residential homes have been, and continue to be, operated in Western society with similar results[66] (although there have not been any operating in the U.S. between the closing of the first Soteria house in 1983 and the opening of the second Soteria house in 2009). After reviewing the results of the studies that looked at the efficacy of such homes, Mosher concluded that "85% to 90% of acute and long-term clients deemed in need of acute hospitalization can be returned to the community without use of conventional hospital treatment"[67].

Mosher's conclusion was further corroborated by the findings of a more recent study that looked at the recovery rates of participants in a program known as the *Open Dialogue Approach* taking place within Lapland, Finland[68]. This program generally does not involve the use of a residential facility, but it does incorporate very similar principles to the Soteria house in that it maximizes the freedom of the members and the expectation of full recovery while also striving to create a social environment that emphasizes safety, empathy, and authentic interpersonal interaction. Five years after initial diagnosis, 86% of the members of this program had returned to work and/or school, and 82% of them were considered to be fully recovered, in that they "did not have any residual psychotic symptoms" and were not using any psychiatric drugs[69].

Chapter 6

Summarizing the Research on Schizophrenia and Recovery

As we have seen, piecing together the evidence regarding recovery from long-term psychosis is no simple and straightforward task. However, we have also seen that there are certain findings that have demonstrated high consistency and reliability across this wide array of research:

- In spite of over a hundred years of research and billions of dollars spent, we still have not found any clear evidence of a biologically-based etiology of schizophrenia, nor have we been able to validate that schizophrenia itself is even a valid construct (there is no doubt, however, that many people suffer from distressing anomalous experiences, what I have been referring to as psychosis, and that these are the individuals who often get labeled as having schizophrenia).

- The use of antipsychotics helps reduce the positive symptoms of psychosis and the associated distressing emotions for many people in the short term (during the first six weeks or so).

- The long-term use of antipsychotics increases the likelihood of the development of a chronic psychotic condition and significantly reduces the likelihood of recovery, as well as carrying the high likelihood of causing other serious physical, cognitive, and emotional impairments. The specific effects of such use clearly vary significantly from one individual to another, but generally speaking, this has been a strikingly consistent and reliable finding.

- Those individuals who are never exposed to antipsychotics have the highest chance of recovery.

- Regardless of the treatment method, it seems that there is always some percentage (although relatively small—apparently about 15%) that is likely to remain in a chronic psychotic condition indefinitely.

- The medical model paradigm, with its associated beliefs of brain disease and terminology such as "mental illness," can significantly increase stigma, fear, hopelessness, and other associated distressing emotions and behavior.

- Residents of so-called developing countries have much higher recovery rates than those in so-called developed countries, and the use of antipsychotics and the medical model paradigm of treatment is *inversely* correlated with recovery rates.

- Residential communities that offer continuous empathic support and freedom, and which minimize the use of antipsychotics, have demonstrated the ability to provide significantly better outcomes for their residents at significantly less cost than what the standard psychiatric model of care has been able to provide. However, these alternative approaches may reduce some personal benefits for many professional caregivers and others in the psychiatric drug industry (e.g., personal income, job security, sense of order and control in the environment, etc.), something that is likely to be a major factor in our mental health care system's resistance to change.

When looking at the summary of the research, it is clear that the medical model paradigm of schizophrenia (and the other related psychotic disorders) has very poor validity and that genuine recovery is surprisingly common, even being the norm in many regions of the world. Yet, in spite of this, there remains the widespread belief in Western society that (a) schizophrenia has been conclusively determined to be a brain disease, and (b) genuine recovery is very unlikely and perhaps not even possible. So why is it, then, that we find such a dramatic disparity between these widespread myths and the actual findings of the research? While there are probably many factors that contribute to this disparity, there is one that may well stand out more prominently than the rest: We may be caught in the grip of a self fulfilling prophecy. Let's take the research we've looked at so far and see how it is that we may have become caught in such a harmful belief system.

First, the evidence strongly suggests that the primary modality that we use in the West for treating psychosis (involving primarily the use of antipsychotics and the insistence that one accepts that one has a "mental illness"/brain disease) significantly increases the likelihood that individuals experiencing one psychotic episode will go on to develop a chronic psychotic condition.

Second, we notice that this treatment is widely prevalent in Western society, with the large majority of those diagnosed with schizophrenia and other psychotic disorders receiving it. Therefore, as would be expected, we find very low rates of recovery and especially of full recovery.

Finally, it is likely that most of those individuals who actually do recover go to great lengths to avoid becoming caught up within the psychiatric system again and therefore are rarely seen again by their former psychiatrists and/or other mental health care workers.* Therefore, many mental health care workers see almost exclusively those who remain in a chronic condition, which creates the illusion of an artificially low rate of recovery on top of an actual low rate of recovery. We are then left with a well established myth that virtually no one fully recovers from schizophrenia, thereby reinforcing our belief that we need to resort to such drastic treatment methods.

Round and round we go, one myth reinforcing the other in a vicious circle—the myth that schizophrenia is a brain disease with no genuine recovery leading to the belief that, in the name of compassion, we must carry on with our harmful treatment methods, even if it requires the forceful coercion of those who "lack insight" that they have a brain disease; and in return, the myth that such treatment is the most beneficial thing that we have to offer actually *causing* widespread brain disease and chronic psychosis and therefore reinforcing the myth that schizophrenia is a brain disease from which there is no genuine recovery (see Figure 6.1). That we have managed to become so wrapped up within this delusional belief system is disturbing enough; but compounding this is the fact that there are a number of players within the health care system who make an enormous amount of money off the current system (the pharmaceutical industry and its many well-paid representative psychiatrists and academics, for example) and are more than happy to perpetuate myths with self serving propaganda and pseudoscience.† It is of no minor significance that since 2008, antipsychotics have become *the single most profitable class of all prescription medications sold within the U.S.*, with prescription sales approaching 15 billion dollars per year[1].

The good news is that some alternative treatment modalities have been showing up in the recent past, and, as discussed earlier, a number of them have shown great promise. The bad news is that, in spite of these promising alternatives, there

---

* Many groups that have organized to provide support to such individuals, such as *MindFreedom International*, the *Freedom Center*, and the *Icarus Project*, are filled with members who understandably share this attitude.

† See Whitaker (2009), *Anatomy of an Epidemic*, for an excellent summary of this and similar issues.

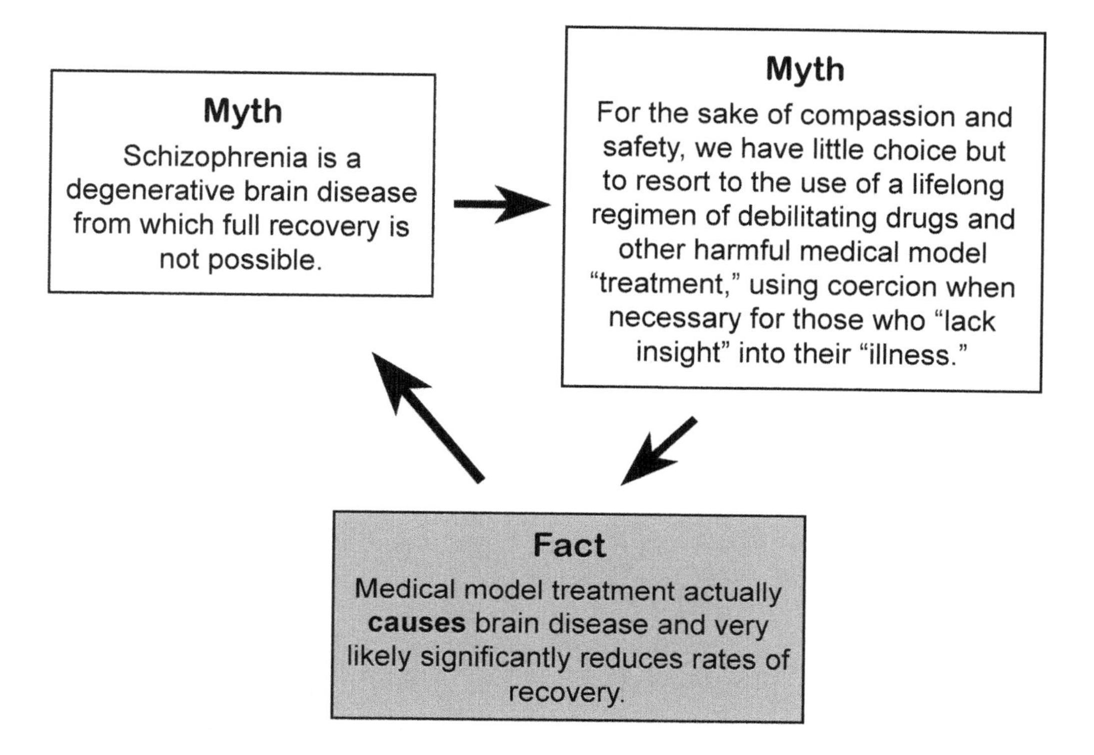

Figure 6.1. The vicious circle of one harmful myth reinforcing the other, leading to the harmful and generally ineffective "treatment" for schizophrenia and psychosis that we find in Western society today.

is still very little sign that the myths of "brain disease" and "no recovery" are losing their strength in mainstream Western society or that the mainstream mental health care system is seriously considering embracing any of these more hopeful alternatives in a serious way. It seems that in order to extract ourselves from the current dysfunctional state of affairs and move in a more hopeful direction, our society must go through a complete paradigm shift in our understanding and treatment of psychosis—coming to an understanding of psychosis that more accurately reflects the research, and developing a treatment model that supports rather than hinders the very high possibility of full recovery that we see in the literature.

Fortunately, we already have a theoretical framework that is much more in line with the research than is the medical model, one that begins with a very different set of assumptions about human nature and offers substantially more hope for healing, growth, and genuine recovery.

Part Two

Alternative Understandings of Psychosis

As we have seen, the recovery research strongly suggests that, when supported in a compassionate and empathic environment, psychosis often (and perhaps even ordinarily) resolves automatically. In addition to this, there is significant evidence that a psychotic episode sometimes provides a breakthrough into profound healing and even psychological and emotional growth.

Silvano Arieti, a renowned clinician specializing in working with clients who have received a diagnosis of schizophrenia, said, "With many patients who receive intensive and prolonged psychotherapy, we reach levels of integration and self-fulfillment that are far superior to those prevailing before the patient was psychotic"[1]. John Weir Perry, another lifelong clinician who served as the clinical director of Diabasis, a medication-free residential facility for young adults suffering from psychosis, said that "85 percent of the clients in Diabasis not only improved, with no medication, but most went on growing after leaving us"[2]. In a recent study conducted by Tooth et al. involving 57 participants who had been diagnosed with schizophrenia and who now identify as being "in recovery," 66% of them describe their functioning as better (and 44% of these as *much* better) than that prior to the development of schizophrenia. In this same study, 62% describe their social situation as better (with 31% of these as much better) than that prior to their development of schizophrenia[3].

A number of scholars and clinicians have suggested that the reason we see these kinds of results is that psychosis may actually be the manifestation of a natural attempt of a psyche to survive and/or heal from an untenable situation or way of being; and therefore, successful resolution of a psychotic episode would naturally entail healing from and/or growth beyond one's former condition[4]. R. D. Laing, a Scottish psychiatrist renowned for his pioneering research on schizophrenia and his clinical work with those so diagnosed, closely studied the social circumstances surrounding over 100 cases of individuals diagnosed with schizophrenia, and he concluded that "*without exception* the experience and behavior that gets labeled schizophrenic is *a special strategy that a person invents in order to live in an unlivable situation* [author's emphases]"[5]. Bertram Karon, a longtime clinician specializing in psychotherapy for those diagnosed with psychotic disorders, stated his belief that any one of us would also likely experience psychosis if we were to have to live through the same set of circumstances as those of his psychotic clients[6].

These individuals, then, who are so often labeled "crazy" may actually be simply doing the best they can to survive extraordinarily difficult circumstances, and when one is confronted with extraordinary circumstances, one often must resort to extraordinary strategies, strategies that may appear completely absurd to those of us who do not understand the full scope of what the individual is struggling with. When viewing these individuals through this lens, then, we can say that there is nothing inherently wrong, biologically or otherwise, with those who suffer from psychosis. They are merely acting as any living organism would in the same situation—they are simply trying to survive, and ultimately aspiring to thrive.

So we see that once we move beyond the very narrow and so far unsubstantiated medical model framework of psychosis, we find that a surprisingly wide array of lines of inquiry have been converging on the prospect that psychosis may be the manifestation of a natural coping/healing/growth oriented process initiated by the psyche. A number of scholars, clinicians, and researchers have generated some compelling models of psychosis based upon this premise, and we will spend the remainder of Part Two looking at some of these. The models I present here are those I have come across that emphasize subjective experience at the deeper levels of experience, and which I believe are the most compatible with the research literature in the field and with the findings of my own studies. It's important to note that while some of these models may appear to differ dramatically from each other, they each emphasize different aspects and/or perspectives of the psychotic process, and so they do not necessarily exclude each other.

Before we turn to look at these models, however, let's first look at the stories of two more participants of my own research so that we can use these stories as references for which to ground these models in actual lived experience. As mentioned briefly in the introduction, over the past several years, I conducted a series of research studies inquiring deeply into the lived experiences of people who had been diagnosed with psychotic disorders and have since gone on to full and lasting recoveries. The final and most thorough of these studies included six participants, and we will have the opportunity to look at the stories of all six of them as we move through the remainder of the book. We've already looked at the story of one participant in the introduction (Sam), we will look at two more now (Theresa and Byron), and we will look at the remaining three stories at the beginning of Part Four.

I want to express my most sincere gratitude to these participants for their willingness to share their journeys of unimaginable trials and tribulations with the rest of us. Every one of them describes a journey filled with heart-wrenching pain, profound healing, and life-changing insights. I find myself deeply inspired by the tremendous courage required of every one of them to have carried on in spite of all that was stacked against them.

Chapter 7

The Case of Theresa

Just as she was entering her 20's, Theresa left her home country of New Zealand on a quest of discovery. But what began as an exploration of distant cultures in faraway lands soon became a profoundly personal journey into the mysterious depths of her inner world. Between the ages of 21 and 23, Theresa experienced two intense periods of psychosis, with a break of about two years between them. She was formally diagnosed with paranoid schizophrenia and manic depression.

The Onset and Deepening of Psychosis

In hindsight, Theresa recognizes that there were a number of experiences in her childhood that worked together to set the stage for her psychotic journey. She believes that one particularly influential factor was the significant sense of isolation that pervaded her childhood, especially after the birth of her younger sister:

> I think I probably had a pretty cushy early few years, you know, until about four or five, and then my sister was born, and that was all okay, but I think all of a sudden, I kind of had gone from .... being the center of everybody's attention to kind of nobody's attention [laughs]. In the course of a normal type of life without other kinds of things layered on top, it probably would have been nothing, but ... it was kind of a bit of a shock, you know, around the time I was starting school which is quite an intense time for little kids, anyway. So I kind of got this sense of disconnection that started right back then, you know, and then I didn't make friends particularly easily at school.

Theresa believes that another important factor was the significant physical and emotional instability that pervaded her middle childhood and adolescence:

> Things started falling apart with my parents' marriage [when] I was about seven or eight, and so there was a lot of conflict and . . . things just started there and just went from bad to worse with their relationship and never really recovered. . . . [Mom] took us off and [then] she was just kind of on the move. . . . I went to more schools than I could remember. [My parents] had a couple of goes at getting back together, so there was all this kind of hope and stuff when they did . . . just that sense of stability, you know, maybe it will come back, you know [laughs]. But before the boxes were unpacked, they were fighting again.

This continuous instability and insecurity further exacerbated Theresa's sense of isolation, creating "layers" of isolation and disconnection that she believes ultimately contributed to the eventual onset of her psychosis:

> So, there was kind of just . . . layers and layers and layers of a similar kind of . . . you know, moving further and further into myself, I think, you know, and less and less able to trust what was going on around me..um..less and less able to trust them, as well. . . . A sort of a disconnection, you know . . . moving schools and not making friends, and not really feeling connected to anything, you know, so . . . it's not difficult to see where those layers and things kind of come from.

Another significant layer to Theresa's deepening sense of disconnection from the world and ultimate vulnerability to psychosis came when both of her parents died suddenly in her late adolescence just a few years apart:

> Apart from those kind of early things, I mean, normally I think I probably could have survived and managed without, you know, developing psychosis later on in my life, but then Mom dropped dead and pretty much just suddenly dropped dead, as well. She had a brain hemorrhage, and it was so sudden that..yeah, our life kind of changed overnight, basically. . . . I was sixteen, so I was still in school, just at the end of school . . . so we moved to live with my father. . . . It was completely miserable for a couple of years until I moved out 'cause we hadn't had a huge amount of contact with him and he was living with a woman . . . who we didn't like, [and] she didn't

> particularly want us around. . . . And then four years after that, when I was twenty, dad dropped dead as well, pretty much. . . . So I think the other stuff by itself might not have needed such an outlet that was quite so extreme, you know [laughs], but the deaths and quite sudden deaths, particularly of Mom, were more than I could probably handle and not have it have to come out in some way.

After the sudden deaths of her mother and father, Theresa believes that another significant layer adding to her disconnection from the world came as a result of having received almost no emotional support for the tragedy of her parents' deaths: "We had no grief counseling of any kind. . . . I think it wouldn't have taken much at that point, but I don't even remember anyone saying, how do you feel, you know. It sort of didn't happen, which means that everything just got pushed sort of further and further down and further and further in." Not being provided with any less harmful way to work with the powerful feelings associated with the loss of her parents, Theresa turned to alcohol: "So I started drinking, which was, you know, which was..fine [laughs]. . . . I mean that helped me sort of skip along the surface of those things."

In retrospect, Theresa recognizes that yet another layer that was forming during her childhood and adolescence was her difficulty with developing a significant sense of self worth: "[I didn't feel] as important as everybody else or everything else around me." She speaks here to her belief that perhaps one role of her psychosis was to develop that sense of self worth: "I seemed to need to break through something to get to the point where I could sort of allow myself to..I don't know..um..to have what I needed somehow."

About a year after her father passed away, at age 21, Theresa followed her desire to travel, eventually ending up on a kibbutz in Israel. Except for a short break to visit her grandmother back in New Zealand when her grandmother became physically sick at one point, Theresa remained in the kibbutz until the onset of her psychosis and subsequent hospitalization. She describes the kibbutz as "a bit like my ultimate heaven": "It was a big family feeling kind of place, communal living, kind of everything that had been missing for me for a long time." Ironically, however, it was this very heaven-like quality that began to overwhelm Theresa and pull her away from consensus reality:

> *Researcher:* You were really able to get your needs for connection met in some way there, it sounds like.
>
> *Theresa:* Well, yeah I did, but . . . I mean, that was kind of the problem, I think. It was all there, but I didn't know that I needed that. . . . I couldn't.

> It was too intense to be able to do that, you know. . . . Because it was so heavenly, it kind of brought to the surface all of the kinds of things that I'd been dealing with, or hadn't actually dealt with. . . . I was having a great time, you know, but I was drinking. I practically just wasn't sober, you know, and so I sort of understand it like, it was all there but it was too much. . . . It was overwhelming and I couldn't cope with it, so I had to kind of suppress it the best way that I could, and then it just kind of bubbled through, you know, and..and just put me over [laughs].

Initially, while living in the kibbutz community, Theresa had feelings of being in Heaven that were not particularly anomalous. Much of these heavenly feelings were a direct result of finding a satisfying community after a difficult and lonely childhood. Over time, however, she began to believe that it really was Heaven. Then the experience of being in Heaven began to alternate with even more powerful experiences of being in Hell. Some experiences in the early stages of this included visual hallucinations on the television: "[I saw] a lot of fiery landscape scenes, kind of classic Hell looking stuff, people transforming into demons, colors changing from normal to red and black. Normal scenes transforming into hellish looking ones. People mutating into horrible looking demonic type creatures." These experiences eventually strengthened, and she began to have similar experiences out in the world, "not demons or devils or anything like that, but just..um..well a lot of fire actually [laughs]."

As Theresa's contact with consensus reality became increasingly precarious, another powerful feeling began to emerge, a feeling "that there was a kind of an ultimate survival somehow": "I was . . . doing things like climbing a high hill behind the kibbutz and working very hard in the kitchens with no breaks, etc. This seemed to be mainly about how hard I could strive. If I could strive hard enough, I could save others from hardships and protect them from having to suffer." These feelings of compulsive heroic striving sometimes combined with the visual images of being in Hell, creating experiences in which she "would have to walk through a hell type thing, like fire."

As Theresa's world became filled with images and sensations of Heaven and Hell, and as she found herself increasingly driven by compulsive heroic striving, she discovered another force that was driving some of her experiences—a desire to have a child, a desire that she later realized was closely intertwined with her desire to have a family. Both of these desires were present prior to her psychosis, and she believes they fed directly into the more anomalous but closely related experiences that occurred during her psychosis. One particularly prominent anomalous experience related to these that arose during her time at the kibbutz was the sense that "if [she] was to have a child, it was to save the whole of humanity," though she was not consciously trying to have a child: "I actually was trying to get pregnant on the kibbutz but not consciously (i.e., was pretty promiscuous and forgetting to take the pill)."

After several months of moving further and further away from consensus reality, Theresa's traveling companion became very concerned and decided to take her to a psychiatric hospital. Unfortunately, Theresa found her treatment there to be more harmful than anything she had yet experienced:

> I got immediately assessed as being deeply psychotic and was put on heavy medication, incredibly heavy medication, that knocked me out for about three days. . . . And when I came to, I didn't really recognize myself at all, and I couldn't think, I couldn't really do anything at all, and [I had], you know, one of those zombie kind of shuffles. . . . [I went] from someone who . . . was, you know, full of beans and had a lot of energy. It was like from one extreme to another. So the trauma of that was probably as bad as if not worse than the trauma of anything that had actually caused . . . the breakdown or whatever, in the first place.

In retrospect, she acknowledges that she did need some kind of strong support, but not the kind of support that the hospital provided:

> The unnecessary bit [was] being pumped full of medication and being knocked out cold for three days. . . . Although, having said that—I mean that was completely unnecessary—I did need some kind of help. I did need a safe place to be because I don't quite know what would have happened if someone hadn't have grabbed me, but I didn't need that extreme level of kind of interference, you know.

Theresa believes that even her friend had been significantly traumatized by the hospital experience: "My friend . . . who's still my friend actually, was as traumatized by the whole hospital experience as I was, and so that's taken her a long time to work out as well."

During her stay in the hospital, Theresa's sense of heroic striving remained to some degree, but the quality of it changed significantly, becoming more personal and more relevant to what was actually taking place in her environment:

> I had some stuff when I was in the hospital about walking through fire and having to survive from the ultimate kind of, I don't know, whatever the most ultimate thing you would have to survive from, I would have to survive from, you know. But I think that was more what was [actually]

> happening at the time. [Being in the hospital] was a traumatic experience, and I did need to survive [laughs].

After being released from the hospital, Theresa returned home, where she lived for a short time with some of her relatives and remained on a heavy dose of antipsychotics. About four to six months later, she moved into her own place and completely stopped taking the antipsychotics. Then, about two years later, she entered her second period of psychosis. In retrospect, Theresa believes that the antipsychotics did play a major role in stopping her psychotic experiences during her first period of psychosis; however, she believes that this did not ultimately serve her well at all, since she feels that they merely interrupted a process that needed to be completed:

> The way I understand it now is that the first [period of psychosis] kind of got interrupted, you know, with the drugs, with the medication, and so the whole process just got stopped, basically. What it was that I was working through and needed to work through just kind of got halted. . . . The purpose that [the psychosis] had, that it was serving, to work through those things that we were talking about before, the trauma and stuff, didn't have a chance to come to a natural kind of conclusion or fruition or whatever, to evolve to where it was needing to be. So I think of it now as if you don't have the chance to work something out, it's gonna come out in some other way, you know.

About a year after returning home, during the period of time that she now believes was essentially a latency period between her two periods of psychosis, Theresa began to attend intensive personal development courses:

> [The courses] were slightly cultish really in a way that you kind of did these weekend workshops and . . . they encouraged you to kind of surround yourself with people who want you to win and all of this kind of stuff . . . and they cost thousands and thousands of dollars. . . . So I got into that and then ended up becoming what I thought was quite connected to the people that were there, and I probably was, but I did isolate myself very much from everybody else.

Theresa recalls that a particular theme began to become more and more prominent in her consciousness during these courses, which was her desire to have a family. She recalled being surprised initially when she noticed the first stirrings of this desire: "Because I think there was kind of grief and trauma associated with it, I had decided I was never gonna have children. I thought . . . I won't bother getting married and I won't have children and that kind of stuff, you know." However, as these courses forced her to inquire more deeply into her longings and aspirations, this longing for a family became undeniable: "[During] one of those courses, one of the early ones . . . you had to make . . . some kind of vision for your life. I think you had to make a promise or something like that, and mine was something to do with having a family or creating a family."

Along with this "strange promise" to herself, however, came an important realization: "That was actually probably what I wanted most, but I couldn't kind of handle it, you know." Theresa believes now that, even prior to her psychosis, she may have been struggling with this core dilemma of wanting a family, yet not feeling that she could "handle it." She now believes her psychosis may have actually been playing an important developmental role in this regard.

The final course was particularly intense, and as it progressed, Theresa felt herself begin to slip away from consensus reality once again:

> The final course that I did was a week long intensive thing where you go off in the bush. It was called breakthrough, break through your mental and physical limits of the things that you think are possible in your life, and that's the idea of it. So I wasn't drinking or doing anything like taking any drugs, but I did use what I now understand to be kind of like mind control techniques really, sleep deprivation and intensive physical exercise and stress and things like that, and I slipped sort of slowly into another mental crisis, I suppose you'd call it, and [soon] became what would be defined by psychiatrists probably as intensely psychotic again.

As Theresa became fully immersed in her psychotic process once again, she found that many of the same feelings and forces that had surfaced within her first period of psychosis came to the foreground again; only this time, her experiences related to her desire for a child had taken a particularly prominent role:

> [This time,] I was actually consciously trying to get pregnant, i.e. had ideas that the creative moment would happen at a certain phase of the moon, with certain writing on the walls of my bedroom, and helped by particular crystals, etc. . . . I completely graffiti'ed the inside of my flat [with] all kinds of curly things [that] artistically expressed all sorts of things to do with creation. I assumed I was channeling a child [laughs]. That's what was going on in my head, you know. . . . And strangely enough, it worked – that is to say, I did get pregnant.

Even though Theresa's experiences related to bearing a child were different in some ways between her two different periods of psychosis, there were significant similarities, particularly with regard to their association with powerful experiences of creation and universal expansion: "I had similar thoughts/feeling both times . . . that I was expanding and expansive to the point that there were no boundaries between me and the rest of the universe and . . . that I could bring about the healing of humanity by bringing all that is good into the spark that would create a child."

The compulsive heroic striving that was so prominent in Theresa's first period of psychosis had also returned in this second period, maintaining a remarkably similar quality:

> The second time was similar in being about physically striving; I had a push bike which I rode all weathers – including long distances thru driving rain etc; walked and then swam out into the sea, walked long distances—aiming to find things to climb (I scaled a really high fence with razor wire on top once and got over it, but was caught by a security guard. I managed to convince him I was lost or something and he drove me home!!). The point of the climbing seemed to be to get to the highest point as then I would be able to see all that was below and take all the suffering I could see away from others.

The themes of Heaven and Hell did not show up as explicitly during Theresa's second period of psychosis as they did during the first. She did, however, have experiences of being watched over by both benevolent and malevolent entities, experiences that could be seen as being somewhat related to her earlier experiences of Heaven and Hell in that they were personified manifestations of good and evil. She describes the experiences associated with being watched over by malevolent entities as "paranoid . . . type stuff." She recalls one such occasion: "There was a van parked outside my place, and I thought that I was being recorded and watched." These experiences alternated with experiences of feeling watched over by benevolent entities:

> I imagined [laughs] there was a huge, great big strong black man living in my wardrobe [laughs] who was there to protect me. It was kind of like being taken care of as well as being kind of watched. . . . [At other times,] a dog came hanging around my flat and I thought, oh, this is kind of like a guard dog, you know. . . . So, sometimes I would get really paranoid and, you know, pulling out the stereo and trying to destroy it and things like

> that; but at other times, I kind of felt like I was hidden away somewhere safe as well. . . . It was sort of like there was a strong protective kind of thing that would sometimes change around. I guess it sometimes flipped over into a paranoia.

Besides the more prominent themes of good and evil, heroic striving, and bearing a child, Theresa experienced a number of experiential themes that were less prominent but still highly significant. One theme that seemed to act as a thread tying many of the other themes together was a feeling of profound interconnectedness. One such experience was a powerful feeling of universal expansion: "[I] felt like I was expanding or believed I was expanding, and there were no limits between me and the universe, and that included being able to kind of move myself anywhere in the universe that I wanted to be, and . . . being able to communicate with every creature." Closely related to this was the belief that she "could communicate with animals" and "could hear people's thoughts." Theresa has some evidence that, at one point, such anomalous communication may have actually occurred:

> When I was in hospital in Israel, my friend came in to visit me and the staff was speaking to me in Hebrew, which is the language that they speak, and they were talking to me, and she got really, really angry at them and said, wait a minute, she can't understand Hebrew. She's from New Zealand. And they said to her, well, she's been speaking to us and understanding us in Hebrew for the last week. What do you mean, we just assumed she knew how to speak it. And that actually happened [laughs]. It's not a delusion . . . but that's one of those things that [my friend] tells people all the time, 'cause she just thinks it's incredible, you know [laughs].

Theresa also had experiences in which this sense of interconnectedness was closely integrated with a sense of profound meaning between things: "Everything was kind of connected. Like I would look at a piece of paper and it would tell me..um..something on there meant that I had to go to this place and..you know."

Another less prominent but significant type of anomalous experience had to do with multiple realities. Theresa recalls occasions during her psychosis in which "several different realities . . . [were] happening at the same time . . . [and] the boundaries weren't clear at all." This kind of experience was particularly vivid during the process of regaining contact with consensus reality:

> The natural regaining of my "normal" mind happened like that as well (and remember this was with no medication). . . . It was like the layers of "other

> realityness" gradually peeled away to reveal a more "grounded (or common sense of) normality" underneath and the more clearly the "normality" came into focus the more I was able to realize what I needed and needed to do - another way to describe it would be like seeing something in a very clear focus . . . so you know that it's real and then it gradually "morphs" into another reality; as one starts to fade the other becomes clearer.

Another common anomalous experience for Theresa was a strong sense of groundlessness, which pervaded much of her experience during both periods of psychosis: "I didn't even feel "physically" connected to the ground most of the time, or "psychically" connected to the planet— and quite importantly (I think) didn't feel connected to another human being through pretty much the whole of the time."

A final type of anomalous experience worth mentioning is Theresa's sense of identifying as alternate characters. For most of her life, even prior to and subsequent to her periods of psychosis, she has had these kinds of experiences, although she generally has not found them to be particularly distressing:

> I don't specifically believe in reincarnation (or not believe in it) but I have very strong feelings, and have had most of my life, that I have some distinct "characters" within my "here and now" being. Aspects of some of these popped up at various times quite clearly during my times of psychosis. One was a "middle eastern" based character. The imagery I get is veils, servitude, lush bedrooms, etc. (almost like being part of a harem of some kind). Another is an Amazonian warrior type character (a man); very tall and strong, travelling long distances barefoot, etc.

Recovery

Even though Theresa experienced a period of about two years between her two periods of psychosis in which she had a relatively solid grasp on consensus reality, she has come to believe that this period of time was essentially a latency period in which there was a general pause in an important healing process that was taking place within her psychotic episodes. She believes that particularly profound healing took place during her second period of psychosis, culminating in a lasting transformation that has provided her with the potential to live a rich and fulfilling life.

Even though Theresa believes that most of the deeper work that led to her recovery took place during the two periods of her psychoses, she does recall one event that took place during the latency period that she feels ultimately played an important role in her recovery process:

> When I came back from the kibbutz, there was a social worker that came to visit me . . . and he had this list . . . it must have been a kind of an assessment list, you know, did you sense you were, you know, the second coming, and do you believe blah blah blah, trying to get a handle, I think, on what kind of psychosis or whatever. . . . But it was a huge relief for me, because I was like, oh my, yes I did feel this, or, no, or whatever, but it said to me that I wasn't completely alone in having these experiences. . . . you know, I'm not a total freak. . . . There was quite a relief in that. That was one of the things that really, really stood out.

After descending into her second period of psychosis, her desire to "channel a child" became a particularly prominent force as well as a very important feature of the terrain she would have to cover as she moved towards recovery. After her experiences related to channeling a child progressed for some time, she did in fact get pregnant: "Before I slipped into, I guess, being in a state that probably people couldn't have related to me at all very easily, I decided that I was going to have a baby [laughs]. Yeah, and so I had a very brief relationship with an old school friend and got pregnant."

Theresa continued to struggle with psychotic experiences and even serious hopelessness for some time after discovering that she was pregnant, but she realizes now that this point represented both a major turning point and an important milestone towards the ultimate goal of her psychosis: "It was the beginning, really, in a way . . . because I mean that was the beginning of creating a family . . . . which is kind of where it was all heading, I think."

But she still had a long way to go. She had finally become pregnant, but she was also homeless and completely alone. She had just been evicted from her apartment, had spent the night in a homeless shelter, and found herself contemplating suicide:

> I was walking across Grafton bridge where people coincidentally quite often throw themselves off, and I thought, okay, considering what's ahead of me, and what's just happened, being dead would be easier than this. . . . But then I had that kind of thing . . . that quite a lot of people describe—well, the situation is as bad as it could possibly get, you know, the only way's up [laughs] . . . so just take the next step type-of-thing.

Not knowing where else to turn, Theresa made the difficult decision to check herself into a psychiatric hospital. Considering the traumatic treatment she received during her previous hospitalization (in Israel), she was very reluctant to risk receiving such treatment again; yet she simply did not know where else to turn for the support she desperately needed. Ironically, rather than receiving forceful interventions, she was immediately turned away: "I needed some help, so I went to Ward 10 of Auckland hospital . . . which was the psych ward, and tried to present myself there, tried to say, look I'm not well and need help . . . but they said, ah, well sorry, if you're well enough to tell us that there's something wrong with you, you're not unwell enough to be here."

Even though her revelation of having "hit rock bottom" on the bridge had given her some sense of hope that things could not get any worse, Theresa was still feeling very stuck. The hospital had turned her down and her recent revelation on the bridge had not provided any particular guidance: "There wasn't really any 'agency' in that – it was just a resignation that I had to put one foot in front of the other." Fortunately, she soon came across a drop-in center run by a group of psychiatric survivors, an encounter that she later came to realize represented another important turning point in her recovery:

> I just wandered in. I was still actively psychotic (i.e. was having uncontrollable disturbing visions/thoughts, but mainly fears for my sanity, fear of being under psychic attack, etc.). I think I was beginning to regain my sanity and realize the situation I was in and was starting to recognize when I was in touch and out of touch with "normal reality." The place had a self contained room—think they called it "emergency accommodation" where I stayed for a week or so. Despite the stuff that was going on in my head, I actually felt safe there. And in all the time I was wandering and psychotic (that time) I hadn't found anywhere I could just stop and feel safe (which was something to do with the striving I think). If there was a point where my recovery started I guess it would have been there. I had the chance to stop and take a breath, gather my strength, and I realized I actually had the strength to face what lay ahead. I found my own internal hero. Then funnily enough I met a real one !!

After being pregnant for about three or four months, Theresa met a man who began to look after her: "I was, I don't know, a little bit strange and a little bit interesting and still pretty..um..still pretty mad, but he kind of took me under his wing, really, and fed me because I was pretty much homeless and pregnant, you know [laughs]." Theresa recalls that a particularly supportive aspect of this man's character was his groundedness: "He actually seemed totally connected to the ground, and also not in even the slightest bit scary, or like he would try and make me do anything I didn't want to do, etc. I was so 'tired' (in every sense of the word) by then, that at that point the ground (with him on it) started to look pretty good."

The stability and care that this man provided were the final important resources that Theresa needed to integrate her psychotic experiences, return to consensus reality, and move in the direction of successfully raising a family: "That quality of just accepting and respecting each other for who we are and how we feel/what we believe, etc., is something that has continued into our relationship and is one of the keys to our success, I think." They remain together to this day, over 25 years later.

In retrospect, Theresa realizes that another important factor in her recovery was in learning to find a balance in her way of connecting with herself and the world. She found that connecting with her body through movement and her creativity through art were particularly important resources, not only in working through the psychosis, but also in her transformation towards a new way of being that is much more sustainable than that which existed prior to the psychosis:

> [Prior to the psychosis,] I kind of lived in my head, really. I mean I didn't really feel. I used my brain to think everything out. . . . I wasn't really very connected to my body, so I've done a lot of work. I've been doing yoga for years and years and years now, and I've done a lot of work on actually . . . just feeling, you know, and expressing my feelings rather than channeling everything through my brain.
>
> . . .
>
> I think [art] was kind of an outlet for all the feelings I was having that I couldn't understand with my mind. Art, for me, comes from a subconscious realm, which cuts through my usual tendency to 'think' things to death. However, there was a drive and a desperation about [my thinking] when I was psychotic – a seeking to understand. So [art] probably did help in some way. Now I've discovered that I can tap into that same creativity, and the more I do that (just relax and let it be) the more "critical acclaim" [my art] seems to get.

Theresa believes that another important resource in her recovery was the attendance of regular psychotherapy sessions that lasted for about six years. Even though she was no longer experiencing any significant psychotic experiences by the time she began attending psychotherapy, she felt that a number of aspects of her journey remained unintegrated, and she found psychotherapy very helpful in tying up some of these loose ends.

◇◇

As of the writing of this book, Theresa is 46 years old. She hasn't used any psychiatric drugs for about 24 years, since coming off of antipsychotics several months after her first period of psychosis, even managing to avoid them completely during her second period of psychosis. She hasn't had any significant psychotic experiences since recovering from her second period of psychosis (over twenty years ago now), although she has had a few relatively minor incidents of nonconsensus experiences and/or beliefs. According to the definitions used in this study, Theresa considers herself as having been fully recovered for a little over ten years. She now works within the mental health field as a consumer advisor, where she offers peer support to those diagnosed with psychosis and other mental disorders. She remains married to the same man who supported her in her recovery over twenty years ago, and together they have raised two children.

Similar to Sam and all of the other participants in the study on which this book is based, Theresa feels that she has gone through a profound and primarily positive transformation as a result of the successful resolution of her psychosis, experiencing far more lasting benefits than harms. We'll explore the details of this in Part Four.

Chapter 8

The Case of Byron

A self-described "child of the 60's," Byron entered his young adulthood with a free spirit and a passionate curiosity for altered states of consciousness. At the young age of 19, however, this intense curiosity led him into realms far beyond his ability to manage, and after receiving a diagnosis of schizophrenia and being locked in a psychiatric hospital for six months, he began the long, arduous process of integrating these experiences and reconnecting with others in consensus reality.

The Onset and Deepening of Psychosis

Byron believes that some of the seeds of his psychosis may have been sown in his childhood—particularly as a result of his close relationship with an eccentric mother and, closely related to this, his own sense of isolation:

> The precursors of [the psychosis] were significant, I think, in terms of my growing up. My mother was an artist and experienced things in kind of a surreal way, and I was very close with my mother, so that affected me a lot. And in adolescence, it was like, you know, now it's time to operate in the world as a functioning member, and I had a hard time with that. . . . [I had] a very troubled adolescence [that] affected me a lot, I held a lot inside. I was pretty intelligent, I was put in a class with, you know, other intelligent people, but . . . I was sort of the outsider with that mix, and I took everything very personally. . . . I had a very hard time with that, and I really didn't want to be there so much.

Soon after leaving school, Byron found himself getting swept up in the 60's movement, and in a very short period of time, he went from feeling very insecure and isolated to experiencing a profound sense of liberation and deep connection, a shift he believes played a significant role in the onset of his psychosis. Byron recalled a particularly powerful incident related to this shift that occurred just two months prior to the onset of his psychosis. He attended the well-known Woodstock music festival, where he experienced profound unitive experiences facilitated by hallucinogens and deep immersion in this once-in-a-lifetime enormous gathering of people all sharing his longing for deep connection:

> I was feeling incredibly ecstatic, waves of bliss coursing through my body. Every cell in my body exploded in bliss. I felt that I was on a heavenly plane of experience, I felt tuned into multiple dimensions simultaneously. Every desire was satisfied, there was want for nothing. . . . I had a transcendent experience. I believe this was a contributing factor to my going off the deep end two months later. Why? Because the experience was so powerful.

Immediately following the Woodstock festival, Byron found himself consumed by the desire to undergo a profound personal transformation. This desire culminated about two months after the festival, when he spent an entire evening in his apartment performing a shamanic rebirth ritual:

> There were a number of events, but then everything sort of came together and I felt that I was gonna die and be reborn. In fact, I was sort of living the experience of death and rebirth, and I did this all night journey which was very intense. I tried to turn myself inside out, trying to die and be reborn. . . . I was like flashing on dying and being reborn in the womb, I was flashing back and forth between these themes to converge to where I would take a rebirth as a new form . . . . And then at the dawn, I felt like I had transcended, perfected going into the future, and I kind of..uh..faded off. Then I walked off the balcony three stories up and broke every bone in my skull. [*Researcher:* Wow. Was that intentional or accidental, when you walked off the balcony?] I have no conscious recollection of that whatsoever.

After his profound rebirth process followed immediately by falling three stories off of his balcony, Byron awoke in the hospital encumbered with a severe head injury and experiencing "extreme altered states":

> I was in the emergency room in a hospital in Boston, and the first thing I remember was seeing my brother walk in and fainting when he saw me..

> tubes hanging out [he gestures toward the right side of his face]..and I was in extreme altered states, and the altered states continued, so I was having all these nonordinary experiences increasing..um..leaving my body..come back in..go out and come back in. . . . [Whether I had suffered brain damage] was never determined. I was extremely fortunate to have retained my functionality. . . . I don't know if my psychosis had aspects stemming from brain damage or not because it all came on before the incident, so my best consideration is it had multiple causations, you know. It was not just one thing.

After waking up in the hospital, a dramatic visionary process began to unfold:

> My initial experience was of..um..everything broke through, and it was like all my sort of..um..angst and questions came pouring out into this powerful energy and I experienced everything in sort of like dream states 24 hours a day. Not unlike some of the psychedelics but much more powerful. It was a full enrapturing experience, if you will, and I experienced a whole lot. There were major themes of death and rebirth and of birth of the Aquarian Age and of the eon. I was always absorbed with the quest for meaning, so some things seemed to click into place big time, but it was all very much three dimensional dreamlike. The television would be talking to me, people on the TV, and some of the dreams were intensely significant, which I remember today.

As this visionary process continued to unfold, Byron realized that a common theme was occurring throughout his visions—in particular, he found himself on a kind of hero's journey in which he worked through various trials and tribulations involving death and rebirth, and other kinds of mythological and archetypal transformations.

Strangely enough, even though these visions occurred long before he encountered Tibetan Buddhism, Byron later came to realize that many of his visions at that time involved themes that closely match those found within Tibetan Buddhism:

> It's really remarkable how many of the visions I had during my psychosis have found context in Buddhist practices. . . . None of that stuff was available at the time, I mean to me, you know, it was all in Tibet, it wasn't in this culture. . . . Most of these contexts fall in with the Buddhist tantra teachings. Over forty years ago, while in the state hospital, I experienced many visions and non-ordinary thought processes such as transforming poisons

> into nectar, liberating suffering beings, performing rituals for world peace, visions of charnel grounds, and much more. The visions I experienced have not gone away; rather they've found context and place within my Buddhist practice.

Byron had numerous visions with themes of transition. Among the strongest and most coherent of these have striking parallels to the Tibetan Buddhist concept of *bardo*, a Tibetan word meaning "intermediate state" and often used to refer to the transitional state between the ending of one life and the beginning of the next. Byron had a number of experiences that he describes as feeling like being "in between" different realms. He recalls one particularly long visionary bardo-type experience:

> I descend through dimensions of energy experiencing symbolic correspondences like astrological planetary energies, colors, chakras, and strong kinesthetic feelings. I move down level after level. As I move down to each particular level, I encounter a person who I have known in this life. This person typifies a level of incarnation. I recognize that in being on this particular level, this person has assumed a certain level of responsibility. Each level is an entire horizontal world existing on a vertical axis. The vertical axis maps onto planet Earth. I descend into the center of the earth. There are female entities in attendance.

Byron now recognizes these female attendants as *dakinis*: "It's a Tibetan term meaning 'sky goers.' These females show up in your life to guide you."

Byron's eagerness for transformation (both for himself and for the world) continued in spite of his serious injuries, and he recognizes now that this enthusiasm resulted in him taking on much more than he could handle. Even though he was cautioned by the dakinis about going deeper and taking on too much, he continued on:

> So, I went further, I went lower. . . . I say to these entities that I want more and want to go deeper. I want it all. [They respond], you're asking for it all, you know. Do you know what you're asking for? And I really didn't. I really didn't, but I was insistent that I wanted to take it all on. I was so insistent, so..I took on all of it, you know. . . . But I cannot handle it all. . . . After asking for it all, there comes a time when I must take on the responsibility and burden of what I have asked for. There is no way I can handle it. Did you ever see *Fantasia Sorcerer's Apprentice*? Mickey Mouse, when he takes hold of the wizard's implements and then conjures up all this stuff, that's

> what happened to me. I couldn't handle it. Or the flight of Icarus. It was very much that.

Another commonly recurring theme within Byron's visions was that of "the transition of the ages," which often tied in closely with "the great battle between the forces of good and evil" and the battle between wisdom and ignorance:

> So I'm living in this whole world along with the current images and symbolism, you know, like the Beatles and Crosby, Stills, and Nash. All the imagery from that all came along, and then people talking to me across the TV, and I'm seeing the great battle between the forces of good and evil, the transition of the age, the Age of Aquarius, and I feel like I'm involved with this, that we'll either go back to the Stone Age or we're going to go into the Aquarian Age with a major shift in awareness, planetary awareness. So I was kind of in that place for most of my psychosis.

A common threat that tied many of Byron's visions together was his experience of himself as a hero. Shortly after being released from the medical hospital, he was taken to a private psychiatric hospital at the suggestion of his therapist at the time: "I thought I was going to a kind of a retirement center or . . . special place where people who had attained the awareness that I thought I had would go." After about three weeks there, he was returned home, but was hospitalized again shortly thereafter. His insurance soon ran out and he was transferred to a state hospital, where his sense of being on a heroic mission intensified:

> They put me in the old-style state hospital, [where] I would end up in a cell, [sometimes] stripped naked in a cell with no bathroom. I was in these extremely altered states, and it was all part of the process. In fact, in the journey from the private hospital to the state hospital, I felt like I was going into the regions of Hell to rescue all the beings. . . . I had experienced heavens and hells, and this had the distinct sense of going down from the heavens into the hells, and then to free the prisoners and hospital patients.

At times, his sense of heroism became so strong that he had visions of literally taking the world upon his shoulders: "I had distinct experiences of being Atlas taking on the world . . . taking the whole planet on my shoulders, handing it off to Heracles and back. All these myths came alive and were personified."

Byron describes experiencing a profound loss of any sense of ground during his psychosis: "My ground was fairly decimated. I had little or no grounding reference points for my experiences. The experiences were like rivers of non-ordinary dream states that carried me along." Along with the sense of losing any semblance of solid footing were experiences of the disintegration of his self. One particularly poignant example of this involves a vision "of being cut into pieces, a very vivid sense of being on a dissecting table with several doctors dicing me to bits." Finding peace with the experience of groundlessness and the loss of solidity would prove to be an important component of Byron's recovery process.

Byron was psychiatrically hospitalized for a total of about six months, and he was heavily drugged with antipsychotics during all of this time, though he stopped taking them as soon as he left. While he does acknowledge that the antipsychotics probably provided some kind of grounding, he did not find the kind of grounding they offered to be particularly helpful and, overall, he considered them to be more of a hindrance than a benefit in his recovery: "The drugs to me were poison, you know [laughs]. I won't take them. I'd rather be in a jail or something. That's why [John] Weir Perry did outstanding work. He had no drugs at Diabasis* house and the staff was instructed to allow people to go through their experience no matter how bizarre it was, 'cause there was kind of a certain element of safety, you know. That's what I needed."

Recovery

As Byron approached the six-month mark following the onset of his psychosis, the intensity of his visionary experiences began to diminish and he found that he began to reconnect with consensus reality in a somewhat cyclical manner:

> I had images of washing to the shores of consensus reality. I mean I was going through these cycles, and the cycles of experience sort of lost their spark, if you will, because these cycles became tiring in a way, and I was heavily medicated, which added to the mix, and it was kind of difficult. And then . . . everything kind of wound down. . . . It wasn't an epiphany of reconnecting to planet Earth because actually, in a way, those places were my real home. . . . It was like everything kind of wound down, and then I was suitable for discharge after about six months.

But after discharge, Byron found himself struggling deeply to find meaning in mundane existence. *Samsara* is a Sanskrit word that refers to the cycles of existence and is commonly used in Buddhism, Hinduism, and other Indian

---

* We will go into more detail of Perry's work in Chapter 10. Also, see J. W. Perry's *Trials of the Visionary Mind* (1999) for an excellent summary of the philosophy and treatment model used at the Diabasis house.

religions. In one definition of *Samsara*, it refers to the cycles of life, death, and rebirth (i.e., reincarnation), but it also refers to the cycles of existence within a given life (wake up, engage with the world, sleep, wake up, engage with the world, sleep . . . ; eat, defecate, eat, defecate . . . ; etc.), and it often implies the challenges of finding meaning within this cyclical existence. Byron recalls his own struggles to find meaning within Samsara:

> So there I was, you know, and now what. Now what. What do I do, you know. Nothing that was part of this world really had any attraction for me, you know, like what do I do, how do I find a reconnection? I didn't want to go back to the state hospital, or go back to that place where everything was unavailable, but I was sort of in an in-between place still. So there I was. What now, what next. I hung out with some old friends, but I couldn't really reconnect, you know. . . . I enrolled in college, I'd gone back to school, but it didn't hold the spark. . . . Samsara actually all my life was very uncomfortable for me. . . . One of the images that came was that it was gonna be a long time before I found my integration, and it then took decades before I really connected with my home, which is Tibetan Buddhism [laughs].

As Byron struggled to find meaning within mundane existence, he came across Ram Dass's book, *Be Here Now*, which inspired him to find a teacher: "That was a major book. I absorbed it, you know. It was like I read it cover to cover right away. It talked to me, and..uh..how do I find my guru, how do I find my teacher?" It took Byron another thirty years to connect with a teacher and a teaching that really resonated with him and finally provided him with the real meaning and stability that he had been longing for: "Over the last ten years . . . I picked up the Tibetan Buddhist practices. . . . I have a wonderful teacher now, born in old Tibet and all that [laughs]. . . . That's how I found my ground after all these years."

Byron believes that his contact with spiritual teachings and teachers has played an invaluable role in his recovery:

> The psychosis alone did not produce the positive changes I experience today. The changes happened over many years conjunct with spiritual practices. . . . I've been very, very fortunate to connect with wisdom teachings from teachers and authentic spiritual masters. The blessings of the masters and the practice of the teachings have made a tremendous difference in my life. Without these influences, I don't think I'd be in a very good place today if I was here at all.

One teaching that has been particularly important to Byron is "the two major principles identified in Buddhism as wisdom and compassion. They are seen as two wings of a bird. One without the other is not sufficient. A bird to fly needs the two wings to work together." A major component of Byron's recovery process has been his struggle to integrate these two principles:

> It's interesting that one of the major vision/hallucinations I had was of a very prominent infinity sign in the sky. One loop was blue, the other red. I understood that I had experienced the one part, perhaps the blue [referring to wisdom]. It was my task to integrate the red loop [compassion] through the subsequent years of my life. I had tremendous visionary insights in my psychosis. I could, however, do nothing with them without grounding in areas of love and compassion. Again, I'm still on this journey.

In the thirty years prior to finding his Tibetan teacher, Byron has practiced a number of different types of meditation, first taking up these practices during a visit to India, and he has found them to be a very important resource in his recovery: "I feel that they grounded me over the long term. . . . It's actually grounding the nonordinary, it's actually working with your own mind and the very subtle influences that come on over time. It really is a transition type of thing, 'cause that was my sort of..um..refuge and resource, the meditative experience."

Over the past ten years, Byron's meditation practice has transitioned to those practices that are associated with Tibetan Buddhism. Even though he has no longer been having psychotic experiences, he finds that these practices continue to play an important role in his ongoing healing and growth: "The Thangkas or the images, the mandalas, the peaceful and wrathful deities, all of them..to me, they're, you know, living archetypes. In the Jungian sense, they're in the collective unconscious, but they become very personal, as well. Mandalas really are forms for understanding the universe, so that's what I'm very involved in, and for me..for me it was a godsend."

Another important factor in Byron's recovery is his having found a loving and supportive wife with whom he has raised several children.

◇◇

As of the writing of this book, Byron is 61 years old and has not used any antipsychotic drugs or been placed in a psychiatric hospital in over 40 years. According to the criteria used in this study, he considers himself to be fully recovered for nearly 40 years, although he adamantly considers his recovery to be an "ongoing journey": "I'm a work in process, an unfinished being," a quality he believes he shares with virtually all of us. He currently works as a spiritual supports facilitator in the state hospital system, finding great joy and meaning in his work.

Like Theresa, Sam, and all of the other participants in this research, Byron feels that he has undergone a profound and primarily positive and healing transformation as a result of his psychotic process. We'll explore this in more detail in Part Four.

◇◇

Let's turn now to look at a number of different models of psychosis that I believe are much more in line with the recovery research than the medical model, keeping the stories of Theresa and Byron in mind so that we can ground these different models in actual lived experience.

Chapter 9

Transpersonal Psychology—Spiritual Emergency vs. Pathological Psychosis

As we make the effort to understand the nature of psychosis, it's important that we expand our horizons to try to understand psychosis as it relates to all major aspects of human experience—not only the physiological and psychological, but also the spiritual. We have already seen that the vast amount of research to date has attempted to understand psychosis from a physiological perspective, a venture that has so far proven profoundly unsuccessful. This lack of success does not necessarily mean that we will never find some significant physiological underpinnings of psychosis, but it does suggest that we may find more fruit venturing down other avenues, with one such avenue being the spiritual.

In most cultures throughout human history, it has been widely believed that those who have anomalous experiences may be contacting and/or interacting with spiritual realms of which the majority of the population is unaware, experiences commonly referred to as *mystical experiences.* However, as the understanding of psychosis has come to be increasingly dominated by the medical model in the West, this belief has been increasingly challenged. One result of this challenge has been significant debate and controversy regarding how and even if we should make the distinction between so-called mystical experiences and so-called genuine psychosis. In the field of psychiatry and even in mainstream psychology, the general stance regarding this issue is quite simple—all such experiences, whether their content appears to be of a mystical nature or not, represent psychopathology. It is likely that most psychiatrists as well as many psychologists would take this conclusion one step further, adhering to a strict belief in the medical model and therefore claiming that the root cause underlying all anomalous experiences is a dysfunctional brain. *Transpersonal psychology*, however, is a branch of Western psychology that has taken a very different stance.

Transpersonal psychology is one of the few branches of Western psychology that has looked closely at spiritual and mystical experiences and has considered such experiences to have validity; in fact, the study of these kinds of experiences is a key defining aspect of transpersonal psychology. Anthony Sutich, one of the founders of transpersonal psychology, described this relatively new field as being distinguished from other branches of psychology in that it emphasizes unitive consciousness, mystical awakening, and peak experiences[1].* Other definitions of transpersonal psychology have since been given, but most of them agree that the most prevalent themes within this field are spiritual experience and practice, states of consciousness, higher and highest potential, experiences beyond the personal self, and transcendence[2].

Since the birth of the transpersonal psychology movement in the early 1960s, numerous models that attempt to explain mystical experiences have emerged within it. This field's exploration of psychosis, however, has been much more limited, and what exploration there has been is often subsumed by the ongoing controversy regarding whether or not mystical experiences and psychotic experiences are fundamentally distinct phenomena. On one hand, I find it disappointing that attempts to make this distinction have so dominated transpersonal psychology's inquiries into psychotic experiences, as I believe that transpersonally oriented inquiries into psychosis have the potential to be far richer than they have been if they were not so constrained by this kind of dichotomous thinking. On the other hand, the reality is that this issue has become a major one, not just in the formal field of transpersonal psychology, but also in other spiritually oriented fields of inquiry within the West, and so if we are to seriously explore psychosis from a spiritual perspective, then this issue does need to be addressed. But before we turn to look more directly at the issue of whether or not mystical experience and psychosis are distinct phenomena, it will help if we first take a look at transpersonal psychology's general understanding of mystical experience.

SPIRITUAL EMERGENCE VS. SPIRITUAL EMERGENCY

In transpersonal psychology, mystical experiences are generally divided into two categories based upon their level of intensity. Spiritual emergence is the condition in which altered states of consciousness and other anomalous experiences may be experienced, but individuals are able to cope with them in a way that does not significantly interfere with other aspects of their lives. Of course, the degrees and intensity of this process will inevitably vary from time to time, but someone

* *Peak experience* is a term introduced by transpersonal psychology founder Abraham Maslow that refers to moments of a profoundly transpersonal nature that typically contain feelings such as tremendous awe, wonder, joy, and interconnectedness.

experiencing spiritual emergence would not say that they felt overwhelmed by the experience. Spiritual emergen*cy*, on the other hand, refers to a sudden and potentially overwhelming plunge into such experiences. Brant Cortright, a long-time professor and practitioner of transpersonal psychology, defined spiritual emergency more specifically as a process in which "the self becomes disorganized and overwhelmed by an infusion of spiritual energies or new realms of experience which it is not yet able to integrate"[3].

It is generally agreed that there are two variables that, when occurring together, tend to open the doorway to a spiritual emergency. The first is any spiritual experience* that is particularly difficult to integrate, and the second is a significant degree of stress. Regarding the first variable, those experiences that are the most difficult to integrate are typically the ones that most directly challenge an individual's personal understanding of the world and of themselves. Regarding the second variable, when a person is particularly stressed, his or her defenses and inner resources are likely to be weakened. According to Cortright, "it may be this very vulnerability or 'thinning' of the person's ego structures that allow spiritual experiences past the usual filtering mechanisms of the psyche"[4]. He said these stressors may be either physical, emotional, or spiritual, or some combination thereof. Physical stressors frequently include near-death experiences, pregnancy and childbirth, fasting, injury, or physical hardship. Emotional stressors frequently include emotional deprivation or loss, experiences that evoke emotional intensity, drugs (especially hallucinogenic drugs), and intense sexual experiences. Spiritual stressors typically involve some kind of intensive spiritual practice, such as intensive meditation retreats or vision quests.

Spiritual Emergency vs. "Genuine" Psychosis

According to researchers Turner, Lukoff, Barnhouse, and Lu, these stressors have been known to not only trigger spiritual emergency but also what they believe is more "genuine" psychosis[5], a point that brings us back to the widely held notion within the field of transpersonal psychology that mystical experiences and psychotic experiences are entirely distinct phenomena. This is an idea that has far reaching implications when it comes to understanding the nature of psychosis, especially from a spiritual perspective, so it is important that we take some time to grapple with this question: Is there a valid distinction between overwhelming mystical experiences (spiritual emergency) and psychosis? And if so, how do we distinguish one from the other?

---

* The term *spiritual experience* here refers to any experience that falls outside consensus reality and/or transcends one's ordinarily more limited sense of self.

If we were to interpret the concepts of spiritual emergence and spiritual emergency using the definitions presented in Chapter One, a spiritual emergence would consist primarily of *non-distressing* anomalous experiences, whereas a spiritual emergency would include significant *distressing* anomalous experiences, the same criteria used to define psychosis. So it seems we are left with two possibilities: Spiritual emergency and psychosis are two distinct and unrelated processes even though they both involve distressing anomalous experiences; or both of these are merely different manifestations of a common underlying process. Many, but not all, transpersonally oriented psychologists subscribe to the former, believing that while the line between spiritual emergency and psychosis is often difficult to discern, such a line does exist. These psychologists typically argue that spiritual emergency is a mystical experience that has the potential for great healing and beneficial transformation when the process is allowed to complete, whereas psychosis is almost entirely regressive and needs to be checked as quickly as possible to avoid an ever worsening spiral into degeneration[6]. The implications of this argument are that the most helpful interventions for each category of experience are essentially opposite, leading to a situation in which it is very important to distinguish one from the other.

A number of different models that attempt to aid in making such a distinction have been put forward by various transpersonally oriented psychologists, including Roberto Assagioli[7], Stanislav and Christina Grof[8], John Nelson[9], Ken Wilber[10], and others. There has been substantial overlap between these different models, however, especially regarding the belief that an important distinguishing factor is the ability of an individual going through such experiences to maintain insight into the process and the ability to distinguish between the individual's own process and consensus reality. Cortright wrote, "There is a better chance that some observing ego* is present in spiritual emergency than in a mental disorder. Many times in spiritual emergency the person is afraid of *going* crazy whereas in psychosis the person *is* crazy and lost in the experience, that is, there is little or no observing ego"[11].

Mike Jackson has put together a table (Table 9.1) that highlights most of the similarities and distinctions between mystical and psychotic experiences that have been suggested to date, mostly from the field of transpersonal psychology[12]. While this table appears to present a picture of two distinct processes taking place, Jackson has pointed out that actual case-study research has shown a very different picture. There has been a significant number of case studies conducted in which there was the attempt to make such a distinction, and in virtually

* The term *observing ego* as used here refers to the capacity to not get completely swept up within one's experience—to question and/or notice any differences between what one is experiencing/believing and what others may be experiencing/believing.

Table 9.1 Suggested Similarities and Distinctions between Mystical and Psychotic Experiences. (Source: Jackson, 2001, p. 170)

SIMILARITIES (found in both spiritual and psychotic experiences)	DISTINCTIONS: [S] Spiritual / [P] Psychotic
CONTENT	
Religious or paranormal content	[S] Sub-culturally based, socially accepted / [P] Idiosyncratic, bizarre, alienating
Belief in personal mission, divine calling	[S] Humility, recognition of personal fallibility / [P] Grandiosity, sense of infallibility
Experience of discarnate entities, "sense of presence"	[S] Benign, recognized entity / [P] Malignant, idiosyncratic entity
Sense of being guided by external power	[S] Volitional control is retained / [P] Involitional
Intense emotional experience	[S] Positive emotions / [P] Negative emotions
FORM	
Hallucinations—visions and voices	[S] Pseudo / [P] True
	[S] Visual modality / [P] Auditory modality
	[S] Mood congruent, coherent, friendly / [P] "First rank", chaotic, critical
Delusions/revelations	[S] Corrigible vs. [P] Incorrigible beliefs
	[S] Comprehensible vs. [P] Bizarre beliefs
	[S] Presence vs. [P] Absence of "insight"
PROCESS	
Duration	[S] Transient / [P] Extended in time
Creative problem solving process: (impasse–insight–resolution)	[S] Spiritual fruits (humility, altruism, creativity) / [P] Mental illness (self-centeredness, inability to function)

every case, it was discovered that the anomalous experiences of the participants could not be neatly divided along the clear division indicated in the *Distinctions* column of Table 9.1[13]. Rather, most of these participants experienced some degree of both categories of experience—both mystical and psychotic. Some expressed alternating between one category and the other; some expressed having experiences that fit mostly in one category while still occasionally or subtly having experiences in the other category; and others seem to have experienced no clear division at all, having experiences from both sides relatively equally and even simultaneously. So in spite of much theoretical speculation to the contrary, what we find when we look at the actual case study research is substantial evidence suggesting that both types of experiences—mystical and psychotic—are merely different manifestations of a common underlying process.

So what can we conclude from this? An advocate of the medical model would most likely suggest that this is evidence that all of these experiences are indeed the result of a common underlying process—a dysfunctional brain. However, as we have already seen, there are several major problems with this conclusion. First, there is a striking lack of evidence of brain pathology associated with such experiences; and second, it is well established that many people fully recover from psychosis and so-called schizophrenia. What appears much more likely, given the evidence at hand, is that all such experiences, whether we deem them mystical/spiritual or psychotic, can be seen as attempts to integrate extra-ordinary experiences and move in the direction of greater health and wholeness (while of course the success of such attempts varies greatly).

So, if we want to remain somewhat faithful to the model and terminology presented by the field of transpersonal psychology while also accommodating the emerging research, I believe we should consider abandoning attempts to draw clear distinctions between spiritual emergency and psychosis, and instead continue to explore the possibility that the substantial variation we see in the manifestation of this process and perhaps even in the recovery outcomes is due (at least in part) to variations in the strength of the individual's observing ego. In other words, perhaps we can say that those experiences that lie closer to the *strong observing ego* end of the continuum would be more appropriately termed spiritual emergency, whereas those experiences that fall closer to the *no observing ego* end of the continuum would be more appropriately termed psychosis (see Figure 9.1). We should also acknowledge that the strength of an individual's observing ego can vary substantially over time.

With this in mind, then, we may still find some guidance by the distinctions made in Table 9.1; but rather than using these distinctions to determine whether or not someone has so called "genuine psychosis," perhaps they can be better

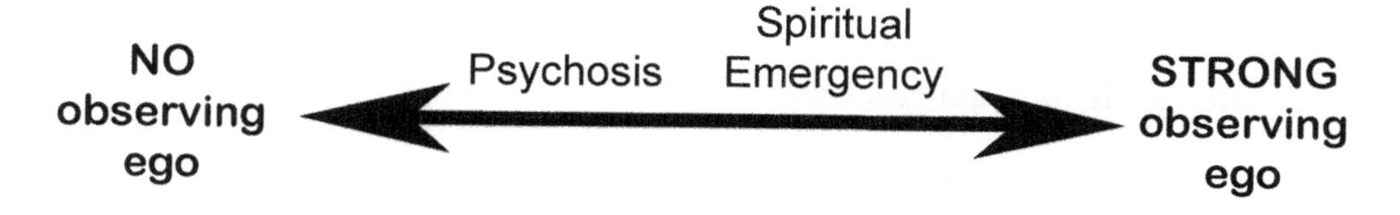

Figure 9.1. The continuum of psychosis and spiritual emergency, where those with a relatively weak observing ego during this process fall more on the "psychotic" end of the spectrum, and those with a relatively strong observing ego fall on "spiritual emergency" end of the spectrum.

utilized in simply offering some guidance in determining an individual's present status with regard to their ability to maintain contact with consensus reality, their level of distress, and their ability to meet their needs on the mundane level.

◇◇

When we reflect upon the stories of Theresa and Byron, we find that the general concept of spiritual emergency as defined within the framework of transpersonal psychology fits their experiences very well, especially if we consider spiritual emergency and psychosis to be merely different points along a common continuum, as illustrated in Figure 9.1:

- In both of their cases, there is clear evidence that their selves became "disorganized and overwhelmed by an infusion of spiritual energies or new realms of experience"[14] that they had a very difficult time integrating, which satisfies the basic definition of spiritual emergency.
- There is also clear evidence that a major factor in the onset of their psychoses was a "'thinning' of [their] ego structures"[15] resulting from significant physical, emotional, and spiritual stressors.
- Regarding the categories of experiences illustrated in Table 9.1, both Theresa and Byron clearly experienced a number of experiences in both categories (both *spiritual* and *psychotic*), lending further evidence to the idea that these two categories are not fundamentally distinct.
- We also see that as they moved through their psychotic processes, their experiences could be seen to move back and forth along the continuum illustrated in Figure 9.1. At times, they were both clearly lost in their experiences with very little observing ego, resulting in a condition generally associated with so called genuine psychosis. And at other times, they both clearly

experienced the presence of very strong observing egos while their anomalous experiences continued to unfold, resulting in a condition generally associated with spiritual emergency.

- Finally, we see evidence that both Theresa and Byron developed a much more resilient observing ego in direct correlation with their recovery, and that they successfully integrated their experiences in a way that allowed them to have a richer, more open, and much more resourced way of being in the world. It appears that these developments resulted in their capacity to keep one foot planted firmly within consensus reality while also maintaining an awareness and experience of the world that is significantly broader than the relatively limiting framework of consensus reality.

Chapter 10

Psychosis as a Renewal Process (John Weir Perry)

John Weir Perry was a Jungian oriented psychiatrist with more than forty years' experience working clinically with individuals suffering acute psychotic episodes (who were typically diagnosed with schizophrenia). He had the rare opportunity to perform many years of deep psychotherapeutic work with people going through these episodes without the use of medication. He concluded that "the process . . . which these millions of [schizophrenics] go through in a way that's usually so very hazardous, isolated and uncreative, is nonetheless made up of the same stuff that seers, visionaries, cultural reformers and prophets go through"[1]. In his own work directing Diabasis, a residential facility designed to support these individuals, "85 percent of the clients . . . not only improved, with no medication, but most went on growing after leaving us"[2]. In his many decades of research and clinical experience in this domain, he arrived at a compelling theoretical framework for psychosis, referring to it with the benign and hopeful term, *the renewal process*[3].

Stages of the Renewal Process

Perry came to the conclusion that psychosis is not the result of damage or impairment, but is actually initiated intentionally by the unconscious psyche (although it is generally uninvited and most likely even unwelcome by the conscious mind of the individual). He said that "when a person finds herself in a state of acute distress, in circumstances that have assailed her most sensitive vulnerabilities, her psyche may be stirred into an imperative need to reorganize the Self"[4]. In other words, the psychotic episode itself is not the main problem, but rather is the attempt of the psyche to address a serious problem that existed prior to it. When one develops a self-image and/or world-image that are unbearably limiting, the psyche initiates a profound reorganization.

During this shift, Perry believed that the deepest levels of the psyche are activated and draw a disproportionate amount of energy in the organismic system, leaving a very low level of energy available to the so-called higher functions (those responsible for relating to the external world). When this happens, a person is likely to find their field of awareness flooded with archetypal *affect-images* (sensory perceptions and emotions that are intricately bound together), and often these very dramatic perceptions of one's inner reality become confused with external reality. This confusion often leads to one of two persistent tendencies—*identification* and *projection*—the very tendencies that often provoke onlookers to perceive the individual as being so "crazy."

With *identification*, an individual may personally identify with the archetypal affect-images and literally believe oneself to be, for example, the Virgin Mother or the Second Coming of Christ. This tends to happen with affect-images that are of a creative, or more desirable, quality. On the other hand, when an individual experiences archetypal affect-images of a destructive, or less desirable, quality, then they are likely to *project* these onto the external world, believing themselves to be stalked by CIA agents, malevolent aliens, or other powerful perpetrators. From this perspective, then, such identifications and projections are not manifestations of senseless madness, but are grounded in very real universal archetypes. The individual is merely overwhelmed and confused by these energies, and with time and the presence of other caring individuals who are validating and empathic, this person is likely to move through this stage and integrate these experiences into a more, rather than less, accurate and helpful construct of the world and of the self.

An important aspect of the renewal process is that in order for such a profound reorganization of the self to take place, a profound disintegration must first take place, followed by a thorough reintegration. During this process of disintegration and reintegration, one's self-image and one's world-image tend to go through a parallel process of dying to old ways of being and being reborn into new ways of being, a process that is rarely linear, often involving a variety of disintegration and reintegration experiences in a more or less unpredictable manner. During experiences of disintegration, one may literally believe that they have physically died or are on the verge of death. They may also have the sense that their very being is on the verge of succumbing to a total annihilation that is even more profound than physical death. This is often a very terrifying stage. During experiences of reintegration, one often has profound experiences of "rebirth and of world regeneration"[5]. Often included within this stage are experiences of messianic affect-images, recognition of the unity of all things, and visions of a new world guided by compassion and love for all beings.

Perry discovered that most of the individuals he worked with who were suffering from acute psychosis and who were allowed to move through it in a supportive way worked through the process to resolution in about forty days. He found that this varied somewhat (with the length of time often being inversely proportional to the intensity of the episode), but the variance was much less than he would have expected. This time period (forty days) has fascinating implications when we consider the frequency with which this same number is used to describe the transformative periods of historical prophets (of Esdras and Jesus in the Bible, and of the world-destruction of the deluge, for example).

The Goal of the Process

When the renewal process is allowed to move through to successful resolution, "a new sense of oneself appears along with fresh interests and motivations in the world"[6]. Throughout the process, "the motivations and capacities that lead to lovingness and compassion"[7] are stirred up, and upon resolution, these qualities become the core values which guide one's life: "This may be experienced both as warmth and intimacy moving into one's relationships, and also as a direct sense of the oneness of all beings—not just a belief or view of how things are, but as the actual experiential realization of it"[8].

Supportive Therapy

Perry suggested that when supporting a person going through a psychotic process, it is important to keep in mind that it is not the psychotic process from which one needs to be healed—this process itself represents an attempt of the organism to heal itself from and grow beyond the limiting self-image and world-image of the person. Rather, the best support we can offer is to support the process itself, rather than try to suppress or reverse it; and the best way to support it is to offer the individual our "*clarita* (clarity) and *carita* (caring)" above all else[9]. Perry's use of the term *clarity* refers to our ability to be genuinely present with the person, and to make the effort to understand and support the person in finding meaning for their experiences. In this regard, it is important to acknowledge that no amount of education or training could ever put us in a position where we know more about the meaning of the person's experience or the experience that they need to have at any given moment than their own psyche. Our job is simply to facilitate this process to the best of our ability. Perry's use of the term *caring* refers to our aspiration to hold the person and all of their experiences, no matter how challenging they may be, with respect and unconditional positive regard.

Implications for Cultural Reform

Perry pointed out that there is an aspect of the renewal process that is crucial for the health of a society. Once an individual has gone through a very profound reorganization of their inner culture, they sometimes emerge with a vision of reorganization for the culture at large that is uncanny in its timeliness and its effectiveness in bringing about just the reform that society needs in order to continue its existence. Perry cited numerous examples of social visionaries and reformers who have done exactly this, especially pointing out a number of Native Americans who played this role as their societies struggled so hard to survive "in the new conditions of white dominance"[10]. This observation, he said, suggests that "the psychic healing process . . . does its work in two principal areas: on one side there is self-healing in the individual persona's renewal process, and on the other, cultural healing in the visionary work of the prophet"[11].

Chronic Schizophrenia

Unfortunately, individuals going through such a process in our society are very likely to be met with fear and invalidation, a response that Perry felt often has the effect of blocking this natural process and leaving the individual in a lost and confused condition indefinitely. In Western society, we generally expend great effort to stop this process in an attempt to return the individual's behavior and experiences to "normal" as quickly as possible. Yet Perry believed that this very attack on the psyche's attempt to heal itself is one of the main reasons why there is such a poor recovery rate in the West, as compared to much more spiritually mature societies in which such visionaries are often met with validation, respect, and even honor. If Perry is correct, then not only is it a very sad irony that our very attempt to support these individuals may actually be one of the main factors in preventing their recovery, but the implications this has for our society as a whole are also quite grave. In a time when our society is in desperate need of the guiding values of love, compassion, and wisdom, suppressing the very process that attempts to renew these values in wounded psyches and in our society at large could be very detrimental indeed.

◇◇

Reflecting upon the stories of Theresa and Byron, we can see that Perry's model appears to fit their experiences quite well:

In both of their cases, we find that their psychoses were very likely initiated directly by the psyche after the development of a condition that was no longer sustainable. In particular, they both found themselves in the position of longing for deeper connection and intimacy with others while at the same time feeling extremely insecure and fearful of such connection.

We also see evidence that their psychoses consisted of a profound reorganization of the self in that they both underwent a profound and lasting transformation as a result of their psychoses.

They both spent substantial periods of time in which their consciousness was flooded with *affect-images* (sensory perceptions and emotions that are intricately bound together). One of the most pervasive affect-images for both of them consisted of overwhelming visions and impulses linked together under the theme of messianic striving. Another theme of affect-images common to both of them was that of being guided by powerful benevolent entities (the strong black man and guard dog for Theresa, and the dakinis for Byron). They also both experienced affect-images of being persecuted by malevolent entities—for Theresa, a prominent form of this was a feeling of being spied upon and persecuted by unknown entities, and for Byron, a prominent form of this was the sense of being dissected by ruthless doctors.

Closely related to these affect-images were the *projection* and *identification* of good and evil forces. We can say that Theresa projected good and evil onto the world in a number of different ways—projecting archetypal evil/destruction with her visions of fire and hell realms, her sense of overwhelming suffering in the world, and her sense of the persecuting spies; and projecting archetypal good/benevolence with her visions of heaven and her sense of protective entities. We can also say that she *identified* with good in her taking on the role of a messiah. We can say that Byron projected good and evil onto the world in a number of different ways—with his many visions of great battles between good and evil, and in his experiences of the ruthless doctors and of the benevolent dakinis. And we can say that he identified with archetypal good in his various messianic and heroic experiences of himself.

Perry suggested that *clarita* (clarity) and *carita* (caring) were important factors in supporting one's recovery, and it's clear that this was true for both Theresa and Byron. Theresa found clarity in her exploration within the development courses, her artwork, and by having the fortune to move through her process with minimal hindrance. She found caring from the psychiatric survivors group and from her future husband. Byron found clarity especially as a result of his own spiritual seeking, and he found caring from many of his spiritual teachers and from his loving wife.

Finally, Perry suggested that the successful resolution of a psychotic process results in the establishment of love and compassion as the core values that guide one's life. There is no doubt that both Theresa and Byron have developed a strong desire and a great capacity to bring these values into their lives and work.

Chapter 11

The Process Paradigm (Arnold Mindell)

Arnold Mindell has developed a theoretical orientation he calls *Process Oriented Psychology* (which he also refers to as the *process paradigm)*, a model based particularly on concepts from Jung's work and on various field and systems theories[1]. Mindell's model provides a unique and interesting perspective from which to look at the phenomena of psychotic and mystical states, to which he collectively refers as simply *extreme states*. Let's first take a brief look at each of the most essential relevant concepts of Mindell's model and then bring them into a more complete description of how extreme states are understood from this perspective.

Key Psychological Components in the Process Paradigm

The core principle in the process paradigm is that all of us have both *primary processes* and *secondary processes* operating within us all of the time. The primary process is the process within our psyche with which we are presently most identified. The secondary processes are those processes within our psyche with which we are least identified, and of which we are likely not even fully conscious (these are closely related to Jung's term, *the shadow*).

The *feedback loop* refers to our ability to take in information from the external environment so that we may modify our own behavior and beliefs accordingly. Mindell suggested that all of us have filters acting on our feedback loop, to a greater or lesser degree, and that all of us have significant blind spots in our perceptions of the world that often serve us in minimizing information that might be painful and/or difficult to integrate into our personal worldview.

The *metacommunicator* is that aspect of ourselves that is able to remain somewhat detached from our experience and comment on what we are communicating, expressing, feeling, thinking, etc. This is closely related to the concept of an observing ego discussed earlier. Mindell suggested that all of us have access to a metacommunicator, but that this access can fluctuate dramatically, even in so-called ordinary people. For example, most of us find it difficult to access our metacommunicator while experiencing intense fear or anger.

The Difference Between "Normal" States and Extreme States

When we perceive someone as crazy or as a schizophrenic, what exactly are the qualities that lead us to make this distinction? Mindell suggested that there are three main differences between being in an extreme state and being in a normal state, and that these differences manifest in behavior that is often easily perceived by onlookers.

One difference is that, in an extreme state, the feedback loop is very limited or even nonexistent. By blocking the feedback loop, the individual is able to completely "filter out signals which oppose his belief in order to preserve and complete the inner story or myth he is working on"[2]. To an observer, this may give the appearance of inappropriate affect and/or sticking stubbornly to a "delusion"; however, while it may be true that the individual's affect and belief system are inappropriate relative to the what is occurring in consensus reality, it is likely that they are very appropriate to what is going on within the person's inner world.

A strongly limited or missing feedback loop is a necessary condition for extreme states, but is not sufficient by itself. The second necessary factor is that the metacommunicator is very weak or even temporarily nonexistent. Whereas a normal person occasionally goes through periods of limited or even no metacommunication (such as when extremely angry, fearful, etc.), someone we label as psychotic goes through much more extended periods without metacommunication. Without a metacommunicator, the primary and secondary processes are free to rapidly exchange places (what Mindell refers to as *flipping*) or to even superimpose upon one another, giving the appearance that the individual's personality has suddenly and dramatically changed and/or is highly unstable. Mindell suggested that we all have the potential to *flip*, given that we all have primary and secondary processes, but that our metacommunicator generally does not allow this to happen. This could explain why even "normal" people can go through dramatic personality changes during times when their metacommunicator is particularly weak, such as when intoxicated with alcohol or drugs or when overwhelmed by strong emotions.

The third factor necessary for the onset of an extreme state is the presence of a strong conflict between the primary and secondary processes. Mindell suggests that virtually all of us have some degree of this kind of conflict within us, and in fact it is this very conflict that lies at the root of most of our suffering. In a relatively mentally healthy individual, this conflict is relatively minimal, and in a so-called neurotic individual, this conflict is generally somewhat stronger. For an individual susceptible to flipping into extreme states, Mindell suggests that this inner conflict is particularly severe, and the individual has been all too successful in avoiding the painful confrontation, and therefore healing integration, between these different processes.

Benefits of Extreme States for Society

Like Perry, Mindell suggested that even though extreme states often wreak havoc on the life of the individual experiencing them as well as on the lives of others closely related to the individual, extreme states play a very important role in maintaining the health of a society. He suggested that the world can be seen as operating like a field that (1) has its own awareness, and (2) "does everything it can to bring itself to consciousness"[3]. In psychotic people, in whom the meta-communicator is weak or temporarily nonexistent, the field finds a channel (the psychotic person) with which to express itself to the rest of the world:

> The field thus informs the general public about its conflicts. This can be formulated in different ways, depending upon belief systems. One might say that the Self wants to know itself, or that God is trying to discover himself, the Anthropos we are living in is trying to wake up, the collective unconscious is trying to express itself or the universe is evolving in such a way as to make us more aware of the meaning of life.[4]

The process paradigm offers us a particularly useful model for understanding the oft-observed irony that society tends to view an individual immersed in an extreme state as ill, while that individual tends, simultaneously, to view society as ill. According to the process paradigm, each represents the shadow (the secondary process) of the other:

> In a given collective, the schizophrenic patient occupies the part of the system in a family and culture which is not taken up by anyone else. She occupies the unoccupied seat at the Round Table, so to speak, in order to have every seat filled. She is the collective's dream, their compensation, secondary process and irritation[5].

◇◇

Returning to look at the stories of Theresa and Byron through the lens of the Process Paradigm, we find that many of their experiences fit well within the concepts of this paradigm:

For both of them, we see evidence of a missing feedback loop at times, particularly during those times when they were almost completely disconnected from what was taking place within consensus reality.

We see evidence of each of them having a virtually nonexistent metacommunicator at those times when they were completely lost in their experience, so absorbed in what they were experiencing that they were unable to question the validity of their anomalous beliefs and perceptions with regard to consensus reality. It's important to notice that both their feedback loops and metacommunicators appear to have varied dramatically over time and even from one moment to the next.

We see clear evidence of a profound conflict taking place between various inner processes for both Theresa and Byron, and relatively frequent "flipping" between them. They both experienced dramatic flips on a number of different levels at different times: between feelings of omnipotence and feelings of powerlessness; between identifying as a supremely confident visionary and being struck down by crippling self doubt; between identifying as a hero and identifying as a victim; between feeling powerful creative energies/impulses and feeling powerful destructive energies/impulses; and between the perception of being taken care of by benevolent entities and being persecuted by malevolent ones.

Finally, we can say that both Theresa and Byron embodied the "shadow" of society quite frequently during their processes, having very clear insights into the various ills, harmful practices, and general suffering within their societies while at the same time being identified by society as the ones who were ill.

CHAPTER 12

Seeing Through the Veil of our Cognitive Constructs (Isabel Clarke)

In this chapter and the next, we look at two different models that have taken a more cognitive approach* to understanding psychosis while still maintaining that psychosis is ultimately a growth oriented process, as do all of the models presented here.

Isabel Clarke, both a researcher and a clinical psychologist with extensive experience supporting people struggling with psychosis, has formulated a cognitively oriented model of psychosis that suggests that psychosis results from an imbalance in the mental systems that we use to make sense of the world. She places particular emphasis on her belief that the distinction between mystical and psychotic experiences is merely superficial—that most experiences from both of these categories arise from the same underlying process[1]. To emphasize her belief in this regard, she prefers to use the single term *transliminal experiences* to capture all such anomalous experiences.

Clarke argued that psychosis, spiritual experiences, and even social movements such as revolutions arise from a common process that involves a shift or discontinuity in consciousness. She proposed that the key to understanding the mechanism beneath this shift in consciousness, along with the associated transliminal experiences, can be found within a cognitive model first put forward by John Teasdale and Phillip Barnard in 1993—the *Interacting Cognitive Subsystem* (or *ICS*) model[2]. In this model, the various modalities of human information processing (such as hearing, sight, language, etc.) are managed by two overarching meaning making systems: the *propositional subsystem* and the *implicational*

* By *cognitive approach*, I mean that these models emphasize cognition—the mental processes involved in making sense of the world via perception, thinking, interpreting, and reasoning.

subsystem. These two systems provide us with two very different yet complementary ways of making sense of the world.

The propositional subsystem represents our logical mind and is capable of fine discrimination. It categorizes our various sensory perceptions of the world so that we can distinguish one object from another. We can think of this subsystem as being primarily concerned with the external world. We store the information and memories derived from this system verbally, so this system is closely associated with our use of language, words and symbols. Clarke suggested that there may be some correspondence between the propositional subsystem and our neocortex, the most recently evolved region of the human brain.

The implicational subsystem, on the other hand, represents emotional meaning and holistic perception. Our view of the world via this subsystem is in direct contrast to that of the propositional subsystem. Instead of being devoted to a relatively dispassionate understanding of the external world, the implicational subsystem is more focused on the experience of our inner world, being directly associated with our emotions/feelings and assigning value to our present experience. It is particularly concerned with determining worth of the self and threat to the survival or position of the self. Also, whereas the propositional subsystem is more concerned with making distinctions between things, the implication subsystem sees the world as an interconnected whole. We store the information and memories derived from this subsystem via several sensory modalities and emotion. Clarke believes that there may be some correspondence between the implicational subsystem and the older and deeper regions of the brain.

Clarke suggested that in everyday functioning, these two systems are in relative balance, but in transliminal (mystical/psychotic) experiences, the implicational subsystem is primarily running the show. When this subsystem is dominant, we experience the world in relatively undifferentiated wholes and are unable to manage fine discrimination; our experience is emotionally charged, often swinging sharply between euphoria and terror, and we find ourselves particularly concerned with the self and threats to the self. Putting this another way, we can say that when we experience the world primarily through the implicational subsystem, we are peering beneath the veil of our discriminating constructs to some degree and experiencing the world in its more raw form.

Clarke suggested that the primary difference between so called mystical experiences and so called psychotic experiences is that in mystical experiences, a return to some balance between these two subsystems occurs naturally and relatively quickly, generally leading the individual to feel more integrated than before the experience; whereas in psychotic experiences, the orderly return does not happen in such a timely manner, and even when a return does happen, the individual often feels destabilized and is easily susceptible to more destabilizing transliminal experiences: "For the person with psychosis, the barrier that makes this sort of experience hard to access for most of us is dangerously loose"[3]. Clarke agreed with other writers such as Chadwick[4] and Laing[5] that it is one's ego strength (the establishment of a stable sense of self) that "will predict whether [a transliminal experience] is a temporary, life enhancing, spiritual event, or a damaging psychotic breakdown, from which there is no easy escape"[6].*

Clarke felt that transliminal experiences are natural and even essential for a healthy society, but that Western society, in its drive for the material and the graspable, has marginalized ungraspable transliminal experiences, and with this marginalization, it has lost its sense of interconnectedness and comfort with mystery and the unknowable. She suggested that the original role of religion and spirituality has been to keep us connected with this holistic, mysterious, interconnected realm. However, by concentrating on the fate of the individual soul rather than on our interconnectedness with a greater whole, she feels that our modern religions have let us down, allowing "the technology born of our ferocious power to discriminate and to bend the material world to our will [take us to] the point where the sustainability of our species is put in question"[7]. Clarke suggested that "without in any way failing to recognize the suffering associated with psychosis, we could respect the experience of those suffering in this way for its connection with the sacred state"[8].

◇◇

Reflecting on the stories of Theresa and Byron through the lens of this model, we can say that, in both of their cases, their implicational subsystems were primarily running the show for substantial periods of time during their psychoses. Recall that the implicational subsystem is primarily involved with perceiving the world holistically, assigning value/feeling to present experience, and determining worth and threat with regard to the self. The propositional subsystem, in contrast, is primarily concerned with discriminatory awareness, assigning categories to our perceptions and experiences, and is closely related to our ability to keep our understanding and experience of the world in alignment with consensus reality. For both Theresa and Byron, it's clear that the propositional

---

* We can see that there is some similarity here with the other models presented previously, in which it is generally believed that the strength of one's observing ego plays an important role in determining one's susceptibility to psychosis as well as the likelihood of successful resolution of such experiences. While the concept of *observing ego* (the capacity to maintain a somewhat detached awareness of one's experience) is somewhat different than the concept of *ego strength* (the stability and well-foundedness of one's sense of self), it is likely that there is significant correlation between them.

subsystem was very weak at times, as their own constructs of their selves and the world often fluctuated wildly. And it's clear that, for both of them, the implicational subsystem was predominant during much of the time, as they had many experiences of extreme threats to the self, holistic perception of the world and their selves, and fluctuation between euphoria (particularly associated with heavenly realms and expansiveness) and terror (particularly associated with hell realms and persecution). Self worth was also clearly a major factor in many of their experiences, especially during the times when they were overcome with feelings of messianic status and heroic striving.

Chapter 13

The Creative Process Gone Awry (Mike Jackson)

Mike Jackson, a clinical psychologist with extensive experience researching the overlap between mystical and psychotic experiences, developed a model based upon a pioneering research study he conducted that provides a compelling framework for understanding what may be happening at the cognitive level during psychosis, while also making sense of the often observed differences between so called mystical and so called psychotic experiences. The study was a qualitative multiple case study involving 18 participants, and the intention was to explore the relationship between benign mystical experiences and psychosis[1]. The participants were separated into two groups— those who were diagnosed with a psychotic disorder in one group and those who were undiagnosed and seemed to have experienced benign mystical experiences in the other. In summary, Jackson concluded that participants in both groups had gone through or were going through a similar "basic, adaptive psychological process, which is also observed in artistic and scientific creativity"[2]. Jackson discovered that a model of the process of creativity developed by Batson and Ventis[3] (which in turn drew from the work of Wallas[4]) fit the findings of his study very well.

This model describes the cognitive processes underlying creative thinking as taking place in four steps:

1. *Preparation and impasse*: First, one becomes aware of a problem and tries to solve it using normal problem-solving strategies, but eventually, finding it impossible to make further progress, arrives at an impasse.
2. *Incubation*: The impasse creates cognitive and/or emotional tension, leading to conscious withdrawal from the problem and a period of incubation.

3. *Illumination*: At some point, a solution suddenly emerges from the unconscious, resolving the impasse, typically by means of a shift in one's personal paradigm.* The illumination often manifests nonverbally, and is often felt to have originated from an external source.

4. *Verification*: The illumination is formulated in rational terms, and its validity is tested empirically. Upon verification, the original cognitive/emotional tension that initiated the process is finally released.

This process corresponds with the epiphanic "Aha!" experience to which most of us can probably relate.

Batson and Ventis suggested that mystical experiences involve this same process, but that they involve existential rather than intellectual problems, typically much higher levels of emotional tension along with cognitive tension, and solutions that involve metaphysical rather than theoretical paradigm shifts[5]. Jackson added to Batson and Ventis's model by suggesting that psychotic experiences can also be the result of this very same process.

Jackson suggested that both mystical and psychotic experiences begin with the same underlying intention of finding a solution to an existential problem, but that in the case of a mystical experience, the process resolves successfully, whereas in the case of a psychotic experience, the process does not. In the case of a mystical experience, we can see the process as acting like a negative feedback loop: significant tension is generated along with a sense of crisis; a necessary shift in the individual's paradigmatic framework subsequently takes place; the tension and sense of crisis fade; and the feedback loop is completed. A psychotic experience, on the other hand, can be seen as a *positive* feedback loop: upon generation of tension and a sense of crisis, a paradigm shift takes place, but for some reason this shift fails to resolve the tension and may even exacerbate it. This may then lead to another unsuccessful paradigm shift, leading to even further tension and the potential for an ever worsening spiral of chaotic paradigm shifts and eventually florid psychosis.

This concept corresponds well to what is often seen in acute psychosis, when belief systems and emotional states can change and alternate very rapidly. Jackson suggested that one possible reason for the lack of successful resolution in the case of psychosis is that the new belief system itself may be highly distressful (such as in the case of persecutory delusions) or the new paradigm may have validity, but if it does not conform to consensus reality within an intolerant society, harm to the individual's relationships with others and/or harm to the individual's confidence in his or her sanity may lead to increased distress.

* *Personal paradigm* is a term that will be used occasionally from here on. It refers to one's experience and understanding of the world and one's self.

It's important to emphasize that the underlying process illustrated in this model is considered to be healthy, adaptive, and actually an essential component of our ability to navigate our way through the world. According to Jackson, it is only a small minority of cases in which this process fails and we see the manifestation of psychotic experiences.

◇◇

Reflecting upon the stories of Theresa and Byron, we can say that their experiences fit this model quite well in that, in both of their cases, their psychotic processes began with an existential problem, which led to an unsustainable resolution (a paradigm shift) that ultimately led to further problems and further paradigm shifts. In both of their cases, we can say that this process finally came to a general completion with the development of lasting and relatively stable transformations in their personal paradigms.

We can say that Theresa began with the existential problem of experiencing a deep longing to connect with others and to ultimately bear a child and raise a loving family while simultaneously having tremendous fear of intimacy and closeness. We see evidence of a series of resolutions/paradigm shifts that provided temporary relief from this problem but which ultimately failed to provide a lasting resolution: using alcohol and drugs to tolerate closeness; taking on the role of a messiah striving against evil/suffering in the world (which we can say allows connection with others in the sense of saving them while allowing her to remain at a safe distance as their savior); and connecting with a profound universal love which she was unable to sustain. Theresa finally successfully resolved this dilemma by going through a very deep healing process (including a profound transformation in her personal paradigm) that allowed her to connect with others again, develop a relationship with a compatible partner, and finally raise a healthy family.

We can say that Byron's existential problem was very similar to that of Theresa*—experiencing a deep longing for connection with others and for a more universal spiritual connection while simultaneously feeling trapped within a highly constrictive and isolated experience of himself. In his story, we also see that his initial resolutions/paradigm shifts were ultimately unsustainable

---

* We can identify other existential problems with which Theresa and Byron may have been grappling, such as finding meaning, freedom, etc., but I believe that the problems I mention here for each of them are likely to have been the ones that were most in need of an immediate resolution. Also, I believe if we look more closely at these various existential problems, we will find that they all share common roots (an idea we will explore in much more detail in Parts Three and Four).

and therefore required further resolutions/paradigm shifts. He began with a powerfully connective experience at Woodstock, but upon not being able to fully integrate this experience, resorted to extreme efforts to carry out a rebirth experience. This in turn led to further overwhelming experiences and further unsustainable resolutions (messianic striving, etc.), until he was finally able to cultivate a spiritual path and practice that allowed him to soften his constrictive sense of self in a much more sustainable way.

Chapter 14

The Life Fear / Death Fear Dialectic (Otto Rank)

The final three models of psychosis that we look at in this part of the book have been formulated by more explicitly existentially oriented* thinkers—Irvin Yalom, Rollo May, and Ernest Becker. Before we look at these models, however, it will help to first take a brief look at Otto Rank's *life fear / death fear dialectic*†, which laid the foundation for the works of Yalom, May, and Becker.

Rank (1884 – 1939) was a protégé of Sigmund Freud for nearly 30 years before breaking away and forming many of his own alternative ideas in the late 1920s, many of which had a strong existential flavor. While he did not offer a particularly comprehensive system for understanding psychosis, he did formulate a model of a core existential dilemma that would eventually become an important foundation for the work of a number of existentially oriented thinkers, such those mentioned above, who would later develop more coherent models of psychosis.

In 1936, Rank published the essay, *Will Therapy*[1], in which he first introduced his idea that at the core of all fear is a dynamic dialectical system comprised of two poles (see Figure 14.1). At one pole lies what Rank referred to as *life fear*, which he defined as "the fear of having to live as an isolated individual"[2]. At the other pole lies what Rank referred to as *death fear*, which he defined as ""the fear of the loss of individuality," the fear of losing one's sense of self by being engulfed by too much merger and connection[3].

* While there are a number of different connotations for the term *existential*, with my use of it here and generally throughout the book, I am referring to that which is related to the most fundamental dilemmas with which we all must struggle as living, conscious beings—maintaining our existence, finding meaning, finding joy, minimizing suffering, etc.

† The term *dialectic* refers to the tension that exists between two opposing forces.

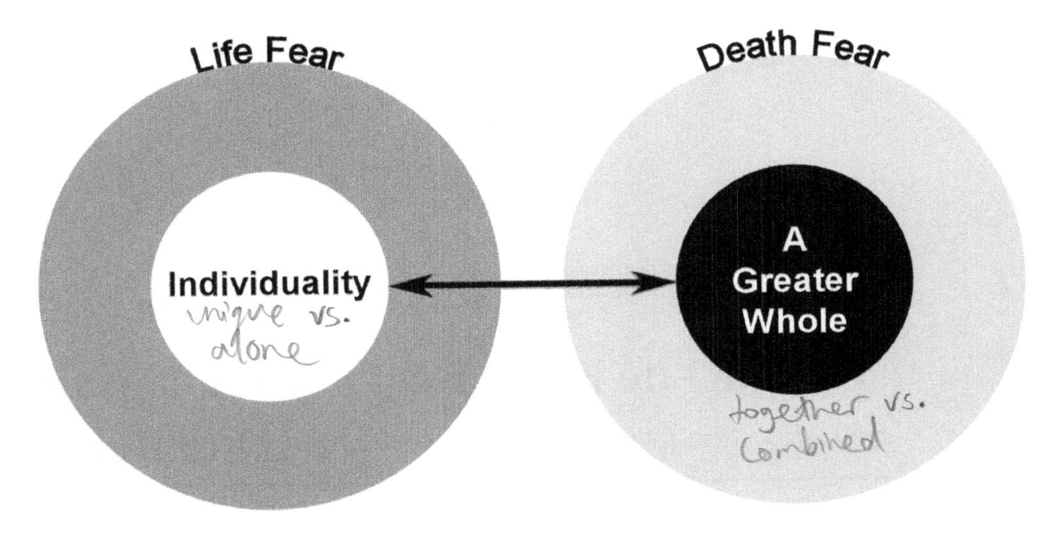

Figure 14.1. A graphical depiction of Rank's life fear / death fear dialectic.

Rank suggested that, as we move through life, we find ourselves perpetually striving towards one side or the other as we seek to find some tolerable middle ground between these two fears. We respond to the fear of isolation—life fear—by moving towards deeper merger and connection; but as we do so, we find that we begin to lose our sense of self and are faced with the threat of self-annihilation—death fear. So, we respond to death fear by moving back towards life and individuation, but as we move in this direction, we once again find ourselves confronted with the fear of isolation (life fear), and are compelled to move once again in the direction of more connection and merger (death fear). Rank suggested that we find ourselves perpetually torn in this way, bouncing back and forth, seeking some middle ground that we can tolerate.

Chapter 15

Overwhelmed by Death Anxiety (Irvin Yalom)

Irvin Yalom is generally considered to be one of the leading contemporary thinkers in Western existential psychology and psychotherapy, having literally "written the book" on the subject (*Existential Psychotherapy*[1]). Nearly half a century after Rank formulated his life-fear/death-fear dialectic, Yalom formulated his own model of psychopathology and psychosis that utilized this dialectic. Yalom suggested that "death . . . is the primordial fount of psychopathology"[2], and that all humans struggle to a greater or lesser degree with the fear of death, even though this struggle is primarily unconscious for most of us. He suggested that when one's strategies are not entirely successful in warding off the fear of death, then one becomes vulnerable to experiencing various neuroses*; and when one is completely overwhelmed by death fear, one becomes vulnerable to developing psychosis. In order to better understand how Yalom suggested psychosis may occur in this way, it will help if we first take a closer look at his model that describes how we all deal with death anxiety, regardless of our degree of contact with consensus reality.

Defining Death Anxiety

Yalom noted that many different definitions for the fear of death have been posited, including Jaspers' *fragility of being*, Kierkegaard's *dread of non-being*, Heidegger's *impossibility of further possibility*, and Tillich's *ontological anxiety*[3]. Yalom felt that it is difficult to come to an umbrella definition that includes all

* While the specific definition of *neurosis* varies somewhat, for our purposes here, we can simply define neurosis as an intrapsychic conflict, and "a neurotic" as someone who experiences a relatively high degree of intrapsychic conflict. More on this in the remainder of this chapter and in the following chapter.

of these variations since we all experience the fear of death somewhat differently, but ultimately established a preference for defining the fear of death as the fear of "'ceasing to be' (obliteration, extinction, annihilation)"[4]. He also made a point to draw a distinction between *fear* and *anxiety*. He borrowed Kierkegaard's distinction, defining fear as the fear of *some* thing whereas anxiety or dread is the fear of *no* thing[5]. In raw anxiety, one experiences the dread of losing oneself, of becoming nothingness. Since no object of this anxiety can be found, we feel utterly helpless in the face of it, and therefore anxiety has the potential to become more overwhelming than any specific fear. Since our fear of death is actually the fear of succumbing to the mysterious nothingness that death represents, Yalom suggested that it is more accurate to use the term *death anxiety* than to say that we have a fear of death.

Coping with Death Anxiety: Two Instinctive Methods

Yalom suggested that beginning in early childhood, we learn to cope with death anxiety using primarily the strategy of denial, which comes in the form of two different core beliefs: (a) we are "personally inviolable," and/or (b) we are "protected eternally by an ultimate rescuer"[6]. He suggested that while most of us use some combination of both of these delusional beliefs, an individual may rely almost entirely on one or the other in extreme cases.

The belief in one's specialness. It is likely that most of us who have inquired deeply into our personal experience would admit that we feel personally inviolable in some way. On a rational level, we admit that of course we will die just like everyone else, but deep down, there is a feeling that death and other such horrible things only happen to others. This delusional belief is what Yalom referred to as *specialness*. Of course, we are all doomed to have the falsity of this belief come crashing down on us one day; in the meantime, however, it does provide us with a very effective strategy for coping with death anxiety. This belief in one's specialness can manifest in different ways in different individuals, ranging from relatively benign manifestations to severely maladaptive ones, depending upon the desperation with which one holds onto it.

The strategy of specialness often manifests in the pursuit of heroism. In its more benign form, we find a relatively healthy individual who cultivates qualities such as courage, self reliance, and a sense of adventure. But as this strategy becomes more extreme, we find what Yalom called the compulsive hero. A compulsive hero is someone who must "seek out and conquer danger as a grotesque way of proving there was no danger"[7]. Yalom used Hemingway's life as an example in which this strategy became so extreme that it eventually led to psychosis. Hemingway's mother said that the first words he spoke as a child were, "'fraid of nothing.'" Yalom pointed out that the irony with this attitude is that one is trying to convince oneself that they are afraid of nothing precisely *because* they are afraid of nothing—*nothingness*, that is. As Hemingway grew older, because he had clung so tightly to his myth of personal inviolability, as the inevitable happened and this myth began to unravel, he completely fell apart. He first fell into a deep depression, then into paranoid psychosis, and he finally took his own life.

Other forms that the strategy of specialness can take are: workaholism, in which one attempts to prop up their belief in specialness by believing that they are constantly progressing or getting ahead; narcissism, which results when "a belief in personal inviolability is coupled, as it often is, with a corresponding diminished recognition of the rights and the specialness of the other"[8]; and the drive for power, whereby one attempts to "enlarge oneself and one's sphere of control"[9] as a reaction formation* against one's own sense of insecurity and limitation.

The belief in an ultimate rescuer. With the strategy of believing in one's specialness, the individual attempts to separate and individuate; with the strategy of believing in an ultimate rescuer, however, the individual takes the opposite approach and attempts to merge or fuse with another. The manifestations of this strategy are somewhat less diverse than those of the strategy of specialness—essentially, there is simply a belief in a supreme immortal other with whom one can merge and therefore defy mortality. This other may take the form of a supernatural figure such as a god or a goddess; or it may take the form of a country, a leader, or even a cause or value system. Yalom believed that, generally speaking, this strategy is more problematic than the specialness strategy. He pointed out that in attempting to merge with another, this strategy ultimately leads one directly into that which it is attempting to avoid—the loss of oneself (i.e., existential death). It leads to the failure to explore one's potentialities, the failure to self-actualize, and it generally results in living a highly limited and unstable life.

---

* *Reaction formation* is a term used in psychoanalytic theory that refers to the attempt to master unacceptable feelings or impulses by exaggerating one's behavior or beliefs in the opposite direction.

A Framework for Understanding Psychopathology and Psychosis

While the strategy of the belief in one's specialness and the strategy of the belief in an ultimate rescuer are clearly very different, Yalom suggested that the majority of us probably use both of these to some degree: "Generally one does not construct a single ponderous defense but instead uses multiple, interlaced defenses in an attempt to wall off [death] anxiety"[10]. He pointed out that rather than being mutually exclusive, these two modes of defense are actually complementary:

> Because we have an observing, omnipotent being or force continuously concerned with our welfare, we are unique and immortal and have the courage to emerge from embeddedness. Because we are unique and special beings, special forces in the universe are concerned with us. Though our ultimate rescuer is omnipotent, he is at the same time, our eternal servant.[11]

Yalom went on to suggest that because of the interplay between these two belief systems, we find ourselves torn between two opposing fears in a dynamic dialectic that closely parallels Rank's life fear / death fear dialectic:

> "Life anxiety" [which corresponds to Rank's *life fear*] emerges from the defense of specialness: it is the price one pays for standing out, unshielded, from nature. "Death anxiety" [which corresponds to Rank's *death fear*] is the toll of fusion: when one gives up autonomy, one loses oneself and suffers a type of death. Thus one oscillates, one goes in one direction until the anxiety outweighs the relief of the defense, and then one moves in the other direction.[12]

Similar to Rank's thinking, then, Yalom suggested that we find ourselves caught in a perpetual oscillation between these two fears, and each of us develops a unique and complex set of strategies in an attempt to mitigate these fears and find some tolerable middle ground (see Figure 15.1).

Yalom suggested that it is in our navigation of this dialectic that a framework for explaining psychopathology can be found. He suggested that a *neurotic* is one who clings especially tightly to either one or the other of these two strategy systems (either the belief in one's specialness or the belief in an ultimate rescuer). This causes them to live a very restricted life, and they subsequently become plagued by *existential guilt*—guilt arising not from a transgression to another but from a transgression to oneself, in failing to live a full, authentic life. Yalom suggested that a *psychotic*, on the other hand, is one who is so tormented by death

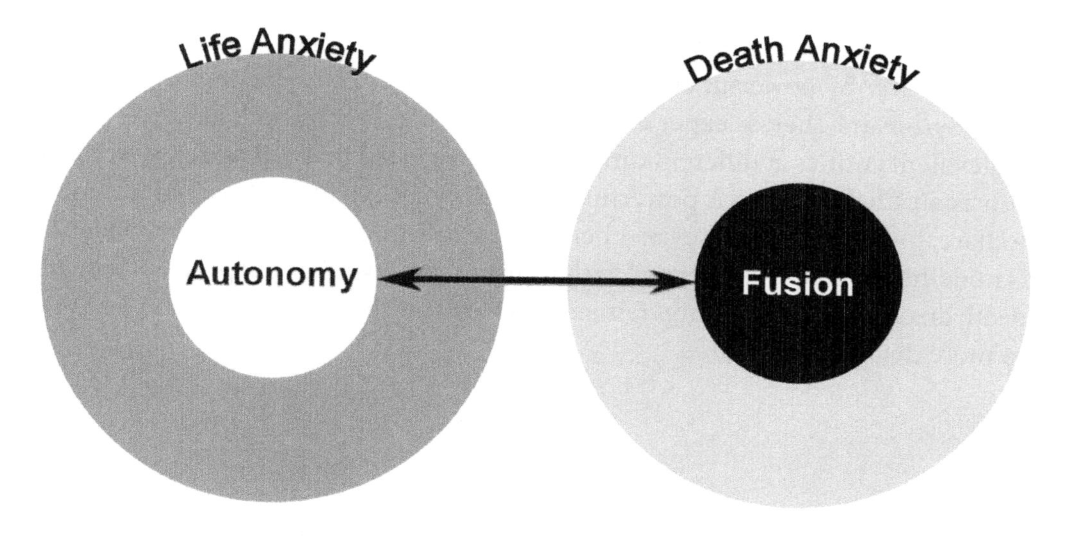

Figure 15.1. A graphical depiction of Yalom's life anxiety vs. death anxiety dialectic. Notice it is essentially the same as that of Rank's, with some subtle differences in the terminology.

anxiety (perhaps because of overwhelming exposure to it at a young, vulnerable age, or because of serious attachment issues with the caretaker as an infant) that she or he clings even more desperately to these defensive strategies than does the neurotic. Such clinging is not only more desperate, but is also much less secure and stable. The psychotic individual may swing dramatically from a sense of extreme specialness (omnipotence and/or compulsive heroic striving) to a sense of extreme merger and loss of boundaries.

◇◇

Reflecting upon the stories of Theresa and Byron, it is quite clear that both of their experiences fit well with Yalom's assertion that psychosis likely entails extreme versions of both types of strategies that we use to stave off death anxiety—the belief in our *personal inviolability*, and the belief in an *ultimate rescuer*.

Yalom suggested that an extreme version of the personal inviolability strategy typically manifests as a form of heroic striving, and both Theresa and Byron clearly experienced a very high degree of such striving. And Yalom suggested than an extreme version of the ultimate rescuer strategy typically manifests as extreme merging with a powerful force or being and a corresponding loss of boundaries; and again, both Theresa and Byron clearly experienced this kind of experience to very high degrees—both in their experiences of a general expansiveness or merging with all, and also in experiencing themselves as being watched over by powerful entities. In the latter case, it's interesting to note that both Byron and Theresa experienced being watched over by both benevolent and malevolent entities at different times. Yalom suggested that this strategy typically consists of believing in a powerful being or force greatly concerned with our welfare, but it may be that even a being or force with malevolent intentions who is nonetheless greatly concerned with us may serve a similar purpose (staving off death anxiety by maintaining tremendous worth of the self). We will explore this in more detail in Part Three.

Chapter 16

When Overwhelming Anxiety Is Insoluble on Any Other Level (Rollo May)

Rollo May (1909 – 1994) was another individual whose contributions to existential psychology are generally held with very high regard. Like Yalom, May also used Rank's work as an important foundation for his own formulation of psychopathology and psychosis[1]. He came to believe that "many forms of psychosis are to be understood as the end result of conflicts and anxiety which are too great for the individual to bear and at the same time insoluble on any other level"[2]. To understand what May meant by this, it will help if we look more closely at what he meant by "conflicts and anxiety."

Defining Anxiety

May spent a great deal of time grappling with the definition and meaning of anxiety, even devoting an entire book to the subject (*The Meaning of Anxiety*[3]). He, like Yalom, drew from Kierkegaard's distinction between fear and anxiety[4], saying that fear is a sense of apprehension in relation to a specific object while anxiety is a vague and apparently "objectless" apprehension, involving a general sense of uncertainty and helplessness[5]. He suggested that a useful approach for understanding the essential nature of anxiety is not to focus so much on the quality of anxiety but to instead ask what it is in our experience that is being threatened. He suggested that fear typically involves a threat to something relatively superficial in our experience (such as physical harm), while anxiety represents a threat towards the core of our personality. In other words, a particular fear that someone experiences is based on the individual's particular security pattern; in anxiety, however, "it is this security pattern itself which is threatened"[6]. May offered the following formal definition of *anxiety*:

> Anxiety is the apprehension cued off by a threat to some value that the individual holds essential to his existence as a personality. The threat may be to physical life (the threat of death), or to psychological existence (the loss of freedom, meaninglessness). Or the threat may be to some other value which one identifies with one's existence: (patriotism, the love of another person, "success," etc.).[7]

The real challenge in working with anxiety is that since it is the very core of our sense of self that we perceive to be under threat, there is no way to stand outside of it—there is no object to confront. When experiencing anxiety, then, we sense that the very structure that maintains our sense of agency is under threat and therefore that our very agency could come crumbling down at any moment. In other words, perhaps the potential to feel such profound helplessness in anxiety arises from the combination of having no object to confront and sensing a threat to our very capacity to confront anything at all.

Normal and Neurotic Anxiety

May proposed that anxiety can be divided into two categories: *normal anxiety* and *neurotic anxiety*[8]. Normal anxiety is (a) proportionate to the degree of threat, (b) does not involve intrapsychic conflict such as repression or neurotic defenses, and (c) can either be dealt with constructively or it naturally leaves once the threat is lifted. With neurotic anxiety, the opposite is true with regard to all of these qualities. The crucial distinction is that with neurotic anxiety, there are intrapsychic conflicts and patterns that interfere with our ability to maintain a clear assessment of the situation and to use our own power effectively. May suggested that neurotic anxiety usually has its genesis in childhood, being especially likely to develop when a child finds she must resort to repressing the awareness of a potentially overwhelming interpersonal threat. For children, the quality of their relationship with their caregiver(s) is a key factor in their core sense of security; therefore, children are particularly vulnerable to being traumatized by the kinds of threats that cause anxiety within these relationships, and their level of anxiety can easily escalate from normal to neurotic.

The Intrapsychic Conflicts within Anxiety

As mentioned above, a fundamental distinction between normal and neurotic anxiety is that neurotic anxiety contains intrapsychic conflicts. It is important to note, however, that May held that neurotic anxiety arises directly out of normal anxiety. According to May, we all enter life experiencing normal anxiety, and

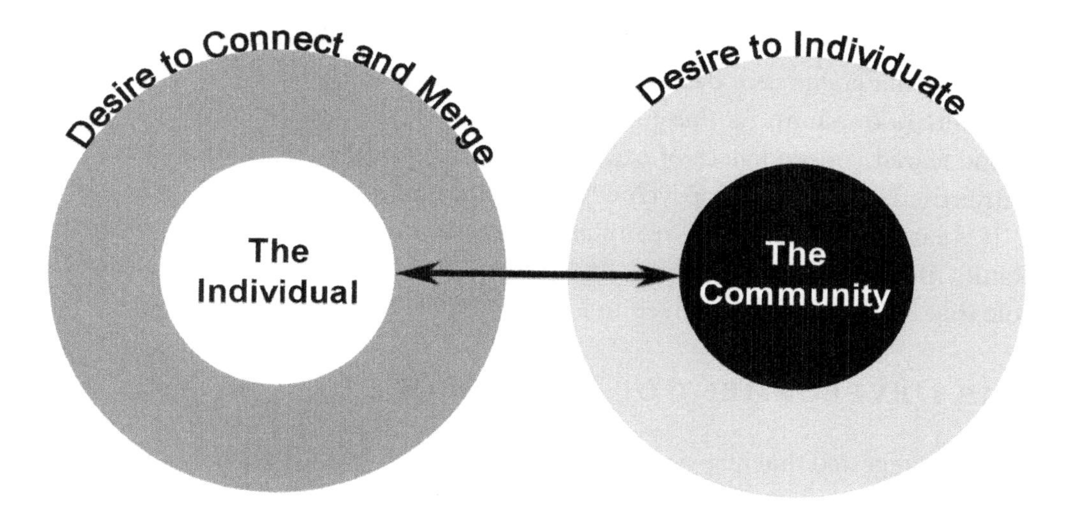

Figure 16.1. A graphical depiction of May's "dialectical relation of the individual and his community." Notice that the essential components of this model are nearly identical to those of Rank and Yalom.

difficult circumstances lead to the exaggeration of this normal anxiety until we experience neurotic anxiety. While some experience more neurotic anxiety than others, none of us are completely free from it. So, the intrapsychic conflicts that are more evident in neurotic anxiety have their roots in a more subtle form in normal anxiety. May suggested that while the particulars of these conflicts may vary significantly, they all share one common root—"*the dialectical relation of the individual and his community* [author's emphasis]"[9]. On one hand, the individual possesses a strong need to individuate, to find one's uniqueness, one's authenticity, one's sense of agency. On the other hand, the individual possesses a strong need to be a member of a community, to experience love and belonging (see Figure 16.1).

May suggested that these two directions of development can be seen as two poles, and according to May, if either pole becomes blocked, psychological conflict and therefore neurotic anxiety results. An individual who develops an imbalanced emphasis on the independence pole develops "the anxiety of the defiant and isolated individual"; an individual who develops an imbalanced emphasis on the dependence pole develops "the anxiety of the clinging person who cannot live outside of symbiosis"[10]. Along with these imbalances develop conflict systems comprised of hostility and repression. Those individuals imbalanced in the direction of independence develop hostility toward those believed to be responsible for the individual's isolation and those individuals imbalanced in the direction of dependence develop hostility toward those believed to suppress the individual's freedom. As these hostilities develop, they are often repressed due to the feared consequences of expressing them, completing what can be a very entrenched system of intrapsychic conflict and neurotic anxiety.

It is easy to see that May's formulation of psychopathology is closely related to Rank's life fear / death fear dialectic, and in fact, May openly acknowledged the role that Rank's work has played in influencing his own.

THE DEVELOPMENT OF PSYCHOSIS

So May suggested that all psychogenetic psychopathology* ultimately arises from intrapsychic conflicts stemming from the unsuccessful resolution of finding both a healthy sense of individuality and a healthy sense of community and relationship. May then went on to suggest that if these conflicts and the associated anxiety become too great for the individual to tolerate "and are insoluble on any other level," then psychosis is likely to ensue[11]. In other words, May suggested that psychosis is nothing more nor less than the means with which the psyche attempts to resolve "otherwise insoluble conflicts and anxiety, at the price . . . of the surrender of some aspect of adjustment to reality"[12].

When reflecting upon the stories of Theresa and Byron, it is clear that they both profoundly experienced the dilemma to which May referred. They were both longing to experience a deep connection with others and/or the world in general while simultaneously being terrified to do so, presumably because they had both become strongly identified with a highly constricted sense of self and were terrified that such intimate connection with others would entail the loss of their autonomy.

* The term *psychogenetic psychopathology* refers to psychopathology (mental and/or emotional difficulties) whose origin is predominantly psychological and/or social in nature. This is in contrast to psychopathology that may manifest as a result of organic brain damage.

Chapter 17

Overwhelming Exposure to the True Nature of the World (Ernest Becker)

Ernest Becker (1924 – 1974) is one of the most highly regarded existentially oriented thinkers of the past century, perhaps being best known for his Pulitzer Prize winning book, *The Denial of Death*[1]. Like Yalom and May, Becker also gave Rank credit for providing an important foundation for his own work, though his own formulation of psychopathology does not map onto Rank's model quite as neatly as does that of Yalom and May.

What lies at the heart of Becker's work is his belief that the fear of death is the fundamental dilemma giving shape to all other challenges in our lives, including all forms of psychogenetic psychopathology and psychosis:

> We might call this existential paradox the condition of *individuality within finitude* [author's emphasis]. Man has a symbolic identity that brings him sharply out of nature. He is a symbolic self, a creature with a name, a life history. He is a creator with a mind that soars out to speculate about atoms and infinity. . . . This immense expansion, this dexterity, this ethereality, this self-consciousness gives to man literally the status of a small god in nature, as the Renaissance thinkers knew.
>
> Yet, at the same time, as the Eastern sages also knew, man is a worm and food for worms. This is the paradox: he is out of nature and hopelessly lost in it; he is dual, up in the stars and yet housed in a heart-pumping, breath-gasping body that once belonged to a fish.[2]

In other words, human beings have the capacity to experience themselves as existing in something like an eternal presence, savoring the sweet illusion of omnipotence and omnipresence (our symbolic self), yet we are trapped in a body that is doomed to decay and die. We find ourselves torn between our seemingly eternal symbolic identity and our impermanent physical body. Becker suggested that this dilemma, though it is generally repressed and unconscious, is the essential motivating factor that drives us through our lives.

Many would argue at this point that if this dilemma were so essential to our experience, we would certainly be more conscious of it. In response to this, Becker suggested that while the fear of our own mortality must be present enough to ensure self-preservation, if we were constantly aware of this fear, especially of the full scope of it, we would be so overwhelmed that we would be unable to function. Therefore, it is essential to our own sanity that we find some way to mitigate this fear, and in fact, Becker has suggested that psychosis is the result of being unsuccessful in this regard. To better understand how Becker believed this failure may occur, it will help if we first look at how Becker believed successful mitigation occurs, which he said we do using essentially just two strategies: the development of our character structure and our perpetual striving towards heroism.

The Vital Lie of Character

Residing at the foundation of our defense against the fear of death is what Becker referred to as the *vital lie of character*[3]. As the illusion of omnipotence that is believed to have filled our experience during our first year of life begins to fade, we find ourselves living in a world at once so awesome, terrifying, and incomprehensible, that if we did not limit, distort, and repress our perceptions, we would be completely overwhelmed and unable to function. We would be overwhelmed not only by external stimuli but also by the thoughts and feelings emanating from within—especially those that make us feel weak, shameful, worthless, and evil. In response to this, we begin to learn at a very early age how to repress our experience on a global level in an attempt to create a "warm sense of inner value and basic security"[4]. Becker argued that other animals are given automatic instinctive programming as the means to achieve this, but we humans have to do the arduous work of constructing our own defensive system, which is our character structure.

This character structure, while providing us with the illusion of security and value, also limits our freedom and expansiveness, and therefore puts us in a very painful dilemma: "Our deepest need is to be free of the anxiety of death and annihilation; but it is life itself which awakens it, and so we must shrink from being fully alive"[5]. We find ourselves in the unfortunate situation of being both afraid to die and afraid to really live; but, according to Becker, this is the high price we must pay for being human.

Heroic Striving

Becker believed that a second line of defense we typically use against our fear of mortality is provided by our perpetual striving towards heroism. In order to fight against the apparent worthlessness and disposability of our physical body, we must "desperately justify [ourselves] as an object of primary value in the universe; [we] must stand out, be a hero, make the biggest possible contribution to world life, show that [we count] more than anything or anyone else"[6]. This stance is apparent in the overt narcissistic demands of young children. That this behavior fades away as we grow into "healthy" adulthood is not a sign that we have fundamentally overcome this stance, but rather it means that, for most of us, we have simply learned to transform our individual heroic strivings into social heroic strivings.

Becker argued that the ultimate function of society is in fact to provide us with the means to negotiate the transcendence of our death, to provide us with *immortality projects*. He described society as a "symbolic action system, a structure of statuses and roles, customs and rules for behavior, designed to serve as a vehicle for earthly heroism"[7]. In this view, cultural differences are merely differences of *hero systems*. Regardless of whether a particular system is considered magical, spiritual, scientific, or civilized, the role is the same: to provide us with a feeling of "primary value, of cosmic specialness, of ultimate usefulness to creation, of unshakable meaning"[8].

Even if society fails to provide an adequate hero system for us, most of us are adept at finding it in other ways—for example, through our family, our work, our goals, or our values—and most of us are able to find more than one avenue for achieving a sense of death transcendence. With more passionate individuals, such as many successful politicians, business professionals, athletes, and even spiritual leaders, such heroic strivings are obvious. In more passive individuals, such strivings are not so obvious but can still be seen manifesting in safer, more mundane ways such as pride in one's profession or the drive to receive recognition for smaller things.

Our striving for heroism, according to Becker, results in a very harmful irony: While our heroic strivings are rooted in our desire to transcend death, and are therefore aimed at destroying evil (which, by definition, is the intentional causing of harm, destruction, and death), such strivings paradoxically generate more evil in the world. Because we each cling so tightly to our own hero-system, depending desperately upon the righteousness of our own system to prop up our fragile illusion of death transcendence, we respond to any other hero-system that contradicts our own as a serious mortal threat. As a result, we often see systems different than our own as evil (be they different societies, religions, governments, value systems, etc.). We justify attempts to annihilate them, and as a result, we end up with a world that is plagued by hatred, war, and violence.

IMPLICATIONS FOR PSYCHOSIS

Becker said that psychopathology is simply "the failure of heroism"[9]. He described a neurotic as someone who has particularly strong awareness of his "creatureliness"[10], of his physical, mortal self; a neurotic's attempt to construct the illusion of immortality, to develop a high sense of self worth, to attain heroic status, has been particularly weak. A psychotic, then, falls further along this same spectrum.

Due presumably to a particularly traumatic upbringing and/or other unfortunate circumstances, a psychotic individual is one who was unable to develop either a secure *seating* in their body (Becker's term) or a secure connection with their culture's hero-system, and therefore is unable to successfully deny the world and the terror of it. Ironically, then, the psychotic is in a position to witness the true nature of the world more accurately than the so-called mentally healthy individual, although they are likely to be completely confused and overwhelmed by what they witness. Therefore, they must rely "instead on a hypermagnification of mental processes to try to secure death-transcendence; [they have] to try to be a hero almost entirely ideationally"[11]. Since this is a poor substitute for the buffers of the mentally healthy individual, we see individuals deemed psychotic often resorting to desperate strategies such as "megalomanic self-inflation" (such as messianic striving) and fabrications of "new symbolic transcendence" (such as anomalous belief systems and perceptions)[12].

◇◇

Reflecting upon the stories of Theresa and Byron, we find that their experiences fit quite well with Becker's assertion that psychosis develops as the result of being unable to deny the fragility of one's existence using the standard strategies of maintaining *the vital lie of character* (a general denial/repression of one's authentic experience) and connecting with culturally sanctioned hero systems:

Regarding the failure of their character structure, in the cases of both Theresa and Byron, we see evidence that their character structures were simply no longer sustainable, that they had become overwhelmingly constrictive and suffocating. Therefore, a desperate need had arisen for these structures to undergo a profound transformation.

Regarding the lack of connecting to their culture's hero-systems, we again see clear evidence that Theresa and Byron had both failed in this regard. In Theresa's case, we see clear evidence of her failure to connect with her culture's hero-systems in her attempt to seek relief from her suffering within a distant culture. Byron showed a similar disconnection from his culture's sanctioned hero-systems with his attempts to seek relief from his suffering in shamanic rebirthing rituals.

As their psychoses unfolded, true to Becker's assertion, it is quite clear that they both experienced a genuine "failure of heroism" to the extent that those qualities of the world and their selves that are ordinarily denied and/or repressed came flooding into their consciousness. They both became frequently overwhelmed by experiences heavily imbued with the qualities of profound suffering, death, and destruction.

We can also say that since they were no longer able to utilize the more ordinary strategies of the vital lie of character and culturally sanctioned forms of heroic striving, Theresa and Byron were both forced to attempt to prop up these strategies almost entirely ideationally (via "the hypermagnification of mental processes"[13]). Becker suggests that such desperate strategies typically manifest as "megalomanic self-inflation" (such as messianic striving) and "new symbolic transcendence" (anomalous belief systems and perceptions). And, of course, Theresa and Byron experienced both types of these experiences frequently within their psychoses.

Finally, we can say that Theresa and Byron were ultimately successful in developing more sustainable methods of dealing with their death anxiety, and from the perspective of Becker's model, we can say that this is essentially what is required to successfully resolve one's psychotic process. In Theresa's case, we can say that she was successful in transitioning from the very fragile immortality projects of messianic striving and the ideational bearing of a child (her attempts to "channel" a child) to the much more robust strategies of developing a real flesh and blood child and family and in finding meaningful work that allows her the opportunity to offer genuine support to others. After all, there are few culturally sanctioned immortality projects as powerful as these: giving birth to a being who is sprouted from our own genetic and psychological material and who may then go on to pass on our name indefinitely; and devoting our energy to a cause that transcends our limited and mortal self.

In Byron's case, we can say that he was successful in transitioning from the very fragile immortality project of primarily messianic striving to the cultivation of a spiritual path that provides him with a way to find a more genuine peace with the fragility and fleeting impermanence of his existence. In addition to this, like Theresa, Byron was also able to develop a meaningful career that allows him the opportunity to offer genuine support to others. In other words, he made the transition from ideational savior to genuine supporter.

Chapter 18

Toward a Paradigm Shift in the Way We View Personal Paradigm Shifts

As discussed earlier, the models presented above are among the most compatible with the existing research, and while the details of each vary significantly from the others, they are not necessarily mutually exclusive since they each emphasize different aspects and/or different perspectives of the psychotic process.

A good metaphor in this regard is the well-known story of the elephant and the six blind men. Having never "seen" an elephant before, each man is said to have taken hold of a different part of the elephant—one grabbed a leg, one grabbed an ear, another the trunk, another the tail, etc. Each man mistook the entire elephant to be merely the part that he had grabbed—the man who had grabbed the tail thought that an elephant must be like a brush; the one who had grabbed a leg thought that an elephant is like a tree, and so on. In actuality, of course, an elephant consists of all of these parts and more. In the same way, we can make the case that psychosis is very complex and multi-faceted, and by opening our minds to many different perspectives, we allow ourselves the possibility of developing a much richer understanding of it.

It is also important to notice that all of these models share as a fundamental postulate the idea that psychosis is a natural process initiated by the psyche and that it is closely associated with a profound reorganization of one's understanding and experience of the world and of oneself—one's personal paradigm, in other words.

So, after distilling the most significant research related to the topic of long-term psychosis and recovery, and after looking at some theoretical models that appear to fit this research well, we find that several interesting shifts happen in the framework with which the topic of psychosis is ordinarily held:

First, contrary to the assumptions made by many of the transpersonally oriented researchers, we find that making a distinction between the categories of psychotic and mystical experiences becomes less relevant and perhaps even ultimately impossible. As Clarke said, the case study research suggests that making such a distinction is "invalid and essentially meaningless . . . . The relationship between psychosis and spiritual does not fit into this type of neat dichotomy"[1].

Second, we find a significant change in the implications of how we can best support people going through such experiences. Recall that a number of transpersonally oriented clinicians have suggested that those having psychotic experiences should receive standard psychiatric care (i.e., medication, invalidation of their experiences, and reversing their process as quickly as possible), whereas those having mystical experiences should receive gentle guidance and encouragement to work through the process. However, if we come from the perspective that both categories of experience arise from a similar growth oriented process, then it should be apparent that regardless of the category to which we attempt to assign an individual's experiences, *all* such individuals should receive support and guidance in attempting to work through their process[2]. One very important implication of this perspective, and one that is well supported by the research, is that serious harm can inadvertently be caused by those offering support when they do not accept the possibility that this may be a natural process. By stigmatizing and invalidating individuals going through such a challenging process, we are most likely hindering their recovery process. Another important aspect of offering support, then, is minimizing interventions that hinder the process, chief among these being involuntary hospitalization and compulsory long-term psychiatric medication use, as discussed earlier.

A third important change the research suggests we make is acknowledging that the process these individuals go through has the potential to contribute greatly to our society. Many authors have argued that by going through what is often a very painful and difficult process in attempting to integrate different realms of experience, loosen ego rigidity, and step outside the box of consensus reality, these individuals may be the very ones who can take on the long-forgotten role that our species needs in order to rediscover a sense of harmony with each other and with our world[3].

A few of the more well-known people who have contributed in this way are: Sir Isaac Newton, preeminent founder of calculus and physics; Joan of Arc, Catholic Saint and national heroine of France; Vincent van Gogh, considered to be one of the most inspirational artists of the past two hundred years; Black Elk of the Sioux, who experienced a *shamanic illness** in his youth that served him

* *Shamanic illness* is a term that refers to an overwhelming spiritual/psychological/physical process occurring in some members of indigenous societies that, (continued on the next page)

in leading his people through a very difficult integration with the European descendants in the U.S. at the turn of the 20th century; Virginia Woolf, considered to be one of the greatest novelists of the 20th century; Carl Jung, the well-known transpersonal psychologist who spent several years struggling with psychotic experiences and emerged with profound insights into the human condition; and John Nash, winner of the 1994 Nobel Memorial Prize in Economic Sciences. Of course, there are many others who are less well known but who have also made significant contributions to society.

It's important to recognize that acknowledging the potential contribution of such individuals is not a romanticization of those struggling with psychotic states. It's clearly evident that such individuals often experience tremendous confusion and inner turmoil for significant portions of their lives. Such acknowledgment merely recognizes that such individuals have likely played valuable roles in human societies since the birth of our species; and when we consider the rigid dogmatism that plagues our societies today, it is difficult to deny that we may be harming not only these individuals but all of society when we deny these individuals the possibility of contributing in their own unique way.

(continued from the previous page) when successfully resolved, can lead to the transformation of the individual into a powerful healer, or *shaman*.

Part Three

Arriving at an Integrative and Comprehensive Model of Psychosis

Now that we have undertaken a thorough review of the literature, looked at a number of different models of psychosis that are in close accord with the research, and have looked in-depth at several real-life accounts of individuals who have experienced the successful resolution of profound psychotic processes, we are finally ready to turn our attention to a model that integrates and simplifies this vast array of information—a model I refer to as the *duality/unity integrative model* (or the *DUI* model* for short). If we return to the metaphor of the elephant and the six blind men to understand how so many different models can all have legitimate perspectives of the psychotic process, then my aspiration in developing this model has been to arrive at a model that looks at the entire elephant (or a much broader view of it, anyway). We will spend the remainder of Part Three going over the details of the DUI model, and then in Part Four, we will apply this model to the recovery stories presented in this book in an attempt to make sense out of the entire process of psychosis, from onset to full recovery.

A very important point that I will emphasize here and elsewhere is that the DUI model refers to *all* human experience as it manifests at the deepest level of our experience (what I refer to as the existential level), not only to experiences we think of as psychotic.

---

* Some people in the U.S. have found that my use of the acronym "DUI" is a little strange, since in some states in the U.S., this is the same acronym used to represent the act of "driving while under the influence" [of alcohol or other intoxicants]. I find that this alternative meaning actually makes this acronym even *more* suitable for this model, however, since the most essential characteristic of this model is the idea that we all find ourselves "driving" through life "while under the influence" of several very challenging existential dilemmas, the details of which will be discussed shortly.

It's also important to mention that, unless you are relatively familiar with both formal Western psychology and the basic principles of nonduality—the cornerstone of most of the major Eastern spiritual traditions such as Buddhism, Hinduism and Taoism—then you will most likely come across many new terms and concepts in this section. You will find that these terms and concepts are explained fairly thoroughly, so there is no need to have any prior knowledge about them; however, some of these concepts can be somewhat challenging to grasp, so you may want to move through this section of the book relatively slowly and with extra attention, having the willingness to stretch your imagination and taking the time to contemplate these concepts and your own personal experience with regard to them. It will be very helpful to regularly refer to the Glossary at the back of the book to maintain familiarity with the terms that are new to you, and it will also be helpful to refer to the "Summary of the DUI Model" on the last two pages of this section to keep a handle on the big picture of this model as you go through the more subtle details.

Chapter 19

The Foundation of the Duality Unity Integrative (DUI) Model

There are three particularly important characteristics of the DUI model that distinguish it from many of the other models of psychosis that have been formulated: (1) It is highly *integrative*—integrating the major components from a number of other models of psychosis and human nature, including the integration of both Western and Eastern understandings of human experience; (2) it is *existentially oriented*—attempting to offer a description and explanation of what is taking place at the root-most level of experience during the psychotic process; and (3) it is primarily *phenomenological**—emphasizing subjective experience rather than attempting to determine whether or not what is experienced is "objectively real†."

The foundational structure of the DUI model is essentially an integration of two different (but what I believe are closely related) existential models—one developed in the West and the other developed in the East. By integrating these two models into one, we arrive at a particularly comprehensive model that captures virtually all of the experiences of the participants in this study as well as

* The term *phenomenological* might seem a little daunting to some readers, but it essentially means simply "pertaining to one's direct subjective experience." This is a very important concept for the study of psychosis and other types of highly subjective experiences.

† Research within modern physics has cast serious doubt on the validity of the classic Newtonian paradigm that assumes the existence of a purely objective reality that exists entirely separate from the observer. In other words, making an attempt to determine whether or not one's subjective experience is in accord with "objective reality" may be based on an entirely erroneous premise. Therefore, as discussed earlier, when exploring psychosis and other topics pertaining to subjective experience, it is probably much more appropriate to discuss whether or not one's subjective experience is in accord with the experiences of others around them (i.e., consensus reality) and leave aside questions of so called "objective reality."

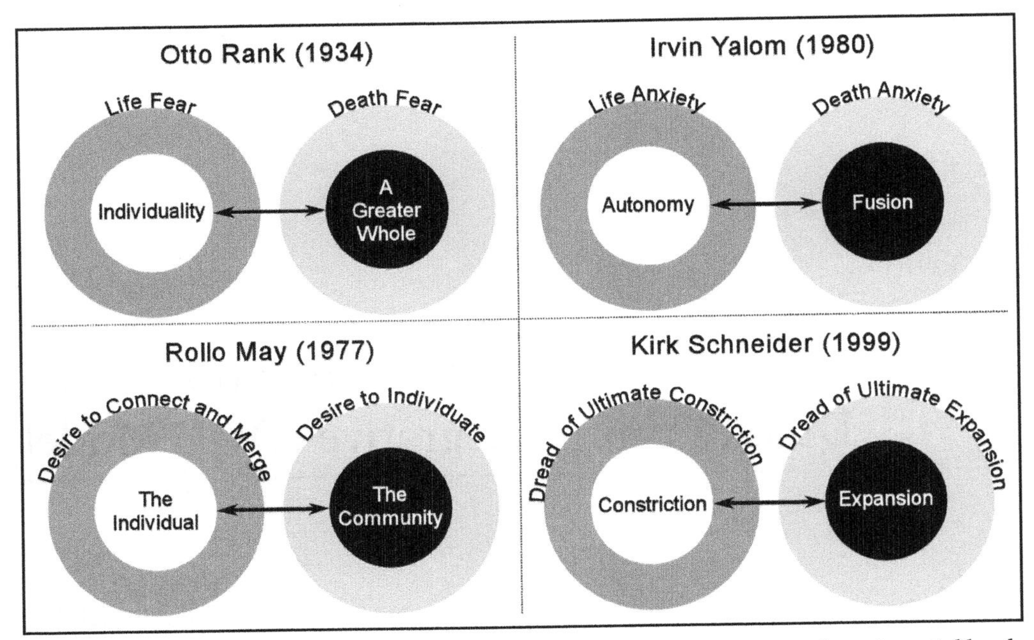

Figure 19.1. Four different dialectic models representing our experience at the existential level. The black and white circles represent the poles of our experience at this level, and the gray circles represent the repulsive field associated with each pole. (*Note: These illustrations are my own highly simplified depictions of these models.)*

human experience in general. This unique synthesis is particularly useful for our task here in that it takes the vast mysteriousness of human experience and makes it more tangible by capturing all subjective experiences within just two categories—(1) our experience of duality, and (2) our experience of the interplay between duality and unity (we will look more closely at what the terms *duality* and *unity* refer to shortly). By simplifying the vastness of human experience in this way, I believe we can arrive at a relatively simple framework for describing and explaining what is taking place at the root-most/existential level of one's being during the psychotic process.

The Western existential model is an existential dialectic that, to the best of my knowledge, was first introduced in 1936 by Rank with his formulation of the *life-fear/death-fear dialectic* (as discussed in Chapter 14). He defined *life fear* as "the fear of having to live as an isolated individual" and *death fear* as "the fear of the loss of individuality"[1]. Since then, several other existentially oriented thinkers in the West have posited similar dialectic models (see Figure 19.1). In 1977, May suggested that virtually all intrapsychic conflict essentially boils down to "the dialectical relation of the individual and his community"[2] (as discussed in Chapter 16). In 1980, Yalom suggested that human suffering arises from the dialectical tension between "'life anxiety' [which] . . . is the price one pays for standing out, unshielded, from nature" and "'death anxiety' [which] is the toll of fusion"[3] (as discussed in Chapter 15). In 1999, Kirk Schneider presented yet another variation on this theme, defining the two poles of this dialectic as *constriction* and *expansion* and suggesting that the root of all of our fears ultimately arise from *the dread of ultimate constriction* (or *obliteration*) and *the dread of ultimate expansion* (or *chaos*)[4].

While this kind of existential dialectic has been introduced and explored in the West for about 75 years now, it is interesting to note that existentially oriented thinkers in the East have been exploring different versions of a very similar dialectic for over 2,500 years. In particular, numerous practitioners of Hinduism, Taoism, and Buddhism have described in great detail a type of existential dialectic that is quite similar to that presented by the Westerners above, but with the significant difference of incorporating the principle of fundamental unity (whereas these Western existentialists have generally placed primary emphasis on the dialectical aspect of existence while remaining more or less agnostic with regard to the existence of a fundamental unity). The basic structure of this Eastern model is sometimes referred to in the West as *dialectical monism* (see Figure 19.2), and it is interesting to note that these and other closely related Eastern traditions, using different contemplative practices as their method of inquiry, have each independently arrived at a model very similar to this one.

While we in the West have been studying the nature of the mind and human experience using formal scientific methods for a little over a century, the traditions mentioned above and other closely related traditions in the East have been formally studying the nature of mind, consciousness, and subjective experience

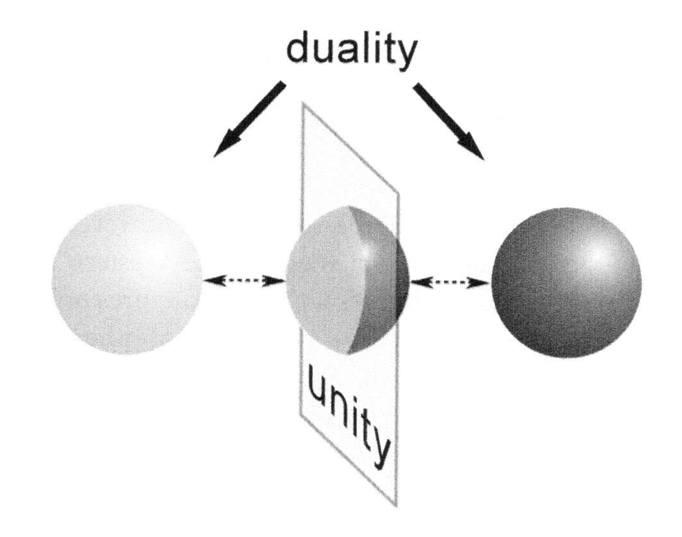

Figure 19.2. A graphical depiction of the common structure of human experience at the existential level as taught within several Eastern traditions—most notably Hinduism, Taoism, and Buddhism. This structure is sometimes referred to in the West as dialectical monism.

for over 2,500 years (and even much further back than that in some cases). Many Western psychologists often discredit the discoveries of mind and consciousness made within these traditions as being religious rather than scientific, but I believe that this has been a serious mistake.

While there may be some limited merit to the argument made by these Western psychologists, especially with regard to a number of dogmatic beliefs that have developed in various lineages of these Eastern traditions, I think that the discoveries made within the present-centered contemplative practices used in these traditions can indeed be claimed to be scientifically derived. This is true particularly if we look at the various forms of mindfulness meditation found within the Buddhist tradition, many of which are clearly forms of unadulterated phenomenological inquiry that would be considered perfectly legitimate methods for studying subjective experience within contemporary Western science (we will discuss this in more detail in Chapter 21). Moreover, these practices have been cultivated and practiced intensely by millions of people over several millennia, most of whom have arrived at very similar results, demonstrating a robustness of scientific inquiry into subjective experience that far exceeds anything we have ever developed in the West.

All of this excellent scientific evidence, however, is routinely discarded by researchers in the West, which I believe has resulted in a very unfortunate setback in our understanding of mind and consciousness. Fortunately, the tide has begun to turn in the last ten to twenty years, and mindfulness meditation in particular has begun to be embraced in the West, both as a valid form of phenomenological inquiry and as a highly beneficial therapeutic tool in clinical practice.

As I went through the process of developing the DUI model, I initially tried to fit the experiences of my participants and those of many others who have struggled with psychosis into a purely dialectical existential model similar to those suggested by the Western psychologists illustrated in Figure 19.1. However, I soon came to the conclusion that the principle of unity as it is used in the Eastern model (Figure 19.2) would have to be incorporated; and interestingly, as soon as I incorporated the unity principle, virtually all of the inconsistencies and missing pieces with which I had been grappling while trying to use a strictly dialectical model disappeared. I did, however, find it very helpful to keep some of the most essential elements of the Western model. The result, then, has been an integration of these two basic structures of human experience at the existential level, along with the integration of other principles and concepts that have been developed in both Western and Eastern inquiries into mind and consciousness—hence the name, the *duality/unity integrative model.*

One final piece about the DUI model that is important to mention at this point is that it incorporates the fundamental existential dilemmas inherent in both the Western and the Eastern models, and it attempts to integrate them. The Western and the Eastern existential models each suggest a slightly different fundamental existential dilemma that lies at the root of human suffering. In the Western models listed in Figure 19.1 above, the fundamental existential dilemma is perhaps most concisely summarized as the need to find a tenable balance between autonomy/individuation on one hand and connection with others and with the world in general on the other hand*. In the Eastern model (Figure 19.2), the fundamental existential dilemma is perhaps most concisely summarized as the need to find a tenable balance between duality and unity—in other words, our desire to maintain the existence of a separate (dualistic) self within a world that is fundamentally unitive. In the DUI model, both of these core existential dilemmas are seen as equally important and very closely related. For the remainder of Part Three, we will continue to explore the various ways that these two dilemmas manifest within our experience, ultimately giving fruit to the vast array of experiences that we all undergo during our journey through life.

---

* Schneider's model deviates somewhat from the other models in Figure 4 in this regard; however, there are strong parallels between his use of the term *constriction* and the others' use of the terms *individual*, *individuality*, and *autonomy*, and there are also strong parallels between Schneider's use of the term *expansion* and the others' use of the terms *community*, *fusion*, and *a greater whole*.

Chapter 20

Our Experience of Duality

The DUI model posits that at the core of our experience, we find a nearly constant dance taking place between our experience of duality (our experience of ourselves as fundamentally *separate* from the rest of the world) and our experience of unity (our experience of ourselves as fundamentally *interconnected* with the rest of the world). We will explore our experience of unity in the next chapter, but we will first devote the remainder of this chapter to exploring our experience of duality.

By "our experience of duality," I am referring to the common experience we have of living in a universe populated by discrete subjects and objects. There is the experience of an "I" (subject) that is separate from everything else (objects), and in turn we generally experience each of these objects as being separate from each other. The DUI model suggests that, within our psyches, there are essentially two primary components that work together to create our dualistic experience of the world—one component can be described as a dialectic that I will refer to as the *self/other dialectic**, and the other component can be described using the concept of *cognitive constructs*. Let's now turn to look at each of these components in turn.

The Self/Other Dialectic

The self/other dialectic consists essentially of two poles—*self* and *other*—set apart from each other in *dialectical tension* existing at our deepest level of experience (see Figure 20.1). This dialectic draws extensively from the similar

* Recall that the term *dialectic* as I have been using it refers to the tension that exists between two opposing forces.

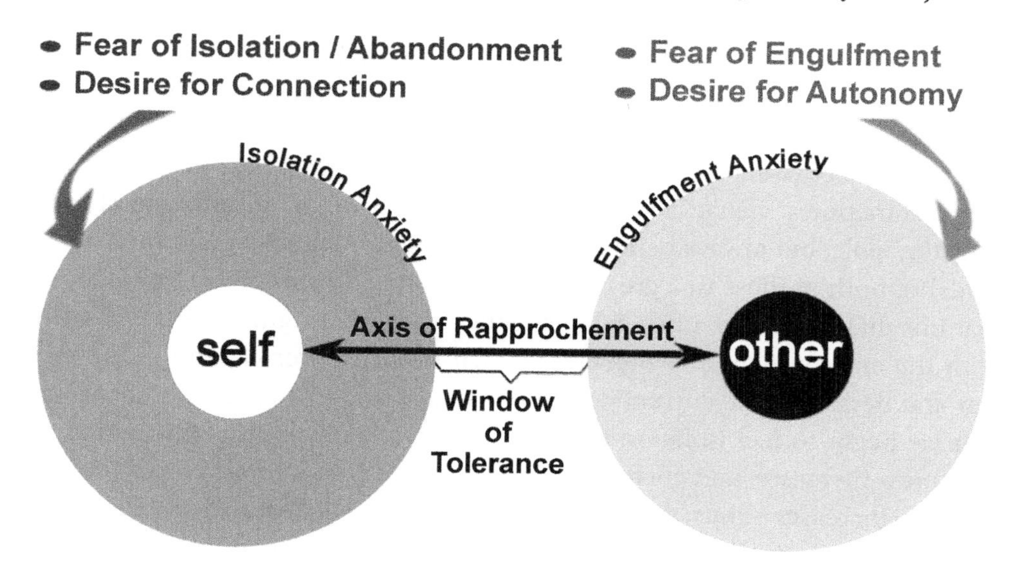

Figure 20.1. The self/other dialectic.

dialectics posited by Rank, May, Yalom, and Schneider (see Figure 19.1 in the previous chapter); however, I have found it helpful to use different terminology and expand upon it in different ways in order to best represent the data generated by this research.

One way that we can divide our experience is to make the distinction between our experience of self and our experience of other, with *other* referring simply to everything that we experience apart from ourselves (other people, other beings, other objects, the world in general). We can also see this as the classic dualistic subject/object split. Most of us generally have the experience of a distinct division between self and other, although many people report having experiences in which this division feels somewhat blurred. The self/other dialectic represents our experience of these two divisions with the two poles of *self* and *other* connected by the *axis of rapprochement*, which represents our experiential location with regard to these two poles (see Figure 20.1). Dialectical tension results from the opposite forces of the two anxieties associated with the two poles. The anxiety we experience in relation to self is referred to as *isolation anxiety*, and the anxiety we experience in relation to other is referred to as *engulfment anxiety*.

Isolation anxiety corresponds closely to what Rank referred to as life fear—the fear of isolation and of being cut off from the greater whole, which includes the corresponding desire for connection. Engulfment anxiety corresponds closely to what he referred to as death fear—the fear of losing our autonomy, of being engulfed by too much merger and connection, which includes the corresponding desire for autonomy. The overall strength of the dialectical tension within the system is directly related to the strengths of these two opposing anxieties.

At any given moment, depending upon the current conditions, we can say that our subjective experience is located somewhere along the axis of rapprochement, sometimes being closer to *self* pole and at other times being closer to the *other* pole, but always being within the experiential fields (and thus anxiety fields) of both of these to a greater or lesser degree. It is important to recognize that this dialectic is dynamic, being affected by ever changing conditions within both the environment and ourselves which result in ever changing degrees of fear and desire. As the current conditions take us closer to the *self* pole, we find that we begin to feel isolation anxiety more acutely. Our fear of isolation and loneliness increases, and corresponding to this, our desire for love and connection also increases—thus, we find ourselves being pushed/drawn in the direction of the *other* pole, feeling compelled to change our current conditions in order to make this shift. As we approach the *other* pole, however, we find that we begin to feel engulfment anxiety more acutely. Our fear of engulfment and of losing our autonomy increases, and corresponding to this, our desire for more autonomy and individuation also increases—thus, we find ourselves being pushed/drawn once again in the direction of the *self* pole. As Rank[1] articulated, and as May[2] and Yalom[3] later reiterated, we spend our lives perpetually being pulled/pushed back and forth between these two poles, continuously striving to find some tolerable middle ground (see Figure 20.2).

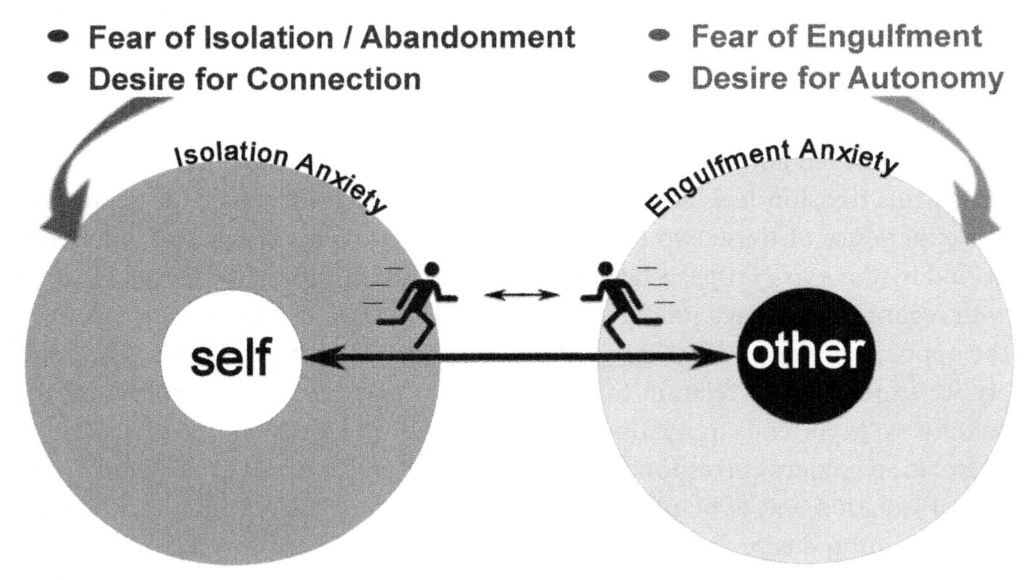

Figure 20.2. A graphical depiction of our constant struggle to find some middle ground between autonomy and connection.

This constant struggle to find some tolerable middle ground between autonomy and connection can be seen quite clearly in children between about one and a half and two years of age, in which they go through a developmental stage that Margaret Mahler, renowned contributor to the field of developmental psychology, referred to as *rapprochement* (from which the term *axis of rapprochement* was derived)[4]. During this stage of development, children can clearly be seen being driven by the desire for autonomy, as they venture out away from their mother (or other primary caretaker) to explore the world, but then upon reaching a certain threshold of anxiety related to their separation, they are clearly seen being driven by the desire for connection and protection as they run back to their mother. Mahler suggested that healthy resolution of this stage of development entails the maintenance of a secure sense of connection with the primary caretaker while simultaneously developing a healthy sense of confidence in one's autonomy. In the DUI model, I am suggesting that even though it is possible to find some sustainable resolution between these two opposing forces quite early in life during healthy development, this dilemma is never entirely resolved. It remains within us constantly throughout our lives, although it may be primarily unconscious.

According to this model, then, our subjective experience at any given moment is determined by essentially three factors: (1) the strength of the isolation anxiety; (2) the strength of the engulfment anxiety; and (3) our position in relation to the two poles (our location along the axis of rapprochement). There actually is a fourth factor that affects our subjective experience, which is our experience of *unity*, but for the sake of clarity, we will postpone discussion of this for now.

One of the most important implications of this model is the existence of what I refer to as the *window of tolerance*, the region along the axis of rapprochement within which the overall degree of anxiety is tolerable. The size and location of the window of tolerance is determined by the relative and overall strengths of the two types of anxiety. If the isolation anxiety is stronger than the engulfment anxiety, then the window of tolerance will be skewed towards the *other* pole (see the top of Figure 20.3); alternatively, if the engulfment anxiety is stronger than the isolation anxiety, then the window of tolerance will be skewed towards the *self* pole (see the bottom of Figure 20.3).

The overall combined strengths of the two types of anxieties determine the size of the window of tolerance. If the combined level of anxiety is relatively high, then one will experience a relatively narrow window of tolerance and will find the capacity to tolerate only a relatively narrow range of experiences (see the top of Figure 20.4). If the combined level of anxiety is relatively low, then one will experience a relatively large window of tolerance and will find the capacity to tolerate a relatively wide range of experiences (see the bottom of Figure 20.4).

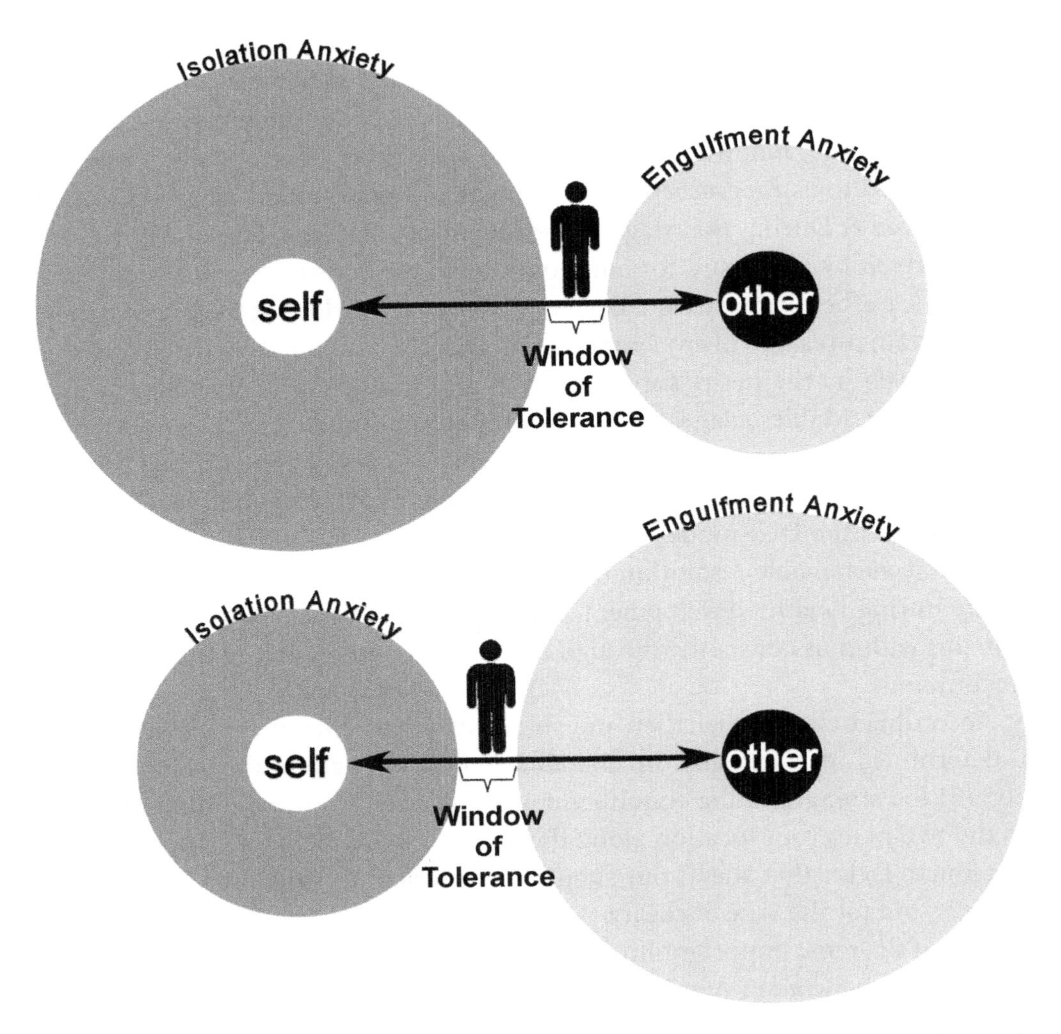

Figure 20.3. Skewed windows of tolerance. (Top) The window of tolerance is skewed towards the *other* pole. (Bottom) The window of tolerance is skewed towards the *self* pole.

The window of tolerance can be relatively dynamic, changing in size and location throughout one's life, and even from one moment to another and in one relationship to another. Research within the field of *attachment theory* suggests, however, that due to conditioning and perhaps innate temperament to some degree, most of us tend to experience a window of tolerance that is relatively stable, typically only changing somewhat gradually throughout our lives[5].

Attachment theory provides a particularly useful framework for understanding the different experiences that result with different variations of the window of tolerance. *Attachment* is defined within attachment theory as "an affectional

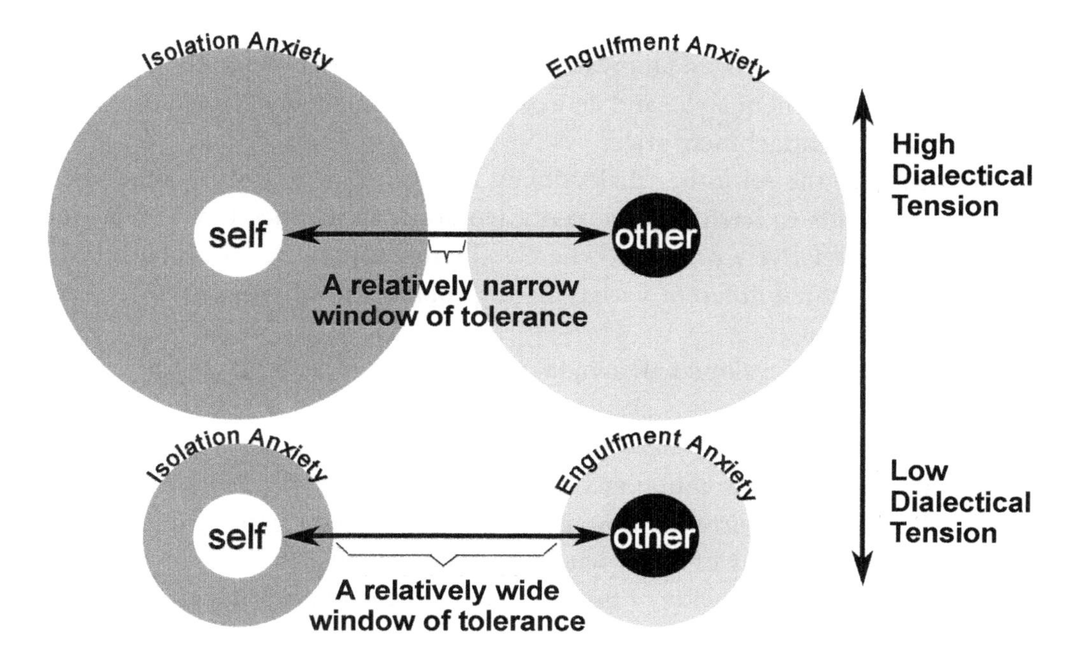

Figure 20.4. The overall combination of the two anxieties determines the width of the window of tolerance.

bond between individuals characterized by a seeing of closeness or contact and a show of distress upon separation"[6]. After over thirty years of substantial research, attachment researchers have generally arrived at the conclusion that there are essentially just four different attachment styles (although there continues to be some debate regarding possible variations of these)—*secure attachment* and three different types of *insecure attachment*.

Secure attachment is defined as "a type of attachment characterized by mild distress at leave-takings, seeking nearness to an attachment figure, and being readily soothed by the figure"[7]. The three types of insecure attachment styles are *avoidant attachment*, which is "characterized by the apparent indifference to the leave-takings of and reunions to an attachment figure"; *ambivalent attachment* (also sometimes referred to as *anxious attachment*), which is "characterized by severe distress at the leave-takings of and ambivalent behavior at reunions with an attachment figure"; and *disorganized attachment*, which is "characterized by dazed and contradictory behaviors toward an attachment figure"[8]. These definitions are based on the observations of infants; however, it is generally believed that they apply fairly well to people of all ages, although upon reaching adulthood, the attachment figure is likely to be a romantic partner rather than a primary caretaker, and the behavior related to each may manifest somewhat differently. In adulthood, an ambivalent attachment style is often referred to as a *preoccupied* attachment style, and an avoidant attachment style is often referred to as a *dismissive* attachment style.

Returning to the self/other dialectic, we can say that the secure attachment style corresponds to having developed a window of tolerance that is relatively wide and also relatively centered, whereas the three insecure attachment styles correspond to three different variations of problematic configurations of one's window of tolerance:

If someone has developed an avoidant attachment style, then we can say that their window of tolerance is skewed in the direction of the *self* pole (see the bottom of Figure 20.3). In other words, they are generally more vulnerable to engulfment anxiety than isolation anxiety and are therefore more likely to modify their behavior and environment in a way that places their experience more on the *self* side of the axis of rapprochement. Like anyone, these individuals are still within the anxiety fields of both the *self* and the *other* poles and certainly still feel the isolation anxiety with its associated fear of isolation and longing for connection. However, since the engulfment anxiety is stronger, they tend to feel more comfortable with isolation than with intimate connection. If such an individual also has a particularly narrow window of tolerance, they may be particularly susceptible to being overwhelmed by engulfment anxiety and therefore cling onto their sense of autonomy with unusual desperation. May suggested that an individual who develops an imbalanced emphasis on this side develops "the anxiety of the defiant and isolated individual"[9]. Some researchers have suggested that particularly strong avoidant attachment styles may be linked to what are referred to in the DSM-IV as *schizoid personality disorder* and *narcissistic personality disorder*[10].

If someone has developed an ambivalent attachment style, then their window of tolerance is skewed in the direction of the *other* pole (see the top of Figure 20.3). In other words, they are generally more vulnerable to isolation anxiety than engulfment anxiety and are therefore more likely to modify their behavior and environment in a way that places their experience more on the *other* side of the axis of rapprochement. These individuals are still within the fields of both the *self* and the *other* poles and certainly still feel the engulfment anxiety with its associated fear of being engulfed and/or losing one's sense of self. However, since the isolation anxiety is stronger for these individuals, they tend to feel particularly uncomfortable with isolation and loneliness. If such an individual also has a particularly narrow window of tolerance, they may be particularly susceptible to being overwhelmed by isolation anxiety and therefore cling onto relationships with unusual desperation. May suggested that an individual who develops an imbalanced emphasis on this side develops "the anxiety of the clinging person who cannot live outside of symbiosis"[11]. Particularly strong ambivalent attachment styles may be linked to what are referred to in the DSM-IV as *histrionic personality disorder*[12] and possibly *dependent personality disorder*.

If someone has developed a disorganized attachment style, then we can say that their window of tolerance is particularly narrow. In other words, both engulfment anxiety and isolation anxiety are so strong that this person can find very little refuge from either one. Such an individual may be able to create a life strategy that works for some time, but their equilibrium is very precarious. Any circumstance that increases either isolation anxiety or engulfment anxiety could easily knock the person out of whatever limited window of tolerance they manage to maintain, at which point they might find themselves oscillating back and forth between the two poles somewhat chaotically (hence the term *disorganized*) until they manage to regain their precarious equilibrium. It has been suggested by some that this kind of attachment style may be linked to what is referred to in the DSM-IV as *borderline personality disorder*[13]. Yalom suggested that such a configuration is likely to make the individual particularly prone to psychosis[14], an idea that appears to fit well with the data of this study and which we will explore in more depth later.

When considering attachment styles, it is important to keep in mind that these merely represent a style and not a permanently fixed condition. People can experience significantly different degrees of both types of anxieties from one moment to the next, depending on a number of factors, including especially the nature of one's relationship with the particular *other*. This said, attachment research does suggest that by adulthood, people do tend to have established a particular pattern of attachment that is relatively persistent[15].

To put these findings within the framework of the self/other dialectic, it will be helpful to make the distinction between movement along the axis of rapprochement and movement and/or change of the window of tolerance. The simplest way to make this distinction is to consider that movement along the axis of rapprochement occurs as a result of our moment to moment perception and interpretation of our current situation, whereas changes within our window of tolerance occur only as the result of fundamental changes within the nature of our very being—in particular, changes regarding our degree of tolerance for the two existential anxieties (see Figure 20.5).

What this means is that we can experience movement along the axis of rapprochement (towards or away from one pole or the other) nearly instantaneously as our perceptions and/or interpretations of our current situation change. Significant changes within our window of tolerance, on the other hand, require a change within our being at a fundamental level, and for most of us, these kinds

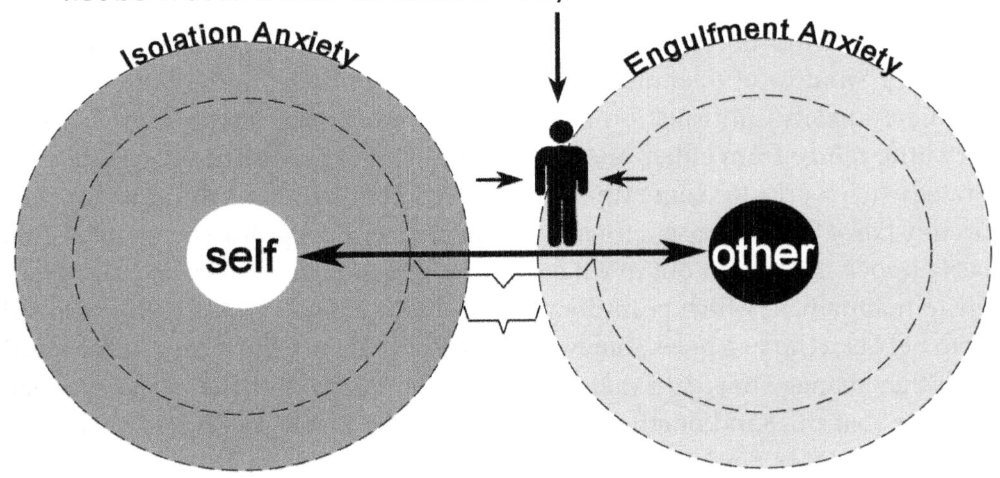

Figure 20.5. Movement along the axis of rapprochement vs. changes in one's window of tolerance.

of changes happen relatively slowly, as is suggested by the attachment research mentioned above. There do seem to be rare exceptions, however, when a change in one's window of tolerance can take place rather quickly—for example, while under the influence of psychoactive drugs, during psychosis, or as the result of experiencing a profound epiphany such as what may result during intensive psychotherapy or deep contemplative practice. I believe there is evidence to suggest that trauma and shame may also be associated with rapid changes within one's window of tolerance, particularly a rapid narrowing (the participant Jeremy's story gives a compelling example of this with regard to shame, as will be discussed in Part Four).

Let's look at a couple of examples to help clarify the difference between movement along the axis of rapprochement vs. changes in the window of tolerance. Imagine, for example, hearing from a trusted friend that your longtime romantic partner is planning to leave you for someone else. It's not difficult to imagine how quickly your experience would shift towards the *self* pole and further into isolation anxiety upon hearing this news (with the myriad related feelings that this might entail—fear, sadness, anger, etc.). In other words, given this situation, most of us would find our experience moving far along the axis of rapprochement nearly instantaneously. Now imagine that you discover several hours later that this was only a false rumor, and that your partner never had any such intention. Your experience would probably return to within your window of tolerance relatively quickly. So notice that (1) such movement can happen relatively quickly, and (2) that it was your *perception* of the situation, upon hearing the rumor and upon later learning that it wasn't true, that ultimately affected your experience, not the actual reality of the situation.

In a second but closely related example, imagine for a moment that your partner did actually leave you. At first, as mentioned above, your experience is likely to move far along the axis of rapprochement in the direction of the *self* pole, maybe even taking you somewhat beyond the edge of your window of tolerance. Let's say that now, due to various reasons, you end up living alone for quite some time without being involved in a significant romantic relationship and even without much involvement in other types of intimate relationships, such as with friends and family. As difficult as this is, however, you somehow manage to tolerate the loneliness. As time passes, your experience changes from mere tolerance of loneliness to some degree of actual enjoyment of your increased autonomy. What happened? Your situation has changed very little during this period of time, and therefore your position along the axis of rapprochement has not moved much, remaining relatively close to the *self* pole. Your window of tolerance, however, *has* moved, expanding in the direction of the *self* pole with a corresponding diminishment of your isolation anxiety until your experience is once again within its bounds. This change in your window of tolerance may or may not have also included expansion in the direction of the *other* pole, and given this particular example, it probably wouldn't have. In other words, your window of tolerance now is likely to look similar to that illustrated in the bottom of Figure 20.3. (We will discuss the difference between the *expansion* of one's window of tolerance vs. the mere *shifting* of it in Chapter 23).

Cognitive Constructs

Our cognitive constructs play a very important role in our experience of duality. In the language of cognitive psychology, the term *cognitive constructs* refers to the particular cognitive framework that each of us develops in order to interpret the sensory data that we take in from the world. In simpler terms, our cognitive constructs are our thoughts, interpretations, understandings, categories, etc., that aid us in distinguishing the various phenomena that we encounter in the world. For example, they allow us to interpret a particular pattern of light waves contacting our eyes as the loving gaze of our mother, or a particular pattern of sound waves contacting our ears as a quickly approaching vehicle. Being able to make such interpretations and distinctions is absolutely essential to our survival. Without this capacity, we would be unable to distinguish a friend from a foe, edible food from inedible matter, a pathway from a cliff, and so on. Without cognitive constructs, we would find ourselves lost in a chaotic sea of unintelligible sensations and uninterpreted perceptions.

It's important to keep in mind, however, that while our cognitive constructs are clearly so essential for our survival, their use also carries significant limitations. Because they are comprised of our own expectations and conditioned interpretations, they clearly distort the world and can interfere with our ability to perceive the world accurately. Metaphorically speaking, then, it could be said that we all look at the world through more or less colored lenses. Furthermore, if we mistake this lens for true reality, we risk becoming dogmatic and inflexible in the face of ever changing conditions. Taken to its extreme, we would only be able to perceive what we expect to perceive, and we would not be able to perceive anything else[16]. Therefore, it is clear that learning to have an appropriate balance in this regard is essential for optimal health and functioning.

Placing the concept of cognitive constructs within the framework of the DUI model, we can say that our cognitive constructs are closely associated with the self/other dialectic in that they provide us with the framework that we use to distinguish self from other, and one type of "other" from another. They also provide us with a framework for interpreting our current situation, which is essentially what determines where our current experience lies on the axis of rapprochement (i.e., in relation to the *self* and *other* poles). The concept of cognitive constructs will be a very important one as we continue to fill in the details of the DUI model and then move on to a more complete understanding of the psychotic process.

Chapter 21

Our Experience of the Interplay Between Duality and Unity

The experiences of the participants of this study and of many other accounts suggest that our subjective experience includes not only the dualistic interplay found within the self/other dialectic as discussed above, but that it also includes a dynamic interplay between our experience of duality and our experience of unity. The evidence suggests, however, that in contrast to the self/other dialectic, the interplay between unity and duality is not dialectical in nature—in other words, it does not consist of two diametrically opposed forces creating tension between them. Instead, unity can be seen simply as the quality of fundamental interconnectedness that exists beneath/within the apparent experience of duality (see Figure 21.1). Speaking metaphorically, we can say that unity is the thread from which all dualistic manifestations of the world are woven.

When discussing the concepts of duality and unity, it is all too easy to fall into abstract philosophical musings, which can then take us very far away from direct subjective experience. So in order to keep our exploration grounded in actual lived experience, let's take a moment to look at how we actually experience the qualities of duality and unity in our everyday world.

Let's take for example the process of a human being ingesting a carrot. This carrot itself is ultimately nothing more than a manifestation of various elements in its environment that all came together to manifest in this particular form. Previously existing relatively distinct forms (other living organisms, rocks, minerals, etc.) initially broke down and decayed until they became the fertile soil and atmosphere from which the carrot takes on life. We can say, then, that these original forms initially flowed in the direction of unity when they underwent decay and disintegration, where the division between "self" and "other," between this form and that form, dissolved to greater or lesser degrees. The manifestation we think of as "carrot" then absorbed these various elements into its own being,

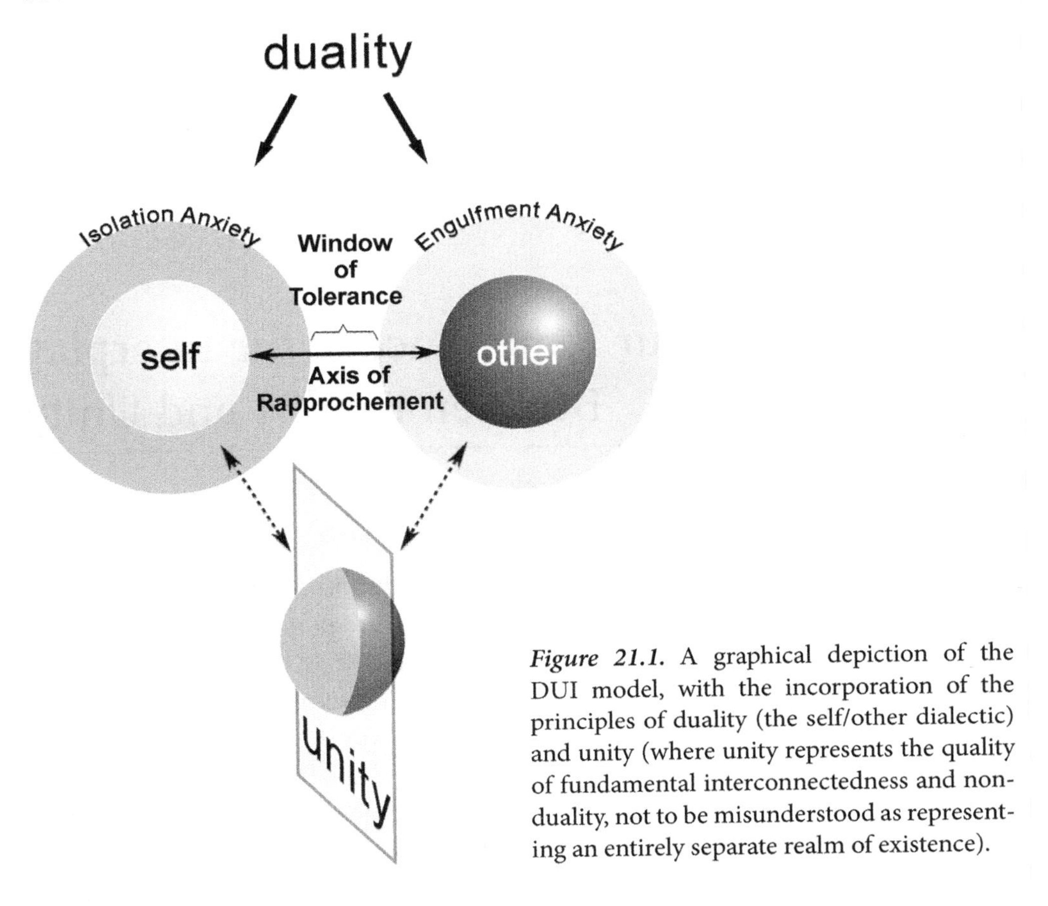

Figure 21.1. A graphical depiction of the DUI model, with the incorporation of the principles of duality (the self/other dialectic) and unity (where unity represents the quality of fundamental interconnectedness and non-duality, not to be misunderstood as representing an entirely separate realm of existence).

soaking them up through its roots and leaves, where it then assimilated them, reintegrating them as a part of itself. We can say that at this point, then, these elements are now flowing in the direction of duality, again manifesting in a particular self/form that is relatively distinct from other selves/forms—this time in the form of a carrot.

Then a day finally comes when a girl comes along (let's call her Mary), plucks the carrot from the soil, and eats it. At this point, then, this distinct self we think of as "carrot" undergoes a profound disintegration in Mary's digestion system, flowing back in the direction of unity as the distinction between self (the carrot) and other (not carrot) dissolves once again. Mary's body then assimilates the various elements of the carrot into itself, and these elements once again flow in the direction of duality, this time being assimilated into the relatively distinct self we think of as Mary. Then, a day comes when Mary's body must also pass on. It eventually undergoes complete disintegration—flowing towards unity—and then eventual reintegration into myriad other distinct forms/selves (worms, plants, etc.)—thus flowing once again in the direction of duality.

On and on this process of perpetual oscillation between duality and unity continues without stopping, not even for even a moment—from the grossest movement of very large and complex forms flowing in and out of existence, to the subtlest movement, in which it is believed that subatomic particles (from which all other manifestations of the universe are comprised) arise and pass away, arise and pass away many trillions of times per second[1] (see Appendix B for more details). For the entire life of our universe, we see an endless procession of forms taking shape, lasting for some time, and then passing away once again back into the unified field from which they came. This is not unlike the procession of waves rolling across a windswept sea. Due to various forces, a wave rises up, setting itself apart from the rest of the sea to some degree, only to eventually fall back into the sea, revealing that it had never actually been anything other than the sea itself. So it is that we find this apparent paradox everywhere we look—on one hand, there is clearly some validity to the concept that there is something that distinguishes one form from another, one self from another; and yet, it is also equally valid to recognize that all forms are nothing more than the most current manifestations in an infinite progression of manifestations arising from and passing away into a sea of unity.

So when we look more carefully at this process taking place on all levels within all forms of the known universe, it is not difficult to see that this constant flow between unity and duality, along with the associated qualities of profound impermanence and interconnectedness, comprise the most fundamental qualities of our universe. And since these qualities comprise the very fabric of our being, there can be little doubt that these must also comprise the very fabric of our subjective experience of the world and of ourselves, even though for most of us, we are rarely directly conscious of them in their more subtle forms.

For further illumination of this phenomenon, let's now turn to look at the findings of one particular contemplative tradition that has devoted literally thousands of years to studying it.

Raw Unconditioned Experience —The Three Marks of Existence

Mindfulness meditation is a method of introspective contemplation that lies at the heart of the Buddhist tradition, and it provides an unusually powerful technique for exploring the fundamental layers of our experience. By referring to the findings of a practice contained within the Buddhist tradition, however, there is the risk that some readers may immediately discount these findings as being dogmatic or religious. If we do this, however, we make the serious mistake of discounting one of the most robust forms of phenomenological inquiry into human experience that has ever been established, one that has produced surprisingly consistent findings for literally thousands of years, and one that is available to anyone, regardless of spiritual beliefs or lack thereof.

It is difficult to find an appropriate category in which to place Buddhism. On one hand, it clearly functions as a religion for many people, especially in the sense that it has the potential to offer a powerful "immortality project" with which to fend off death anxiety, to use Ernest Becker's language[2]. It contains many different lineages, many of which seem to be filled with dogmatic beliefs and rituals, the same as can be said about any of the other major religions. On the other hand, Buddhism had a very different beginning from that of most other religions in that its founder was a common man (Siddhattha Gotama) who never claimed to be otherwise and who acted very much like a scientist.

Gotama spent many years in the earnest pursuit of a deeper understanding of the nature of our suffering in the hopes of finding a way out of it. According to the Buddhist literature, he did finally succeed in working through his own suffering and in teaching others how to do the same (at which point his name was changed to the *Buddha*, meaning "the awakened one"). Whether or not one believes that Gotama actually managed to succeed in this regard, he clearly succeeded in developing a very powerful method of deep and penetrating inquiry into present subjective experience, a method he referred to as *Vipassana*[*], and what is often referred to in contemporary Western society as mindfulness meditation.

While there are different variations of mindfulness meditation, the essential steps are generally the same and are surprisingly simple (though not necessarily easy). First, one develops the capacity to maintain relatively steadfast concentration on a single object (most often using one's breath as the object of concentration); then, one directs this power of concentration onto the subjective experience of one's being (the experience of one's mind and body), making a determined effort to maintain steadfast awareness and firm equanimity in the face of whatever arises.

From a scientific perspective, the fact that such a practice is the very keystone of the Buddhist tradition suggests a very important distinction between Buddhism and most of the other major religions. If we were to put aside the various myths, ceremonies, rituals, beliefs, etc. that we often find within the different lineages of Buddhism, what remains is a tradition spanning over 2,500 years during which time millions of people have practiced this highly disciplined form of phenomenological inquiry; and those who have had the discipline to take this inquiry far enough have arrived at the common experience of three qualities that

* *Vipassana* is a Pali word that has been translated as "clear seeing" or sometimes literally as "before the eyes," implying an emphasis on direct present experience in contrast to the use of reasoning or blind faith.

appear to lie at the foundation of human experience—what are referred to as the *three marks of existence**.

According to these practitioners, raw unconditioned experience is comprised of essentially three qualities: (a) impermanence—the experience of all manifestations of the world, including the self, being in a state of constant change, constantly fluxing and flowing (known as *anicca* in Pali, the language used in the original Buddhist scriptures); (b) interconnectedness—the experience of all dualistic forms of the world, including the self, being merely different manifestations of a common unified whole (also sometimes referred to as *emptiness* or *no-self*, as in the idea that no individual entity is ultimately comprised of a separate and permanent self; *anatta* in Pali); and (c) the suffering that results from our desire to maintain a secure and permanent self in a world that is so fundamentally impermanent and interconnected (*dukkha*). In the DUI model, this suffering is synonymous with the dialectical tension found within the self/other dialectic, which we'll discuss in more detail later.

What we find with these three marks of existence is a powerful framework for understanding the interplay between duality and unity: the ultimately nondualistic "no self" conditions inherent within the first two marks of existence—fundamental impermanence and interconnectedness—and the suffering that is inherent within our need to maintain the survival of a self (a dualistic entity) within these nondualistic conditions.

Although it is somewhat tangential to the discussion here, it is very interesting to note that the field of modern physics has been developing an understanding of the fundamental nature of the universe that is very much in alignment with the findings of intensive mindfulness meditation practice†. They have both generally come to the conclusion that the fabric of the universe and of our very beings consists of an interconnected and impermanent sea of activity arising from the

---

* See the *Resources* section in the back of the book for some books and articles that go into more detail about the specific findings of mindfulness meditation, both from the perspective of contemplative inquiry as well as from the perspective of Western scientific methodology.

† I consider the research within the field of physics to be somewhat tangential to the discussion here because it (and most other research within Western science) typically uses *positivistic* rather than *phenomenological* inquiry. Positivistic inquiry emphasizes objective observation rather than direct subjective experience, typically using a combination of objective measurements and linear reasoning to generate hypotheses and theories. Ironically, the field of physics, widely considered the most pioneering of the Western scientific pursuits, has arrived at a theoretical paradigm that brings significant doubt to some of the foundational assumptions of positivistic inquiry. In particular, the assumption of the possibility of purely "objective" observation has been profoundly discredited, as it has been repeatedly demonstrated that subject (observer) and object (observed) cannot ultimately be separated. This paradigm shift in the field has suggested the importance of incorporating more phenomenological inquiry into scientific pursuits.

interplay between the conditions of unity and duality. See Appendix B for a brief discussion of the most relevant such findings within modern physics.

TRANSLIMINAL EXPERIENCES

So we can say that when we peer beneath the veil of our cognitive constructs, we experience the world in its raw form without the distortion created by our elaborate framework of expectations, assumptions, and interpretations. And when we experience the world in this way, we find ourselves immersed in the direct subjective experience of the three marks of existence. Following in the footsteps of Clarke[3] and other researchers with both cognitive and transpersonal orientations, I will refer to all types of such experiences as *transliminal experiences*.

Parapsychologist Michael Thalbourne originally coined the term *transliminality,* meaning literally "going beyond the threshold," to refer to the process whereby unconscious material crosses the threshold into consciousness (from the subliminal to the supraliminal)[4]. I find that this term fits the concept I am trying to illustrate here quite well; however, I believe I'm using it with a slightly different emphasis than what Thalbourne had in mind. In particular, I am referring to *transliminal experiences* as those experiences that are unitive or reflect the interplay between unity and duality, and which therefore cross the threshold of our cognitive constructs into conscious experience.

All such transliminal experiences are generally considered to be quite rare—for most of us, our cognitive constructs are quite sound and stable and do a very good job of ensuring that these kinds of experiences do not happen. However, in unusual circumstances, one's cognitive constructs may be unusually unstable and/or weak, and the possibility of having transliminal experiences is correspondingly increased. This can happen as a result of being under the influence of certain psychoactive substances (especially hallucinogens), during intensive mindfulness meditation or other practices that alter one's state of consciousness (such as sensory deprivation, sleep deprivation, fasting, etc.), or when the validity of one's present cognitive constructs is significantly challenged (for example, when experiencing a profound epiphany or paradox).

As will be discussed in more detail later, having such a raw unconditioned experience of the world at this most fundamental level can represent a particularly severe threat to our sense of self by bringing directly into our consciousness the fundamental qualities of impermanence and interconnectedness, which so profoundly threaten the experience of our self as a separate, solid, and stable entity. Our psyche apparently goes to great lengths to avoid this threat, primarily using our cognitive constructs to do so. Even when someone does have a glimpse of the transliminal, in most cases it seems that their psyche almost immediately attempts to integrate this experience into their cognitive constructs in a way that minimizes its threat to their sense of self. The research suggests that successful integration of transliminal experiences into one's cognitive constructs can result in genuine growth, whereas the failure of integration can lead to an escalating degree of instability, suffering, and possibly even psychosis[5]. In the context of the DUI model, the successful integration of transliminal experiences would lead to a lasting diminishment of one's dialectical tension, which includes a greater sense of wellbeing and maturity, whereas the failure to integrate transliminal experiences would lead to escalating dialectical tension along with increased suffering and the possibility of psychosis (more on this in the next two chapters).

Successful integration of transliminal experiences seems to occur when a genuine realization of the fundamental principle of unity is directly incorporated into one's worldview (cognitive constructs) to a greater or lesser degree—in other words, when one is able to further integrate the paradoxical coexistence of our fundamental nature as both dualistic and unitive. This is likely to result in the softening of one's self/other split (reduced dialectical tension), which is associated with an expanded window of tolerance and a corresponding reduction of negative feelings and an increase in positive feelings (discussed in more detail in the next chapter). While such successful integration may happen immediately upon the occurrence of a transliminal experience, there is the possibility that this process may take some time, leading to escalating dialectical tension and an ongoing attempt at integration until it is finally successfully integrated. When the psyche cannot successfully integrate a significant transliminal experience, the dialectical tension may continue to escalate until the point of overwhelm and psychosis (see Figure 21.2).

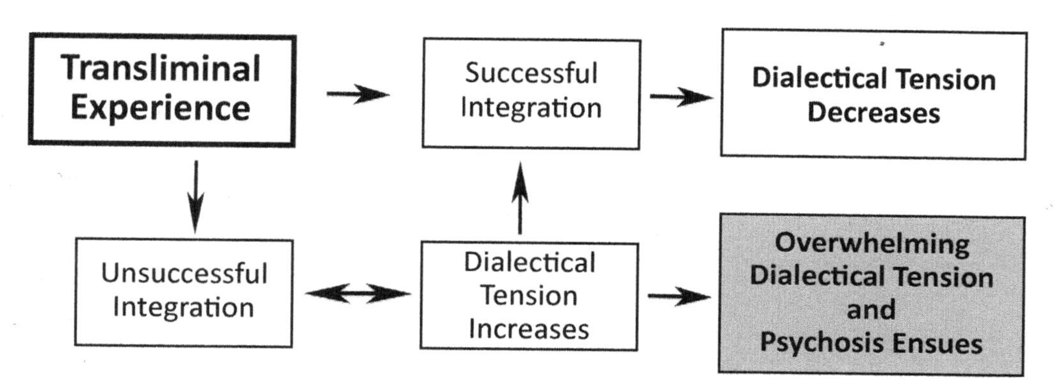

Figure 21.2. A diagram illustrating the possible outcomes resulting from a significant transliminal experience.

The data generated from my own studies and from other research suggest that transliminal experiences are an important component of psychosis, and in fact may even be an essential defining feature of it[6]. For yet undetermined reasons, it seems that there are certain individuals who have a particularly unstable set of cognitive constructs, and the evidence suggests that this may contribute to an unusually high susceptibility to psychosis due to their inability to keep the lid on powerful transliminal experiences.

We will continue to explore the implications that transliminal experiences and unstable cognitive constructs have for the development of psychosis and recovery. Let's now turn to look in more detail at our feelings, and how the DUI model offers a particularly cogent framework for describing our entire spectrum of feelings.

Chapter 22

A Description of the Full Spectrum of Our Feelings

We have discussed the main structural components of the DUI model—the self/other dialectic paradoxically coexisting with the principle of fundamental unity—and we have also discussed some of the most important aspects of our experience of these components—our perpetual dance between the two anxieties of the self/other dialectic, our cognitive constructs, and transliminal experiences. Now, in order to add important depth and detail to this model so that we can better understand how the structure it describes gives rise to our actual conscious experience, we need to look more closely at our spectrum of feelings. How does the dialectical struggle between self and other, and the delicate balancing act between duality and unity, give rise to the myriad feelings that we experience? To better understand this, it will help if we turn once again to look at the three marks of existence—those most fundamental qualities of experience.

Upon further contemplation of the three marks of existence, we can see that interconnectedness is the mark/quality that most closely represents the condition of unity; impermanence is the mark/quality that most closely represents the interface between unity and duality—the continuous activity of dualistic manifestations arising from the condition of unity, changing form, and passing away into the condition of unity; and suffering is the mark/quality most closely associated with duality in that it results from our ongoing struggle to maintain a secure sense of self in the midst of a world that is so fundamentally impermanent and interconnected. In order to understand more fully how these three fundamental qualities so profoundly shape our experience and our spectrum of feelings, it will be helpful to again draw from the findings within the Buddhist literature, in which a distinction is made between *dualistic feelings* and *unitive feelings.*

DUALISTIC AND UNITIVE FEELINGS

In the Buddhist literature, the Pali term *kilesa* is used to refer to any of the feelings we experience as a result of attempting to maintain a sense of self (a sense of duality) in a world in which the underlying unitive qualities make this so precarious. This concept fits the framework we are constructing here very well, but for the sake of clarity, I will use the term *dualistic feelings* rather than *kilesas*.

Within the context of the DUI model, dualistic feelings are any of the feelings directly associated with the self/other dialectic. As already discussed, we find ourselves perpetually torn between diametrically opposed fears and corresponding desires related to our need to simultaneously maintain autonomy while also experiencing connection with others and the world around us, and these fears and desires, along with all other related feelings, are considered dualistic feelings. From this perspective, then, we can see that all dualistic feelings essentially fall within two categories—*craving* (those associated with desire) and *aversion* (those associated with fear), although we find that these two categories give rise to a wide array of feelings (for example, hatred, animosity, anger, jealousy, envy, fear, lust, greed, longing, sadness, despair, etc.).

It would seem that the best we can hope for within this dialectical experience, then, is temporary appeasement of the fears and temporary satisfaction of the desires—in other words, the relative ease we feel while residing within our window of tolerance. It is clear to many of us, however, that there is the potential to experience much more than this—profound peace and joy, unconditional love, and deep compassion for others. Where do these kinds of feelings fit in with our experience of duality? According to the findings contained within the Buddhist literature, they essentially do not. It is suggested that these kinds of feelings require a direct experience of the quality of fundamental interconnectedness, or unity, and so are available to us only to the degree that we are able to relinquish the illusion of strict duality (our self as fundamentally separate from the world). Therefore, we will refer to these kinds of feelings as *unitive feelings*.

The unitive feelings are arguably much less diverse than the dualistic feelings, and they are commonly divided into just four types of feelings, what are sometimes referred to as the *four immeasurables*—unconditional love, compassion, sympathetic joy*, and equanimity. When we directly experience our interconnectedness with all things, we naturally discover a sense of unconditional love,

* Just as *compassion* refers to feeling connected to the suffering of others and/or the suffering inherent in life, *sympathetic joy* refers to feeling connected to the joy of others and/or the joy inherent in life. This is distinguished from the pleasure derived from the temporary satisfaction of a desire, which would be considered a dualistic feeling.

compassion, and sympathetic joy for all beings—since we recognize that others are not ultimately distinct from ourselves, we find that we have these feelings for others as easily and as naturally as we have them for ourselves (when we have a relatively healthy sense of self). As the experience of our self as a fundamentally distinct and separate self fades, the associated anxieties of the dualistic self/other split naturally fade and we discover a profound sense of equanimity.

Within this framework, unitive feelings are considered to be the ultimate ground of our experience since they correspond to an understanding (an experiential understanding rather than merely an intellectual one) and integration of the most fundamental qualities of experience—the three marks of existence. So unitive feelings never actually leave us but are merely "covered over" by the dualistic feelings that manifest within the dialectical struggle of duality. Based on this framework, then, all of our subjective feelings are comprised of some combination of feelings from both of these categories, as an ever changing mosaic of dualistic feelings covers and interacts with the underlying unitive feelings to a greater or lesser degree.

Our Experience of Good and Evil

One particularly interesting implication of this model of our spectrum of feelings is that it provides a relatively simple framework for understanding our experiences of good and evil. Most major religions and systems of morality, and I suspect most people in general, would consider the unitive feelings as defined here (unconditional love, compassion, sympathetic joy, and equanimity) to represent the highest qualities of goodness; and likewise, most major religions and systems of morality, and probably most people in general, would agree that if we become overwhelmed by the feelings referred to here as dualistic feelings, then we are likely to act in ways that are generally considered evil. In other words, if we become overwhelmed by greed, lust, hatred, animosity, jealousy, despair, fear, etc., then we become much more likely to act in ways that cause harm to ourselves and/or others. In the Buddhist tradition, there is no reference to good and evil in quite the same way as it is typically described in other religions or systems of morality. Rather, there is simply wisdom (*dhamma* in the language of Pali) and ignorance (*mara). Wisdom* refers to the experiential understanding and integration of the three marks of existence, and *ignorance* refers to the lack thereof. These conceptions of good and evil are particularly pertinent to the research here in that all of these participants had profound experiences of good and evil, and this model appears to fit their experiences particularly well (the details of which will be discussed in Part Four).

Terror and Euphoria within Transliminal Experiences

If we have a direct experience of the raw nature of the world, there is the likelihood that we will experience profound terror as we experience our sense of self being undermined at the most fundamental level. Yet, we have also seen that the diminishment of our dualistic sense of self provides us with the capacity to experience liberating unitive feelings. How can such apparently similar experiences lead to both feelings of abject terror and feelings of profound liberation and euphoria? There is a way to explain the propensity for both terror and euphoria within transliminal experiences that is in close accord with the findings of research within the field of cognitive psychology[1], the findings within the Buddhist literature, and the experiences of the participants within this study.

As discussed earlier, we can think of a transliminal experience as bringing into full consciousness the core existential dilemma that exists within our being all the time—the need to maintain a sense of self (a sense of duality) within a world in which the qualities are so fundamentally nondual. If we happen to have such an experience strong enough to undermine our cognitive constructs, it is likely that we will experience profound terror to a greater or lesser degree as we experience our sense of self as a stable and secure dualistic entity being undermined to a correspondingly greater or lesser degree. We can understand this sense of profound terror as being literally our direct experience of the very high dialectical tension generated when our psyche recoils from the threatening unitive experience, desperately attempting to maintain the dualistic sense of self by exaggeration of the dualistic split which necessarily entails this increase in dialectical tension (see Figure 22.2).

On the other hand, if our psyche is not immediately overwhelmed by the transliminal experience, then we may have a prolonged experience of very powerful unitive feelings and possibly even a sense of euphoric liberation as the dualistic split softens to a greater or lesser degree and tension within the self/other dialectic correspondingly diminishes (what transpersonal psychologist Abraham Maslow referred to as a *peak experience*[2]; see Figure 22.1).

So, as briefly touched upon earlier, there are several different possible outcomes that may result from the psyche's attempt to integrate a transliminal experience: One possibility is that our psyche may be able to successfully integrate the transliminal experience without ever being significantly overwhelmed. This will likely lead to a lasting expansion of one's window of tolerance to a greater or lesser degree and more or less follow the sequence illustrated in Figure 22.1). Another possibility is that our psyche may come close to being overwhelmed by

Figure 22.1. A successfully integrated transliminal experience.

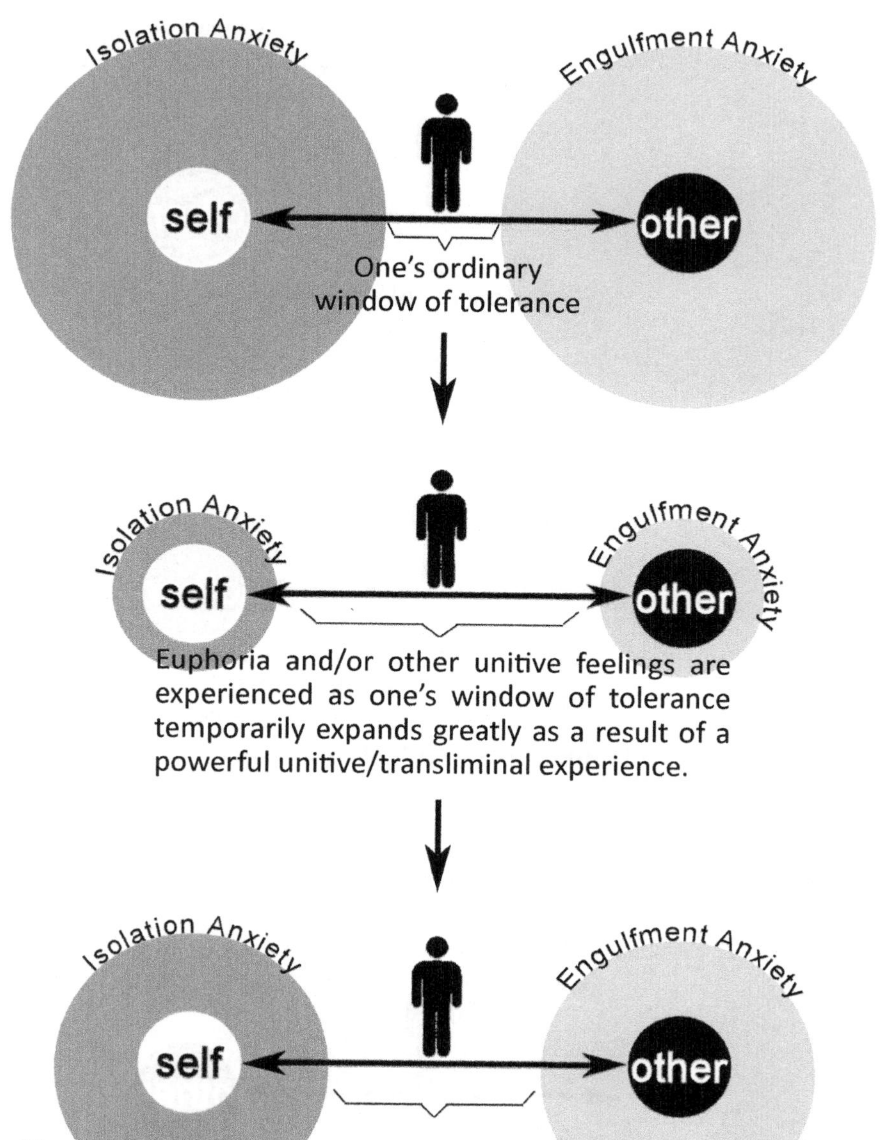

Figure 22.2. A recoil response to a transliminal experience may occur if one's psyche is not able to integrate the experience.

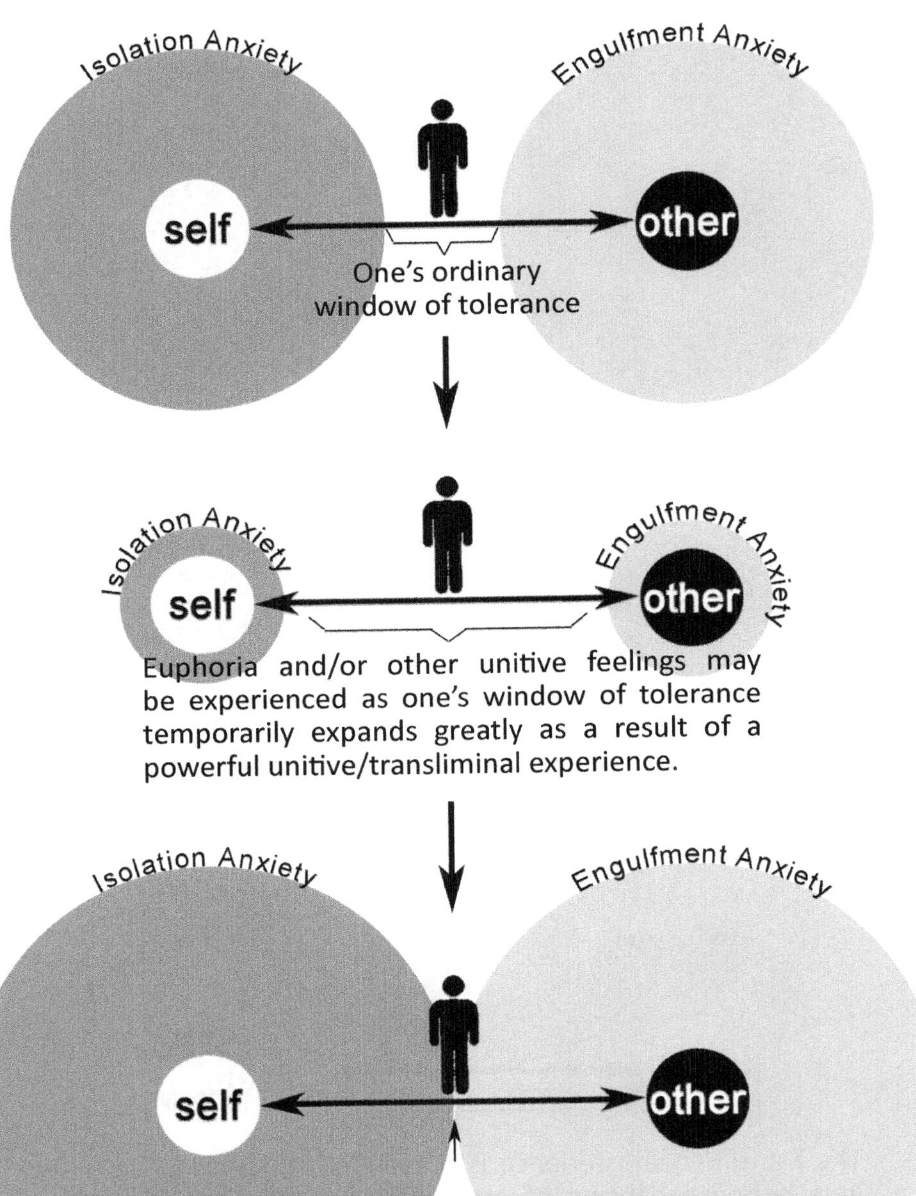

the transliminal experience, perhaps wobbling back and forth to some extent between high and low dialectical tension before ultimately integrating the experience. This will likely also result in a lasting expansion of one's window of tolerance. A final possibility is that our psyche may not be able to successfully integrate the transliminal experience at all, leading to ever increasing dialectical tension within the self system until the point of complete overwhelm, at which point psychosis is likely to ensue (see Figure 22.2). By definition, this sequence will pass through an initial phase of profound opening to unitive experiences; however, depending on how quickly the recoil response occurs, the experiencer may or may not be consciously aware of any liberating and/or euphoric feelings before proceeding to feel the terror of overwhelming dialectical tension.

Defining Our Personal Paradigm

If we combine the above framework for understanding the full spectrum of our feelings with the concept of cognitive constructs discussed earlier, we now have a relatively complete framework for defining one's *personal paradigm*. We can say that one's personal paradigm consists of both how one *understands* oneself and the world—one's cognitive constructs—and how one *experiences* oneself and the world—one's spectrum of feelings. Essentially, it is changes within one's personal paradigm (both one's cognitive constructs and one's spectrum of feelings) that is so closely associated with both personal growth and the process of psychosis. In personal growth, one's cognitive constructs incorporate the unity principle to a greater or lesser degree without threatening the existence of the self, leading to a gradual expansion and centering of one's window of tolerance, which leads to more unitive feelings and less of the extreme dualistic feelings. In psychosis, as we will discuss in more detail in Part Four, one's personal paradigm (both one's cognitive constructs and one's spectrum of feelings) undergoes extreme and rapid fluctuations as the psyche tries desperately to bring one's dysregulated self system back into balance.

Chapter 23

The Fundamental Roles of the Psyche

So we have now established the basic *structure* of the self at the existential level—the dualistic self/other dialectic existing paradoxically within the unitive qualities of fundamental impermanence and interconnectedness. We have also established our basic *experience* of the self—the polarity of fears and desires within the self/other dialectic that gives rise to all dualistic feelings, the unitive feelings that exist beneath the dualistic feelings and which naturally arise in consciousness upon diminishment of the dualistic feelings, and the veil of cognitive constructs that maintains the sense of a dualistic self while protecting the self from being inundated with unitive (transliminal) experiences that it cannot integrate. The final essential ingredient that needs to be addressed when developing a comprehensive model of human experience at the existential level is the *agency* of the self—that aspect of our being that contains the wisdom and the power/agency necessary to maintain our existence and ensure that our most essential needs are met.

Psyche is a common term used to refer to this agency of the self, so for the sake of clarity, I will also use the term *psyche* in this way, as I have already been doing to some extent. I suggest, then, that the psyche has primarily two functions: (1) survival of the self, and (2) growth, or maturity, of the self—attempting to move the self in the direction of ever increasing health and wholeness. And in order to carry out these functions, the psyche is guided by a profound source of wisdom, a wisdom sometimes referred to as *organismic wisdom*.

Organismic Wisdom

By using the term *psyche*, I am not suggesting that the psyche represents some kind of permanent soul or self in its own right. Recall that, according to the DUI model, all manifestations are profoundly impermanent, and that ultimately there is no such thing as a static, permanent self. I am suggesting, rather, that the psyche is merely that aspect of the ever-flowing movement of mind and being within any living organism that maintains a constant direction—namely, the direction of perpetually striving for survival and growth.

To understand this more clearly, it is helpful to think of the continuous flow of mind as being analogous to the light emanating from a light bulb. When we look at a light bulb, we typically have the illusion that the bright filament is static and non-changing—it is merely switched on. But in actuality, there is a constant flow of photons (light particles) streaming continuously from it. The light being emitted from the bulb does not remain constant, not even for a moment. What does remain constant, however, is the fact that there is a continuous flow of photons moving in a consistent direction—from the filament outwards. In the same way, then, we can see that while our entire mind and being is continuously changing and flowing, profoundly impermanent and ultimately devoid of anything we can call a solid, fixes entity, there is a consistency in the direction of this flow—namely, the movement towards continued survival and growth of the organization of experience that we think of as the self; and it is this constant movement, and the energy and volition within it, that we can think of as the psyche.

An important question worth considering when discussing the psyche is, what is it exactly that guides it—from where does the psyche draw the incredible wisdom necessary to maintain the survival of something as complex as a living organism, and particularly as complex as a human being? While, again, we want to be careful not to get drawn into speculative philosophical discussions, it is important to take a moment to recognize that the psyche is undeniably guided by a profound and mysterious wisdom, what I am referring to here as *organismic wisdom*.

Organismic wisdom is so common that we see signs of it everywhere we look, and yet the source remains mysterious and elusive. Embryos form, seeds develop, wounds heal, damaged ecosystems return to a balanced climax state. Whereas machines break and need to be fixed, living beings and living systems *heal*, and the act of healing can only come from within the living being or system itself. A doctor can set a bone, a psychotherapist can bring to light an unresolved trauma, but then she or he must get out of the way and allow the individual organism the freedom to do the actual healing. As mysterious as the source of this wisdom and power may be, we cannot deny that it exists within all living organisms and indeed within all living systems; and perhaps we can even say that the psyche itself is nothing more nor less than organismic wisdom, being merely the manifestation of it within a single organism.

As we continue our exploration of the psychotic process, it's important that we not lose sight of the fact that organismic wisdom is constantly operating within all living beings, even within those who are stricken with the most severe physiological or psychological distress, which of course includes people struggling with psychosis.

MAINTAINING SURVIVAL OF THE SELF

Within the context of the DUI model, I suggest that the psyche sustains the existence of the self by maintaining a tenable balance between duality and unity. On one hand, the psyche must ensure that enough duality (i.e., dialectical tension) is maintained so that the self is not flooded with overwhelming unitive/transliminal experiences that it cannot integrate. In other words, an adequate degree of dualistic self/other split must be maintained in order to maintain a secure sense of self. On the other hand, the psyche must ensure that the degree of duality (the dialectical tension) does not become so strong as to result in the loss of one's window of tolerance, which results in intolerable suffering and therefore also threatens survival of the self (see Figure 23.1).

Another way to put this is that the psyche must ensure that the self maintains enough of a sense of duality to continue to exist, but it must also ensure enough connection with unity so that it does not experience the intolerable suffering and existential annihilation of being cut off from its source. Before discussing the strategies that the psyche uses to maintain this equilibrium, let's first turn to look more closely at how our experiences of duality and unity have the potential to overwhelm us and represent existential threats to the self.

Threatening experiences of duality. In what is probably the most common way that a significant existential threat to the self occurs, someone experiences a significant degree of dialectical tension (a high experience of duality) when a situation occurs that takes them beyond the edge of their window of tolerance. In other words, one becomes inundated by either one or both of the two existential anxieties—isolation anxiety and/or engulfment anxiety. There are essentially two ways that such a threat can reach the level of complete overwhelm: (1) if we find ourselves outside of our window of tolerance in either direction (either towards the *self* pole with corresponding overwhelming isolation anxiety or towards the *other* pole with corresponding overwhelming engulfment anxiety)

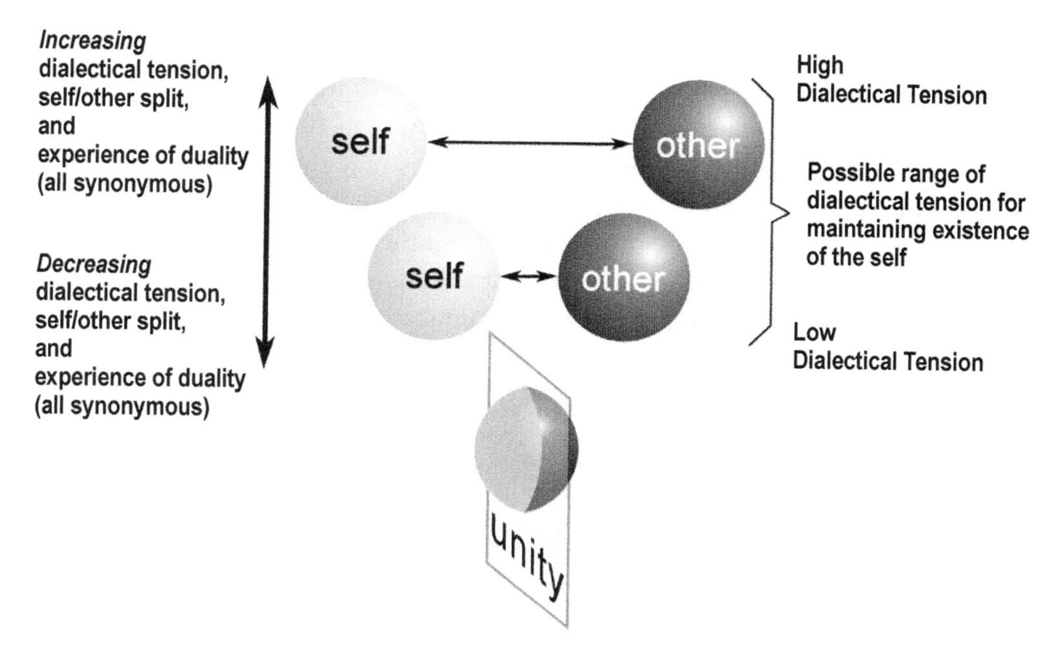

Figure 23.1. The psyche attempts to maintain the dialectical tension within a range that allows for the ongoing existence of the self.

(*Note:* In this graphical depiction, change in dialectical tension is represented by a change in the distance between the self and other poles, whereas in most of the other graphical depictions used in this book, change in dialectical tension is represented by change in the sizes of the two spheres that represent the isolation and engulfment anxieties. The emphasis in this diagram is on the correlation between dialectical tension and the degree of the experienced dualistic self/other split—the higher the dialectical tension, the greater the experience of a self/other split, and the lower the dialectical tension, the less the experience of a self/other split.)

and are unable to return our experience to within our window or tolerance for whatever reason; or (2) if our window of tolerance has essentially disappeared. Both of these can take place very gradually or happen within just a few moments, depending upon the particular circumstances. If someone experiences such an overwhelmingly high degree of dialectical tension and are unable to resolve it, then psychosis may ensue.

While experiencing the existential anxieties to such a degree that they entail an overwhelming threat to the self is fortunately quite rare for most of us, we can say that since we constantly face the existential dilemma of needing to maintain some balance between autonomy and connection (rapprochement), we are constantly within the fields of both of these anxieties to some degree all of the time, even though we may not always be explicitly conscious of them. Therefore, since we exist constantly within the fields of both of these anxieties, we all continuously feel a threatening experiencing of duality to a greater or lesser degree (again, not necessarily consciously), an idea that is in accord with the works of existentially oriented thinkers Becker[1], May[2], and Yalom[3], as well as the Buddhist literature. This also implies that the two existential anxieties play the very important role of warning us when we are approaching a degree of dialectical tension and/or an imbalance within the self/other dialectic that significantly threatens the existence of the self, compelling us to act in a way that will maintain our experience within our window of tolerance and thereby maintain stability within our self system.

Threatening experiences of unity. The second way that one may experience a significant existential threat to the self (and what is probably the least common way) is as the result of a sudden and/or relatively intense experience of unity—a transliminal experience.

If we take a moment to contemplate what it would be like to experience the world in its raw sensory form—as a sensorial experience of an utterly groundless and dynamic sea comprised of dualistic manifestations (including one's very own being) continuously manifesting from and returning to this unified sea—it is not difficult to understand how such an experience can be perceived as a severe threat to the self. As living beings striving to maintain some sense of having a solid and secure self within a solid and secure world, it is difficult to imagine a more severe threat than the experience of the world at this level. Those who have experienced this have often described it as if one's entire being were disintegrating at the most profound level, which often evokes a terror even more intense than the threat of bodily death. Most people who have a strong transliminal experience, however, will probably not be directly conscious of the raw nature of the world in this way, since it is likely that the psyche will almost immediately attempt to integrate this experience into one's cognitive constructs in a more tangible, less threatening form.

Recall that a transliminal experience has the potential to contribute to our growth, resulting in a significantly increased sense of wellbeing when successfully integrated. However, if we are unable to integrate a transliminal experience—whether because it is particularly strong or because we are clinging too tightly to our dualistic sense of self, or some combination of the two—the experience may result in such a severe threat to the self that the psyche may recoil sharply away from the unitive experience, exaggerating one's experience of duality and corresponding dialectical tension in a desperate attempt to maintain a secure existence of the self. Therefore, an overwhelming experience of unity is likely to ultimately result in an overwhelming experience of duality (overwhelming dialectical tension; see the bottom image in Figure 22.2).

The psyche's strategies for maintaining existence of the self. In the face of continuous threats of overwhelming experiences of duality and unity, the psyche must work hard to maintain existence of the self by maintaining equilibrium on two levels: within our experience of duality (maintaining a tenable balance between the two existential anxieties of the self/other dialectic), and between our experience of duality and our experience of unity (maintaining a tenable overall degree of dialectical tension). In simplest terms, then, we can say that our psyche maintains the necessary equilibrium required to maintain existence of the self by simply maintaining our moment to moment experience within our window of tolerance. So in order to do this, our psyche essentially needs to regulate just two factors—our moment to moment experience and our window of tolerance.

As mentioned above, the two existential anxieties associated with the two poles of the self/other dialectic can be seen as a warning system letting us know when our experience is moving toward the edges of our window of tolerance, or in other words, into territory that threatens the existence of the self. We can say, then, that the experience of discomfort that we feel when approaching the edges of our window or tolerance as well as the compulsion we feel to act in a way that brings us comfort once again are simply the conscious experience of our psyche's effort to maintain survival of the self. For most of us, this warning system is very effective most of the time and we find that our experience very rarely ventures outside of our window of tolerance; and when it does, we are usually quick to act in ways that bring it back.

The psyche has essentially just two ways of bringing our experience back to within the comfort of our window of tolerance after it moves outside of it. We can either change the current aggravating conditions in some way, or at least our perception of these conditions, until they return to our window of tolerance; or we can make adjustments directly to our window of tolerance, bringing our window of tolerance into alignment with the current conditions.

Generally speaking, it is far more likely that we will make the appropriate adjustments to the current conditions since it is usually far easier to do this than to adjust our window of tolerance. For example, if we've been spending a lot of time alone and find ourselves approaching the edge of our window of tolerance on the *self* side of the dialectic, and therefore experience excessive isolation anxiety, the first impulse for most of us is to reach out and connect with someone—in other words, to directly change the aggravating conditions. We then feel the desire to call a friend, visit a relative, or perhaps even connect with an animal or even nature as a way to bring our experience back towards the *other* pole until we find ourselves once again comfortably within our window of tolerance. Alternatively, if we have been spending a lot of time with others and find ourselves approaching the edge of our window of tolerance on the *other* side of the dialectic, therefore experiencing excessive engulfment anxiety, we typically feel the desire to spend some time alone. We can then take whatever measures are necessary to do that, again directly changing the aggravating conditions and this time bringing our experience back towards the *self* pole until it is once again comfortably within our window of tolerance.

If a situation occurs, however, that places our experience outside of our window of tolerance, but we find that we are unable to make adjustments directly to the aggravating conditions, then our psyche will most likely resort to attempting to adjust our window of tolerance until it comes into alignment with these conditions. The adjustment of our window of tolerance involves a fundamental change in the nature of our being, rather than merely a change in our situation (or in our perception of our situation). In particular, such an adjustment involves the willingness to cultivate tolerance and equanimity in the face of the current challenging situation. This adjustment also typically includes some modification of our cognitive constructs as we come to understand and experience the world in a way that is more compatible with this new set of conditions.

Assuming that this new situation is not particularly far outside of our window of tolerance and that we are willing to have the patience and openness necessary for this kind of change, then it is likely that our window of tolerance will naturally undergo the necessary adjustment. This adjustment may involve the shifting of our window of tolerance in the appropriate direction, the overall expansion of it, or some combination of the two. If our window of tolerance is expanded, then we typically experience this as spiritual or emotional growth, and our overall sense of wellbeing correspondingly increases (see the top of Figure 23.2). If our window of tolerance is merely shifted without being expanded, then we find ourselves more comfortable within this new situation but will probably not feel any overall benefit to our wellbeing and may even find that we have developed some additional limitations (see the bottom of Figure 23.2).

Our psyche may be forced to take a more desperate strategy if we find that conditions have taken our experience outside of our window of tolerance and we are unable to make either of these more ordinary adjustments—either directly changing the aggravating conditions or changing our window of tolerance. If this occurs, our psyche may be forced into significantly destabilizing our cognitive constructs in order to have extra leverage in bringing our experience and our window of tolerance back into alignment with each other. This type of more extreme strategy will still involve one or both of the two methods discussed above—changing the aggravating conditions and/or adjusting our window of tolerance; however, the strategy used to effect these changes will be significantly

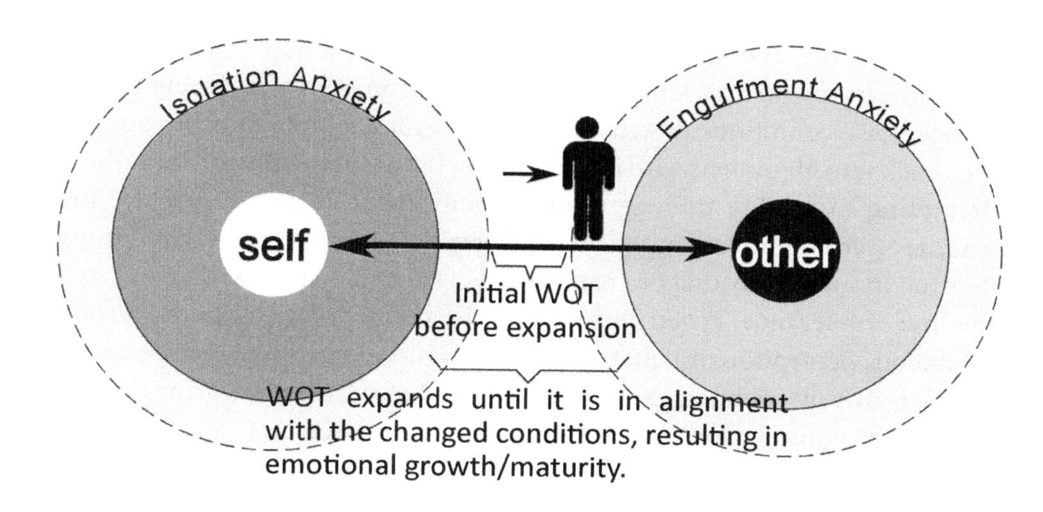

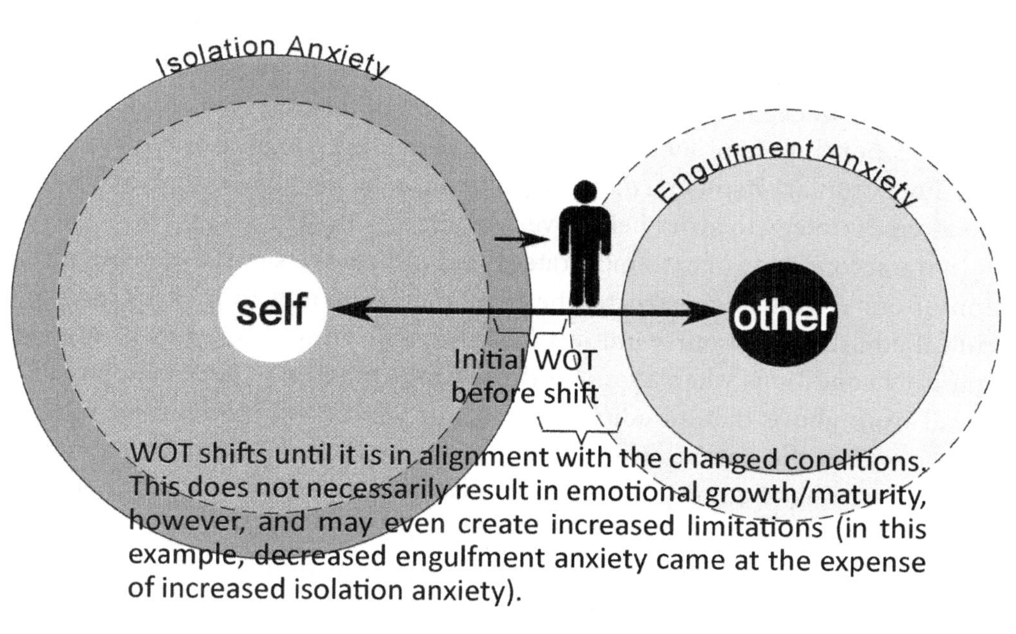

Figure 23.2. The difference between the expansion of one's window of tolerance (WOT; top) versus merely the shifting of it (bottom). These figures represent two different variations—expanding vs. shifting—of the psyche's strategy of adjusting one's WOT to bring it into alignment with changed conditions (in this example, the changed conditions took the person's experience beyond their WOT in the direction of the other pole). Both variations are effective in bringing one's WOT back into alignment with the new conditions; however, expansion results in psychological and/or emotional growth and maturity, whereas shifting without expansion does not necessarily lead to growth/maturity and may even increase limitations in some ways.

more radical and more risky since it involves the significant destabilization and alteration of our cognitive constructs.

Regarding the strategy of changing the aggravating conditions by destabilizing our cognitive constructs, since, for whatever reason, it does not seem possible to change these conditions directly, our psyche still attempts to bring our experience back into alignment with our existing window of tolerance. But rather than attempting to modify the aggravating conditions directly, the psyche instead modifies our cognitive constructs until our *perception* of these conditions has changed in such a way that our experience of them returns to within our existing window of tolerance. When successful, this strategy often takes the form of an anomalous perception or belief that provides more relief than distress.

One relatively common example of this kind of strategy is hearing the voice of or having visions of a loved one who has recently passed away*. Such a death has the potential to result in overwhelming isolation anxiety for the one left behind without the possibility of adjusting the aggravating conditions to reduce this anxiety (bringing the loved one back to life). If the individual is unable to tolerate the process of properly grieving the loss (which would presumably entail the combination of expanding/shifting one's window of tolerance and/or developing other relationships in order to bring one's experience back towards the *other* pole in a more ordinary manner), then the psyche may be forced to resort to this kind of extreme strategy to avoid being overwhelmed by existential anxiety.

In the second type of extreme strategy that utilizes the significant destabilization of our cognitive constructs, the psyche may attempt to make a relatively radical adjustment to our window of tolerance to bring it in line with the aggravating conditions when a more ordinary adjustment does not seem possible. Recall from above that an adjustment of our window of tolerance requires a change within the nature of our being, which typically involves at least some minor loosening of our cognitive constructs along with developing an increased capacity for tolerance. However, in order to maintain contact with consensus reality and also maintain relatively secure stability within the self system, this kind of adjustment generally must be relatively slow and paced. Thinking about

* Considering such an experience to be an anomalous experience does not nullify the possibility that such spirits may actually exist outside of consensus reality. However, such experiences are clearly not considered part of consensus reality, at least within mainstream Western society, and so regardless of whether or not such spirits actually do exist, being able to have such an experience would most likely require the significant destabilization of one's cognitive constructs, at least for most of us in the West. In a number of other societies, the existence of such spirits is consensually accepted, and such experiences of them are correspondingly much more common. Therefore, for members of these societies, such an experience would not necessarily require a radical destabilization of one's cognitive constructs.

this in more practical terms, most of us would agree that we can rarely develop a significant and lasting sense of peace with a difficult situation overnight. It is possible, however, to find ourselves so far outside of our window of tolerance (or perhaps find that our window of tolerance has disappeared altogether) that we are simply unable to tolerate the gradual adjustment of our window of tolerance necessary to maintain relative internal stability.

In this kind of situation, when no other option seems possible, our psyche may be forced to attempt a radical shift of our window of tolerance by using a high degree of destabilization of our cognitive constructs. By destabilizing our cognitive constructs in this way, we can say that the psyche is intentionally "cracking open the lid" to allow some unitive/transliminal experiences into our consciousness in order to expand our window of tolerance in a relatively rapid manner, with the intention that these experiences will be fully integrated and return our self system to a state of balance. Even when these experiences are successfully integrated, however, they will almost certainly carry with them a profound shift in our personal paradigm, since both our cognitive constructs and our window of tolerance would have been significantly altered.

Transpersonal psychology offers a useful framework for understanding this type of strategy, considering it to be either a peak experience or a spiritual emergency, depending upon the intensity and difficulty with successful integration. The term *peak experience* refers to occasions when this sudden "cracking the lid open" to transliminal experiences results in primarily unitive feelings and is able to be integrated relatively quickly, usually resulting in a relatively rapid and lasting expansion of one's window of tolerance. The term *spiritual emergency*, on the other hand, refers to occasions when both unitive feelings and existential terror may be present at times, and when the process of integration may be somewhat more precarious and last significantly longer.

As discussed in Chapter Nine, the main difference between spiritual emergency and what I am considering more "genuine" psychosis is that in spiritual emergency, a relatively strong observing ego remains present throughout the process, meaning that the individual manages to maintain some thread of connection to consensus reality, even though this thread may be very thin at times. In a more genuine psychotic process, the individual's connection to consensus reality may be completely lost at times and the process of integration is even more radically unstable and precarious (more on this in Part Four). It's important to reiterate that these different kinds of experiences—peak experiences, spiritual emergency, and psychosis—are not fundamentally distinct experiences, but rather their difference is primarily a matter of degree, particularly with regard to the ease and success of the integration of transliminal experiences.

One good example of the successful use of this kind of strategy involves the case of a woman who found herself in an existentially untenable situation with what I believe may have been a virtually nonexistent window of tolerance as the result of a very traumatic childhood. In very early adulthood, her entire personal paradigm (her beliefs and experiences of herself and the world) underwent a profound transformation that clearly involved a radical shift within both her cognitive constructs and her window of tolerance, a modification that has remained relatively stable to this day, nearly forty years later. Within her new belief system, she experiences herself as a very powerful being on a messianic quest, and she also experiences herself as belonging to a very benevolent and supportive family existing outside of consensus reality*. It is clear that she has also integrated the unity principle into her cognitive constructs to a significant degree, showing evidence that her window of tolerance had widened significantly. Her life has become an inspiring model of compassion and sympathetic joy for all beings, and the unitive principle is incorporated in some way or another into virtually all of her beliefs and behavior. This belief system has clearly been successful in staving off both engulfment anxiety (with the reinforcement of experiencing herself as a very powerful and confident individual) and isolation anxiety (with the sense of being looked after by such a benevolent and powerful family). To this day, she continues to do quite well, having a relatively high sense of wellbeing and meeting her needs at least as well as the average person whose personal paradigm is more in alignment with consensus reality.

We see significant evidence that this woman's psyche was forced to use the strategy of radically expanding her window of tolerance via the radical destabilization of her cognitive constructs. It is clear that her cognitive constructs were dramatically altered and have since restabilized, and that along with this, her window of tolerance was expanded significantly and probably also centered somewhat, which is associated with her having developed much more equanimity with her circumstances and her also having developed other unitive feelings to a significant degree. We can also say that this strategy was highly successful, in that she has clearly integrated this new configuration in a way that has led to lasting stability in her self system.

---

* As mentioned in the previous footnote, just because such beliefs and experiences are clearly anomalous relative to consensus reality (at least within mainstream Western society), this does not nullify the possibility of their having some degree of validity. After all, we all perceive ourselves and the world through the significant limitations of our own cognitive constructs, so I think it's important that we be skeptical of any tendency to declare with any degree of certainty that one individual's set of cognitive constructs is any more or less accurate than that of anyone else.

In the mainstream understanding of psychosis, these kinds of strategies involving a great degree of destabilization of one's cognitive constructs are generally considered to be a form of psychosis. I think it's important, however, that we seriously reconsider this perspective, especially when the evidence is clear that such strategies, though somewhat radical, are often highly successful in staving off overwhelming existential anxiety and maintaining relative equilibrium within the self system (as with the examples given above). While these kinds of strategies often involve anomalous experiences, as long as these experiences are not significantly distressing, as long as one's cognitive constructs are able to become relatively stable in their new configuration, and as long as one can continue to function in the world and meet their needs in a relatively effective and reliable manner, then I don't think it's appropriate to consider such strategies to be manifestations of psychosis. Examples abound of individuals who hear voices or have unusual belief systems but are able to function normally and have a relatively high sense of wellbeing[4], suggesting that these kinds of strategies can be perfectly valid strategies for maintaining a tolerable existence within particularly challenging conditions. It is important to acknowledge, however, that such strategies clearly do entail the significant risk of making one more vulnerable to both overwhelming unitive (transliminal) experiences as well as overwhelming dualistic experiences, both of which can lead to a more genuine psychotic process, as discussed earlier.

Regarding increasing one's risk to overwhelming transliminal experiences, since the psyche is intentionally "cracking open the lid" to such experiences in order to bring about radical changes within one's cognitive constructs and/or window of tolerance, it is easy to see the risk of accidentally "opening the lid" a little too far, allowing transliminal experiences into consciousness that the psyche is unable to integrate; and, as discussed earlier, unintegrated transliminal experiences are likely to lead to escalating dialectical tension and possibly even psychosis.

Regarding such extreme strategies increasing one's risk to overwhelming dualistic experiences, these strategies sometimes serve as only temporary solutions in staving off overwhelming existential anxiety, possibly taking away the person's desire to change the actual aggravating conditions. Since they may temporarily bypass the need to develop genuine tolerance and equanimity, they are often unsustainable in the long run.

One particularly interesting finding that emerged in my own study—especially in the cases of Trent and Theresa—was that psychoactive substances can be seen as a quick and effective method of facilitating these more radical strategies (particularly the strategy involving a radical change in one's window of tolerance) in that they can alter one's cognitive constructs to a significant degree and can also rapidly shift and/or expand one's window of tolerance. However, it is likely that the use of psychoactive drugs in this way carries an even greater risk than when this strategy is initiated naturally by the psyche. First, such use provides much less regulation with regard to how widely one is opening one's consciousness to potentially overwhelming transliminal experiences; second, sustaining the window of tolerance in its shifted and/or expanded configuration nearly always requires the continued use of the psychoactive substances; and finally, because the use of such substances is so effective in the short term, their use makes it much more likely that an individual will remain in an unsustainable situation far too long, creating the vulnerability of being completely overwhelmed by the existential anxieties when and if this strategy ultimately fails.

It's important to note that while resorting to anomalous perceptions and belief systems in order to maintain equilibrium within the self system might seem extreme, there is evidence that virtually all of us use a milder form of this kind of strategy. Yalom suggested that there are essentially two anomalous belief systems that we all use to a greater or lesser degree to fend off the threats to the self from the two existential anxieties—those who are more comfortable with the *self* side of the dialectic, and who therefore struggle more with engulfment anxiety, are likely to cultivate the belief that they are "personally inviolable," which often takes the form of some kind of heroic striving; and those who are more comfortable with the *other* side of the dialectic, and who therefore struggle more with isolation anxiety, are more likely to cultivate the belief that they are "protected eternally by an ultimate rescuer"[5].

Yalom suggested that most of us use some combination of both of these strategies, even those of us with a well centered and relatively wide window of tolerance, and that they are both important in maintaining a relative sense of security in the midst of our precarious existence. A problem arises, however, when a significant imbalance develops (when one's window of tolerance becomes significantly skewed to one side) in that the stronger the imbalance, the more heavily that person is likely to rely upon the strategy associated with that side, and the more extreme that strategy must become in order to be effective. The implication of this is that the more strongly one relies on just one strategy or the other, the more vulnerable one becomes to having that strategy undermined and thereby finding oneself very far outside of one's window of tolerance and particularly vulnerable to an overwhelming threat to the self.

For those using a relatively strong strategy of personal inviolability (whose window or tolerance is therefore significantly skewed towards the *self* side of the dialectic), their very sense of self depends upon the maintenance of an inviolable self. Therefore, they become more vulnerable to the fear of engulfment and may become overly sensitive to experiencing intimate connection or authority as an overwhelming existential threat to the self. For those using a relatively strong strategy of being protected by a powerful other (whose window of tolerance is therefore significantly skewed towards the *other* side of the dialectic), their very sense of self depends upon the maintenance of connection with others. Therefore, they become more vulnerable to the fear of isolation and may become overly sensitive to experiencing any kind of rejection, abandonment, or even diminished contact from others as a significant existential threat to the self. Such a strong imbalance and/or narrowness of one's window of tolerance could risk putting the psyche in the position where it cannot deal with a powerful existential threat in an ordinary manner and must instead resort to the extreme strategy of significantly destabilizing one's cognitive constructs to deal with them (as discussed above), thereby risking one's departure from consensus reality in various ways—via nonconsensus beliefs and/or perceptions—or possibly even initiating a full-blown psychotic process.

Following are some examples of extreme strategies that can arise as the result of having a particularly skewed and/or narrow window of tolerance:

Someone with a window of tolerance skewed strongly towards the *self* pole, and who therefore has already naturally established a significant degree of the "personal inviolability" strategy, may find themselves needing to resort to beliefs of omnipotence or even taking on a messianic status in order to stave off overwhelming engulfment anxiety.

Alternatively, someone with a window of tolerance skewed strongly towards the *other* pole, and who therefore has already naturally established a significant degree of the "protected by a powerful other" strategy, might find themselves literally perceiving this powerful "other" (perhaps in the form of a voice or a vision), and one that is not necessarily benevolent, in an attempt to stave off overwhelming isolation anxiety.

Someone with a very narrow window of tolerance, even if it is well centered, may be vulnerable to both types of overwhelming existential threats (both overwhelming isolation anxiety as well as overwhelming engulfment anxiety), and therefore may be forced to resort to either or even both of these kinds of extreme strategies. The example I gave above of the woman who successfully incorporated a highly anomalous belief system provides an example of someone who resorted to both kinds of these extreme strategies—believing herself to be on a messianic quest while also believing that she is looked after by a very powerful family existing outside of consensus reality.

As mentioned above, these kinds of extreme strategies might be successful for some time, and possibly even manage to stave off overwhelming existential anxiety for the person's entire life. However, they do require the significant instability of one's cognitive constructs, at least initially, making one more vulnerable

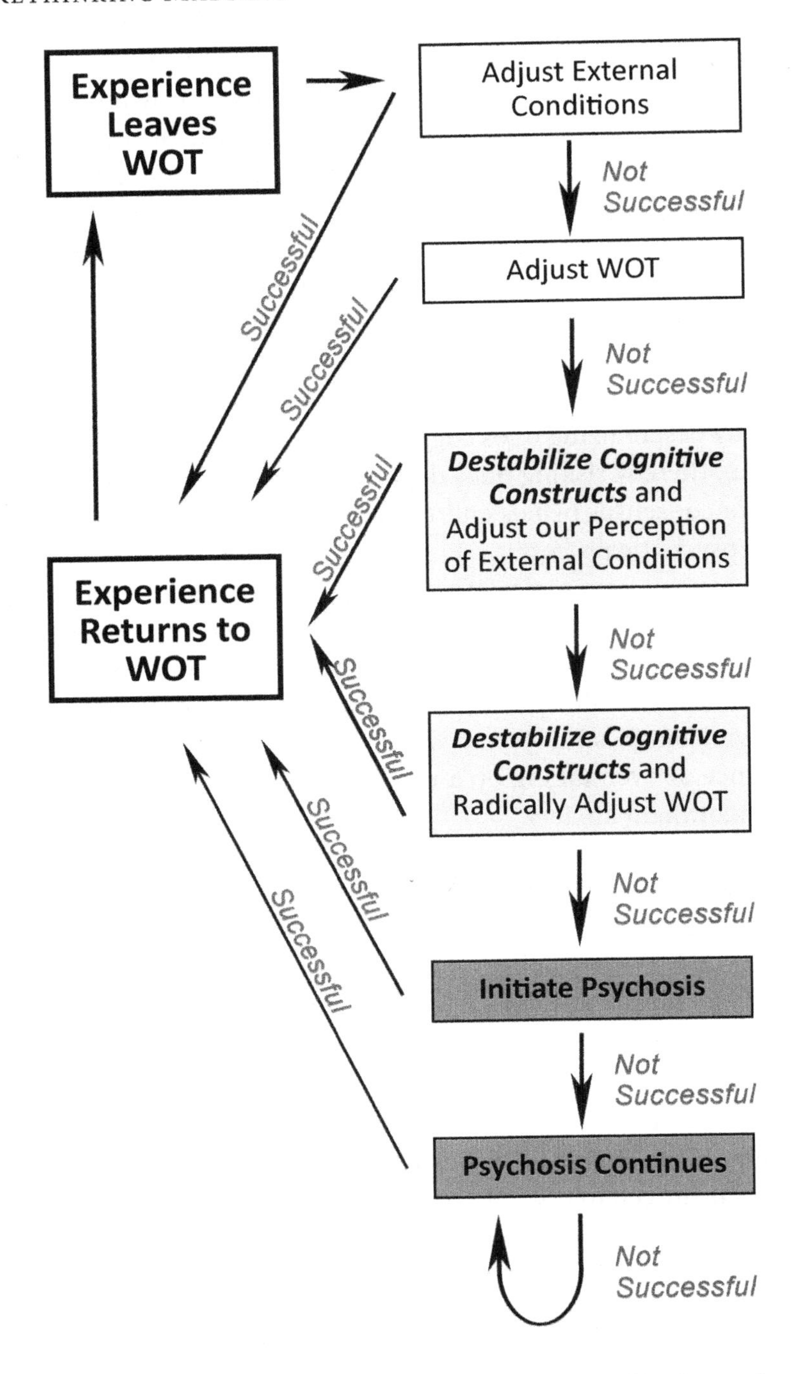

Figure 23.3. A diagram listing the succession of strategies employed by the psyche to maintain one's experience in alignment with one's window of tolerance (WOT). The top strategy is the one mostly likely to be employed first, with each successive strategy coming into play upon the failure of the previous one. Each successive strategy generally carries greater costs, so this particular sequence is likely to be the most common; however, individual cases may vary, and more than one strategy may be used simultaneously.

to overwhelming transliminal experiences, and they often do not address the underlying problem of an unhealthy configuration of one's window of tolerance, often dealing with the symptoms of the problem and not the root cause. Therefore, there is the high likelihood that such strategies may ultimately serve only as temporary solutions, ultimately requiring a significant and lasting adjustment of one's window of tolerance, either via the more benign route of significant but manageable personal growth or via the more haphazard route of a full-blown psychotic process.

While we will save the details of the psychotic process for Part Four, it's important to mention here that, according to the DUI model, psychosis can be seen as the last resort strategy that the psyche uses to maintain existence of the self when faced with an overwhelming existential threat to the self that it cannot deal with in any other way—in other words, when all of the strategies mentioned above are ultimately unsuccessful. See Figure 23.3 for a diagram outlining the various strategies that the psyche uses to address existential threats to the self and bring one's experience and one's window of tolerance back into alignment with each other.

Growth of the Self: Towards an Optimal Personality

The experience of these participants and of many others suggest that not only does the psyche play an essential role in maintaining the existence of the self, but that it also plays another similarly essential role. Reflecting upon what it is that really drives us, most of us would agree that we not only have a deeply rooted desire to maintain our survival, but that we also have a strong desire to thrive. By *thrive*, I am referring to our innate desire to move towards growth and health, especially with regard to our capacity for and tolerance of an ever broader range of experience. We see clear evidence of this in any healthy person and in any healthy organism. So we can say that the role of the psyche (the agent of the self) includes not only survival of the self but also growth of the self. Schneider suggested a particularly useful concept that can be incorporated into the DUI model to accommodate this growth-promoting role of the psyche—the *optimal personality*[6].

Schneider posited that what defines limitation and dysfunction is our dread of the polarities of our experience*, which leads to extreme counter-reaction and

---

* As mentioned earlier, Schneider (1999, 2008) posited a dialectical model of human experience that defines the poles as *constriction* and *expansion* and the fears associated with these poles as *the dread of ultimate constriction* and *the dread of ultimate expansion*. His use of constriction and expansion offers a penetrating inquiry into (continued on the next page)

the possibility of becoming fixated at one pole or the other, what corresponds to the development of a particularly skewed window of tolerance in the self/other dialectic. Schneider said that what defines our freedom, then, is our ability to *center*, which refers to our capacity to face our entire range of experience with awareness, resilience, and the confidence that we can survive all of these experiences.

Schneider suggested that if we can muster the courage and willingness to face our present experience with equanimity, regardless of how painful that might be, we find that it's possible to develop a sense of mastery within our experience. We learn that we can develop the ability to return to a more tolerable middle ground after having extreme experiences, and as our confidence builds in this regard, we find that we can continue to expand the range of tolerable experiences along the continuum. Therefore, being centered does not mean literally remaining fixed within a central position between the dialectic poles of our experience; it means remaining centered in the sense of being able to maintain an overall integration of both poles while also maintaining the ability to shift from one extreme to the other. As we develop this capacity, we move ever closer to what Schneider refers to as the *optimal personality*—we find that our enhanced capacity to be attuned to an ever greater range of experiences leads to an increasingly rich, healthy, and fulfilling life.

Placing the concepts of *centering* and the *optimal personality* into the context of the self/other dialectic, we can say that as we develop an increasing capacity to center—to face with equanimity an increasingly wider range of the spectrum of desires and fears associated with our relationship with self and other—we increase our window of tolerance, which corresponds to Schneider's concept of growth in the direction of the optimal personality. We find that the range of experiences that are tolerable to us widens, we feel the anxieties of isolation and engulfment less acutely, and we find that even when we do have experiences that take us outside of our window of tolerance, we are able to return relatively quickly. In other words, we find that both our equanimity and our resilience strengthen.

Returning again to the concept of a growth-promoting role of the psyche and the corresponding experience of our conscious desire for growth and health, we can say that the psyche is not only attempting to maintain survival of the self via the desire/will to maintain our experience within our window of tolerance, but that it also continuously strives to maximize the width and centrality of this window. In other words, we can say that the fundamental role of the psyche is

---

(continued from previous page) the core existential nature of being; however, I have found that the use of *self* and *other* creates a more suitable framework for capturing the majority of experiences reported by the participants of this study.

to survive and to thrive—to keep the self alive and, conditions permitting, to continuously move it towards ever increasing growth, health/wholeness, and the optimal personality.

One important implication of this movement towards the optimal personality is that as one's window of tolerance opens in both directions, one's experience of both poles begins to change—both isolation anxiety and engulfment anxiety correspondingly diminish. As a result, one is able to experience a richer, fuller, healthier relationship with oneself and with others. As we gain more comfort and confidence with the *self* side, we are likely to find increasing feelings of self connection, self worth, personal power/confidence, autonomy, and a greater sense of ease with aloneness; and as we gain more comfort and confidence with the *other* side, we are likely to find increasing feelings of romantic love, affection, respect, consideration, empathy, and friendship. Such an individual is more comfortable "in their own skin," experiencing the appreciation of aloneness more than the despair of loneliness; and they are also more comfortable within intimate relationship, enjoying a sense of deep connection with others without the fear of losing one's autonomy. In other words, to use the language of Gestalt therapy, one is able to experience genuine *contact* with others rather than either isolation or confluence (in which one loses connection with oneself and becomes somewhat lost within the other)[7]. In the language of attachment theory, we would say that such an individual is *securely attached* to a greater or lesser degree.

When we incorporate the concepts of dualistic and unitive feelings into the concept of the optimal personality, we find some interesting implications. As one's dialectical tension and therefore experience of duality is diminished, unitive feelings are freer to rise into consciousness (see Figure 23.4), offering a depth and richness to the experiences associated with the self/other dialectic and also increasing the full spectrum of available feelings and experiences. As unconditional love, compassion, sympathetic joy, and equanimity become an increasing part of one's experience, not only do personal relationships begin to have greater depth and intimacy, but one also begins to experience a more profound sense of connection with all (what is often referred to as spiritual growth). To use the language of Martin Buber, well-known Jewish mystic, one begins to experience more *I-Thou* relationships in contrast to *I-It* relationships[8].

A final important point to make about the process of growth is that it is generally a relatively slow process. It involves the lasting expansion of one's window of tolerance, and this kind of change typically takes time. In contrast to the individual who experiences terror when their veil of cognitive constructs is suddenly lifted, during movement towards an optimal personality, such transliminal experiences are generally quite subtle and are able to be integrated without overwhelming the individual. Such a process may happen so gradually that one never need experience the sudden undermining of one's cognitive constructs at all—during such a process, the cognitive constructs and one's window of tolerance have time to shift gently as new material is integrated. In the language of transpersonal psychology, this could be seen as the process of spiritual emergence in contrast to spiritual emergency or psychosis.

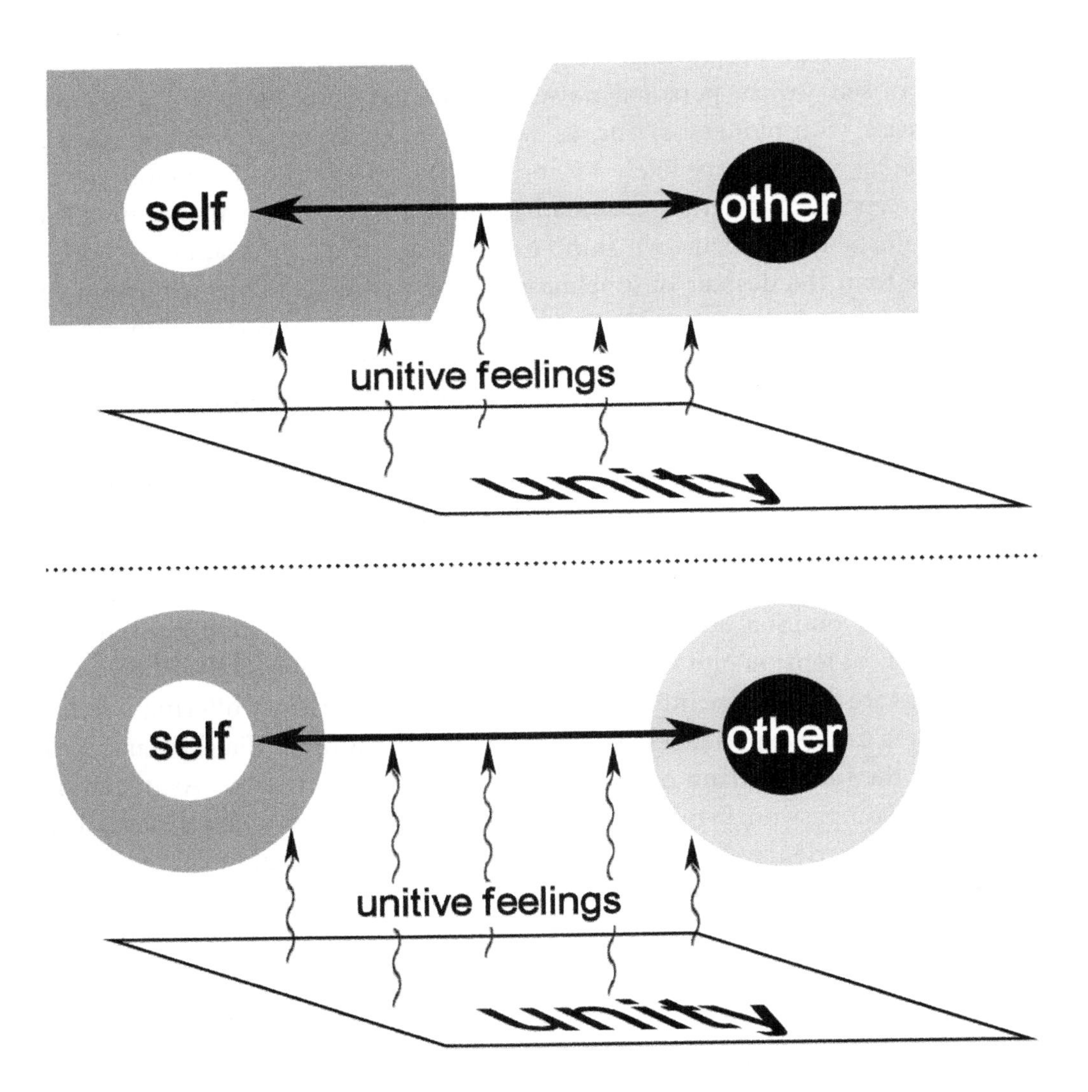

Figure 23.4. When dialectical tension is high (top), very few unitive feelings are able to make it into our conscious experience. As dialectical tension decreases, we naturally experience more unitive feelings (bottom), which then add depth and richness to the dualistic feelings of the self/other dialectic.

Summary of the Duality/Unity Integrative (DUI) Model

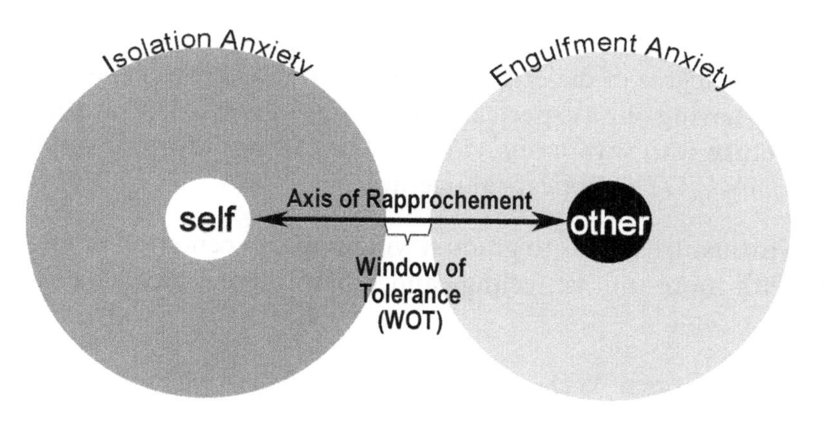

Increasing Dialectical Tension

- and -

Increasing Experience of Duality and Dualistic Feelings

Movement along the axis of rapprochement results from changes in our moment-to-moment perception of our current situation. These include all types of perceptions, including those of both internal and external events and those of events that may or may not be in accord with consensus reality.

Changes within the WOT involve changes within the actual nature of our being, particularly with our degree of tolerance for the two existential anxieties.

Decreasing Dialectical Tension

- and -

Increasing Experience of Unity and Unitive Feelings

Key Points of the DUI Model:

- We have essentially two core existential dilemmas with which our psyche continually grapples:

  (1) The need to maintain the survival of a dualistic self within a world that is fundamentally nondual—in other words, maintaining a tenable balance between duality and unity by maintaining stability within the overall degree of dialectical tension.

  (2) Within our dualistic experience (the self/other dialectic), there is the need to maintain a tenable balance between autonomy/individuation and connection with others.

- When one's experience along the axis of rapprochement leaves the WOT, the psyche will utilize whatever strategy is necessary to bring one's experience and one's WOT back into alignment with each other (see Figure 23.3).

- An overwhelming degree of dialectical tension can initiate psychosis. This can be caused by having one's experience far outside of one's WOT without being able to return it to within one's WOT in any other way, or it can be caused when one's WOT disappears altogether.

- Genuine growth/maturity leads to gradual widening and centering of one's WOT along with more unitive feelings and less of the extreme dualistic feelings.

- While expansion of one's WOT is generally beneficial when it can be integrated, when expansion occurs too rapidly to be integrated (an overwhelming transliminal experience), this can represent a severe threat to the self and result in a recoil response, leading to very high dialectical tension, which may then initiate psychosis.

Part Four

Making Sense of Madness, From Onset to Full Recovery

After having gone over the basic outline of the DUI model, we're in the position to take the final step in our exploration, using this model as a framework for exploring what takes place at the root-most (existential) level of experience during the psychotic process. But before we do this, let's first turn to look at the remaining three stories of the six participants of the study on which this book is based (recall that we have already looked at the stories of three of these participants—Sam, Theresa, and Byron).

Chapter 24

The Case of Cheryl

At the age of 25, Cheryl was a young woman filled with compassion and a strong desire to ease the suffering in the world. However, she also carried with her a heavy burden in the form of very painful feelings and an inner turmoil that began many years earlier. Up until this point in her life, she had managed to keep these feelings somewhat at bay and to generally move forward with her life; but now, they had finally risen to the surface, threatening to tear apart her sanity and send her reeling into an abyss of overwhelming despair. She went on to experience a period of psychosis that lasted a little over half a year, including a two month period in which she suffered what she describes as "extremely severe psychosis." She was hospitalized twice during this period and formally diagnosed with "major depressive disorder, recurrent, moderate to severe, with psychosis."

The Onset and Deepening of Psychosis

In retrospect, Cheryl has come to recognize that the painful belief system that became so dominant during her psychosis had roots extending all the way back to her early childhood. She feels that she suffered from mild to severe depression for much of her childhood, adolescence, and early adulthood, until the onset of her psychosis at age 25. She recalls that her parents were not abusive, and in fact they played a significant role in her recovery. She does believe, however, that there were some significant shortcomings in their parenting skills, which may have played a role in the development of a painful belief system that ultimately came to full force during her psychosis:

> When I was little, I think they were a little immature to be parents and didn't deal with their own emotions well and they would project them on me a lot. Like I remember when I was four, my mom asked me if she was fat and I said yes. I didn't even know what that meant, I just thought, well, she's a lot bigger than me, you know. She got real mad and stormed out and slammed the door, and my dad said to me, how could you do that, that's so terrible, you know. . . . When I was young, my mom was always yelling at me, we didn't get along, and then my dad was always ignoring me. . . . They loved me, they were never ever abusive to me, there was never anything like extremely traumatic. . . . They had their own unresolved issues. . . . They didn't, you know, teach me how to love myself, and I never did.

Just before the onset of her psychosis, Cheryl went through a series of very stressful incidents, most of which contributed to an increasing sense of isolation:

> I was having a real hard time, because I had moved to be with my boyfriend, but then we broke up like a week later, and then my friend moved away, like I didn't know anyone. I wasn't getting along with my roommates, I was unemployed, I finally got this job but it was this awful job with juvenile delinquents, and they were all really mean to me, and it turned into this like traumatizing experience.

Cheryl was on the verge of being overwhelmed by her tremendously isolating situation when she had a profound experience of connection. With the intention of developing a set of skills that would help her to support others in her work, Cheryl decided to attend a spiritual counseling course. The course was designed to teach the students how to communicate with "positive spirits" and make psychic readings. The course went reasonably well—perhaps a little too well: "After the class, I started hearing these voices. They said they were my spirit guides, you know. They said they really cared about me, and I was like, this is great because I have no friends and I [laughs] broke up with my boyfriend and everyone hates me, so at least these things in my head, they like me, right [laughs]?"

Cheryl felt a sense of "mild euphoria and liberation" during the initial stages of contacting these "spirit guides." She also felt a sense of hope and the belief that "life was going to make sense again if only [she] listened to them and followed their guidance." It was not long, however, before things dramatically changed. Cheryl believes that this unusual yet profound experience of connection and support coming right on top of her extreme experience of isolation opened the door to her psychosis.

After some time, the quality of the voices changed from that of primarily offering support to becoming more directive with her: "They started telling me all these weird things, you know, like, oh, you need to go here and do this, and if you go here, you'll meet your future husband, and you have to go rescue this person from here." She initially followed these voices' instructions without much thought, and they were relatively harmless at first, consisting primarily of attempts to help her find her future husband. However, the instructions soon became increasingly unusual and other people began to take notice: "[The voices] actually . . . told me that I needed to go to my job in the middle of the night where I worked with juvenile delinquents because I had to save the life of one of the students, and I did that. I think they thought I was like a child stalker or something, you know, but . . . I think it eventually came out that I was just insane [laughs]."

The voices continued to offer these unusual instructions, but after repeatedly failing to provide any success for Cheryl, they took a decidedly malevolent turn: "When it didn't work out because it wasn't real, that's when these voices started to get really mean and say, well, you know, you're a failure anyway, and we hate you, and God hates you, and you're like worse than Hitler, and [laughs] you should go blow your brains out."

From this point on, the voices continued to berate Cheryl with the most malicious language imaginable: "[The voices] told me all these things like God hates you, and I wasn't even religious, but they said, you know, God hates you, you're such a terrible person that even God who has never hated anyone before hates you now. He hates you so much that he wants to send everyone to Hell. . . . Now God hates everyone in the world because he hates me so much."

Cheryl soon found herself spiraling into an unbearably painful belief system that was in accord with the messages of these voices: "So . . . the delusions turned into, you know, you're the most horrible person and really this was the central thing was that I'm so terrible that everyone in the world is going to go to Hell for ever and ever because of me." As Cheryl's belief system of her utter terribleness escalated, she experienced other closely associated unusual beliefs: "I thought people were going to kill me. I thought that somehow I had failed this imaginary husband and that he hated me now and he was gonna come kill me. Then, when I went to live with my family, I thought that my dad was gonna kill me a couple times. I ran away once . . . because I thought that my dad was gonna kill me, which [laughs]..my dad was never gonna kill me." . . . And:

> I thought that my life was like a reality show that only I didn't know about, but in my reality show, it was about how I was the most horrible person in the world and everyone was in on it but me. And the voices said that my recent ex-boyfriend that I had just broken up with was the star of the

> show, and he secretly hated me and found the idea that he cared about me disgusting and laughable. And meanwhile, these voices told me that he had secretly been dating an earlier . . . girlfriend of his the whole time. Then they told me that my family was helping the two of them plan and prepare for their wedding and everyone was invited but me.

Even though Cheryl's experience of utter terribleness (both in the messages of the voices and in her own escalating belief system) was the core of her psychosis, some other types of anomalous beliefs also emerged. In the early stages of her psychosis, while the voices were still benevolent, Cheryl went through a brief period where she believed that she "had been given this revolutionary theory":

> *Cheryl:* I thought that somehow I had in my mind that because time is an illusion, that somehow people that were all present in the world at the same time were all past lives of each other somehow. . . . I just had this idea and I started thinking that I could figure out that, oh, this person is the past life of this person.
>
> *Researcher:* I see, so different past lives of the same [person] might actually be coexisting.
>
> *Cheryl:* Yeah, that was my idea.

Cheryl also suffered from an overwhelming sense of terror and groundlessness throughout a significant portion of her psychosis:

> I would say I was in a constant state of extreme terror (and groundlessness) for at least two months. Some moments were worse than others. When I entered the crisis unit the first time, I was wailing in terror until a staff member threatened me that I had better stop it or else. . . . Another time I remember feeling completely "dead," "not me," "not human," at one point, as if there was nothing left of me but terror and pain. I gave my brother a hug to try to "feel human" again. I think this helped a miniscule amount, although it probably scared my brother, as I know I was saying something to the effect of "Help! I'm not human! Help! I'm so scared!"

Throughout the most intense stage of her psychosis, Cheryl found herself experiencing profound turmoil and confusion between the voices, her own beliefs, and what was happening within consensus reality; and she was desperate for any kind of relief, even if it had to come by suicide. She describes one particularly poignant example of this chaotic mixture of confusion and despair:

> The voices had been telling me to kill myself, but it was like I..at that time, I really wanted to, anyway, you know, so . . . it was kind of both. With the voices telling me that you're so terrible and you need to do this, well you started to feel like, yes, I need to do this, so at that time it was *my* feeling, but..also I had been hearing voices that said this, so . . . . It just kept getting worse and worse and then at one point I saw a bottle of pills and I was gonna take them but I was so out of it that I started to put my shoes on because I thought, okay, I'm gonna go to Hell now, I need to put my shoes on. And then I realized, wait a minute, I don't need to put my shoes on to go [laughs] to Hell. And then I thought, no, I shouldn't kill myself here in my house, my family will be crushed. I should, you know, do it later, and ultimately when I did try to kill myself, I went to the beach and tried to kill myself on the beach. But then I wandered back home when it didn't work, thinking, oh, well maybe when I get home, maybe my dad will kill me then.

Cheryl managed to work through her psychosis with only two relatively brief hospitalizations in a crisis unit, and she felt that the psychiatric support she received was "not helpful at all." When staying in the crisis unit, Cheryl was originally placed "on very heavy drugs," which presumably included some combination of antidepressants, antipsychotics, and perhaps others, and she recalls experiencing some severe side effects: "They made my heart feel very abnormal and I was scared I was going to have a heart attack. They also gave me visual trailers (i.e., totally slowed my brain down to where I would be looking at something but still be seeing the thing I'd looked at before that)."

Frightened by these effects, she managed to find a way to reduce her intake of the drugs: "These things (esp. the heart issue) scared me, so I started tonguing my meds (hiding them under my tongue to spit out later), a trick I had learned from my previous clients (when I was a [mental health] tech)." She found that one of the drugs seemed to reduce the voices, but this surprisingly resulted in exacerbating her psychosis: "There was one medication that made the voices harder to hear (not sure what any of these were; they never told me) but I thought what the voices had to say was important (listening to them gave me some sense of control, I guess) so I strained even harder to hear them, an effort which brought me further away from reality."

After leaving the crisis unit, she was maintained on small doses of two different psychiatric drugs—Celexa (an antidepressant) and Seroquel (an antipsychotic), which "was a compromise between what [she] wanted to do (take nothing) and what [her] family and doctors wanted (more drugs)," and she carefully weaned herself off of them about six months later. Overall, Cheryl believes that the use of these drugs were "mostly irrelevant" to her recovery: "There may have been a benefit I wasn't conscious of (like healing from the extra sleep—I sure had been desperate for sleep!) I had a couple of health and weight problems that started around that time (which have since been resolved) that I believe may have been related to the drugs, but I can't prove it."

As already alluded to, Cheryl struggled tremendously with hopelessness and suicidality at times during her psychosis. The combination of the constant assault by the voices and her painful sense of terribleness and self-hatred were overwhelming. Shortly after her release from the crisis unit, she tried to commit suicide using sleeping pills, was subsequently returned to the crisis unit, and she again tried to kill herself inside the unit, resorting in desperation to trying to cut her wrist with the under wire from her bra.

Recovery

After several months of falling ever more deeply into despair and the certainty that the utter enormity of her own terribleness had doomed everyone in the world to the fate of everlasting hell, a surprising incident occurred that would be the first in a series of incidents that would set Cheryl well on the path to recovery. After the failure of psychiatry to offer any real support for their daughter and feeling very desperate, Cheryl's parents made the decision to take her to an alternative healer—a practitioner of *BodyTalk*, which Cheryl describes as an integrative form of healing that attempts to balance the various systems of our beings (energetic, psychological, and physiological) by "prompting the body energetically to heal itself":

> The session with [the BodyTalk practitioner] didn't really help, but she was like, well I'm gonna go see *my* practitioner tomorrow at one, and I was like, okay, whatever, I was still really psychotic and . . . I'm thinking to myself, we're all going to Hell, so [laughs] you do what you want but it won't work. So, the next day, all of a sudden I had the weirdest experience. I had heard all these voices constantly for like two months, and then all of a sudden these voices inside me screamed, "No," like an extended "NOOOOoooooooo," and it seemed like they were leaving and they didn't want to, and all of a sudden it got really quiet inside of me. I looked at my watch, and it was like ten after one, and right away, I got like really terrified because I thought, well, these screaming voices are parts of me that are leaving and they're gonna go see my healer and her healer and they're gonna tell them what a horrible person I really am, and then my family's gonna find out and they're gonna be heartbroken, and finally they are really gonna kill me so they can save [laughs] themselves from me. . . . I [had just been] starting to think that

> maybe they did love me and now they're gonna kill me 'cause now they're gonna figure it out [laughs]. . . . So, finally my BodyTalk practitioner called and she sounds really upset, and I'm like, oh god, it's true. She said, "Cheryl, [my practitioner] and I worked on you today and it was a very interesting experience," and she said it in like this really weird voice. Then she said, "[My practitioner] discovered that you had 64 negative attachments, and she removed them for you. She pulled them through herself and sent them to the light or something, and she was like coughing and choking, and I've never seen anything like it in my life," and I was like, what, okay [laughs].

At first, Cheryl had some doubts regarding the effectiveness of this "exorcism":

> I didn't really give this much thought at the time because it didn't immediately eliminate my delusions. . . . I had this sense up until that point that it was like there were monsters inside of me, and then when that happened, I was like, well maybe there were, but I didn't really give it much thought because I still thought, well, I'm still the most horrible and evil person, everybody still hates me, I'm still going to Hell.

But then it soon became apparent that the quality of the voices had changed significantly:

> [Prior to this experience,] there had been these extremely elaborate abusive stories that just went on and on like 24 hours a day, where I was like not even sleeping for [laughs]..for like a week, you know. And I remember at one point there was a word that one of the voices used, and I didn't know what it was, and I looked it up in the dictionary and it was a real word, so it was like really real and really constant and really elaborate. Then after this happened, I was still hearing voices but they were kind of like less distinct from my own consciousness, maybe like my own brain. I kind of, and this is my theory, I don't know, but I think it was kind of like all of a sudden it was so quiet that my own brain couldn't stand it 'cause it wasn't used to it. . . . This was like a lot simpler, it was like voices that would say, die, die, die, you know, like really simple stuff. These new voices were kind of like rehashing the old stuff, but just like really simplistically.

With this shift in the quality of the voices, they began to lose their hold on Cheryl and she began to take them less seriously. Cheryl now recognizes this as having been a very important turning point in her recovery.

A second turning point happened not long afterwards with a different psychic healer. Cheryl's ex-boyfriend arranged to have her visit a psychic who ironically practiced the same modality as that taught in the spiritual counseling course that had played such an important role in the onset of her psychosis. When Cheryl's voices first began to turn malevolent, she had sought advice from the teacher of that course, but the teacher had only dismissed her experiences: "She said, 'Oh, that's a different thing, that has nothing to do with my *class*, you need to go take some medication.'" But this psychic took a completely different approach with Cheryl, validating her experiences but explaining that she was "like a child running out in heavy traffic and . . . getting squashed":

> [She said,] "you don't really know what you're doing and you're getting all of these negative energies that are tricking you and manipulating you." She told me, stop worrying that everyone's gonna go to Hell because of me. The world has way too much love in it to be harmed by anything that I can think or say or do. And it was like all these other delusions and hallucinations have been proven wrong one by one, and then it got to this point where this particular lady was challenging the essential part of it, and she was coming at it from the angle that I guess I needed to come at it from.

These words had a profound impact on Cheryl:

> So I'm talking to her, and it was like everything that my family and friends and everyone had been telling me seemed like it hadn't made any dent at all, and then when she was telling me this, it like all coalesced at once, and I was like, oh my god [laughs], I'm mentally ill, and I was like totally stunned. I was like, I mean my jaw was probably open, I was like, wow [laughs]..um..wow, you know, like I mean, I..I can't tell you how stunned I was. Everything all came together at once.

The psychic's words paved the way for Cheryl to make some sense out of the intense turmoil and confusion of the previous several months:

> There was a gradual process where, as I got better, I was able to look back on what happened and make more sense of it than I did at the time that it originally happened, but I think that . . . there were two things going on with me and one was that I was in touch with actual..um..you know, negative energies that were manipulating me, . . . but also that I had psychological problems that were there either way, that I was like super, super

> depressed, and these two things matched up together. . . . I would say that the psychological problems and these negative energies were really feeding off of each other. . . . It's like they were all a big glob of the same experience.

This new understanding and ability to make some sense of her madness in turn paved the way for a renewed sense of hope and determination to blaze a trail towards genuine recovery.

Immediately after her discussion with the psychic, Cheryl took the first important intentional step in her recovery. She made the determined effort to reclaim her power and stand up to the voices:

> I thought, how am I gonna do this, you know. . . . I didn't really entirely know, but I just took it step by step and the first thing I said was I sort of made this little announcement in my head like, okay, everybody out, you can talk all you want, I don't care, but I'm not listening to you, nothing you say is true. . . . I made a conscious decision that I was gonna ignore everything. And after that, it was easier than I thought it would be, I guess because I hadn't quite yet adjusted to the fact that it had already slowed down [due to both the earlier "exorcism" and now the lesson from the psychic], but then . . . . when I started to say, I'm not listening no matter what, that kind of helped them go down some more, and then the rest of my recovery was worrying about loving myself and forgiving myself.

At this point in Cheryl's recovery, the voices had eased considerably, but she was still haunted with the stubborn core belief that she was simply not lovable. Surprisingly, the most potent challenge to this painful core belief came from the family cat:

> Everybody was trying to tell me that they loved me, that I was lovable, and it was like I wanted to believe them, but at first I still..I didn't believe it. I remember that it was like three a.m. one night and I couldn't sleep and I was like, well, it seems there are these real spirit guides that the psychic says that I have, these nice ones that she can talk to but I never talked to, you know. . . . I said, well, if you're really out there, if it's really true that I'm lovable, can you give me some kind of sign. And . . . as soon as I said that, the family cat came into my bedroom and jumped on my bed and started rubbing up against me and purring, and I thought, huh, well *she* just loves me no matter what.

This experience with the cat began to crack the foundation of Cheryl's painful core belief of being unlovable and her own self-hatred, but she realized that she still had a tremendous amount of work ahead of her: "I had this huge challenge of forgiving and loving myself. . . . Okay, the voices weren't real, I still..like I just tremendously hated myself." She felt unsure how to proceed at this point, but she soon came across the book, *Return to Love,* by Marianne Williamson. The teachings in this book offered her a shift in perspective and some tools that allowed her to begin the difficult work of learning how to forgive and love herself:

> I came across the book in Barnes and Noble, and it just sort of leaped off the shelf. . . . I thought that because of what these voices were saying, and like I said, I wasn't really religious before, but from what these voices were saying, I thought that God was gonna punish me. Then this book talked about God as being the essence of love, and it taught me that I was okay the way I was and I didn't have to earn love. I didn't have to be a certain someone or do a certain thing or accomplish something or whatever in order to be lovable. And I thought, you know, I used to think, well, I can't be forgiven, there's a line where you can't forgive, and this book taught me, no, there isn't, you can forgive no matter what, and you just gotta start over and start loving yourself.

With some guidance and a renewed sense of hope, Cheryl became determined to start chipping away at her hatred and anger. She began to practice the intentional generation of love for herself and others, facing the continuing onslaught of hateful thoughts:

> I decided that I would start visualizing love coming out of my heart and surrounding me, and that gave me feelings of self love, so I kept trying to do that. . . . I kept reverting to the feeling, no, you deserve to die, I should kill myself, but I found that when I didn't love myself, I didn't love others, and I felt like these other people who were so nice to me, they really deserve it even if I don't, and I kind of forced myself to love myself for the sake of my family. . . . So I just kept practicing this, you know, I kept like being, god, I'm the worst person ever and ever, but no, no, no, but then I would start to feel like, grrrr, everyone is awful, I hate everyone, the world is bad, but then I thought, no, no, no, I have to be loving towards them, I have to start by loving myself, okay, I love myself, I love myself.

Cheryl feels that another important factor in her recovery was the persistent love and care from her family. For much of her psychosis, she was unable to believe that anyone could really love her. However, she realizes now that her family's persistent love and care for her eventually broke through this belief system and played a crucial role in her recovery:

> My family..everyone kept trying to tell me that they really did love me and it kind of seemed pretty obvious meanwhile, but . . . the voices said, oh, well they think they love you but their higher selves don't love you. . . . But meanwhile, I was starting to think, when I tried to kill myself and they were sobbing, and then when I lived and they were so ecstatic, I'm like, huh, well maybe [laughs]..maybe that's not true that nobody loves me.

One final essential factor that Cheryl believes was crucial in her recovery was her belief that genuine recovery really was possible. She had studied psychology in college and had worked in the mental health field prior to her psychosis. She said that, like most people in the mental health field, she was exposed to the belief that once someone is diagnosed with a severe psychotic disorder, full recovery is not possible and they must remain on debilitating drugs for the rest of their lives. Cheryl is extremely grateful that, during her undergraduate studies, she came across the research by the World Health Organization[1], Loren Mosher[2], and others who validated the possibility of full recovery from schizophrenia and other long-term psychotic disorders. She credits this knowledge for providing her with some spark of hope for her own recovery, even in the darkest times:

> As soon as I realized, oh, I'm like not well, I'm having a mental illness type of experience, like I remembered, okay, but I can get better because . . . these other people have gotten better . . . but I didn't know how, and there was no professionals who were telling me how. I mean, the professionals that I worked with were like so unbelievably maddeningly clueless [laughs].. but I thought, well, I'm gonna have to figure it out. My main belief is that, you know, here we don't believe that people can get better and so they don't, and I feel that in addition to love and loving myself and other people loving me and figuring everything out, the most important thing was that I had the idea that maybe I could get better. . . . Whereas if I hadn't had that prior knowledge that maybe I could get better, I feel like maybe I really would have finally succeeded in killing myself because it was like I thought, well, I just can't keep living and staying this sick. . . . I was seeing psychiatrists but they weren't helpful at all. I followed my inner guidance and I followed my,

> you know, spiritual guidance and whatever as to..step by step, how do I get better. . . . I just feel so grateful that I had had that message of hope, because hope and love really are the two things that saved me.

◇◇

As of the writing of this book, Cheryl is 34 years old and, according to the criteria used in this study, she considers herself as having fully recovered within a year or two after the onset of her psychosis, and having maintained her recovery for the approximately eight years since. She returned to working as a mental health care worker after her recovery and has continued this work ever since, offering her unique and invaluable experience to the service of those who are struggling with severe mental disorders.

As was the case with all six of these participants, Cheryl feels that she has undergone a profound and primarily positive transformation as a result of her psychosis, with far more lasting benefits than harms. This is with regard to both her personal paradigm and her general sense of wellbeing and ability to meet her needs.

Chapter 25

The Case of Trent

In spite of being raised in a very troubled family, Trent somehow managed to hang on to his sanity throughout his childhood and adolescence. By the time he reached the age of 24, however, his precarious grip on consensus reality was failing and he soon found himself spiraling into a chaotic and confusing world filled with overwhelming forces of good and evil, highly persuasive and unusual belief systems, and profound despair. Trent's first episode was his longest, lasting about six months, after which he went on to experience about seven more significant psychotic episodes over the next six years. During this period of time, he was diagnosed with both catatonic schizophrenia and manic-depressive disorder*.

The Onset and Deepening of Psychosis

Trent refers to the onset of his psychosis as his "mental collapse," and he feels that the single largest factor leading up to this collapse was having been raised in a highly "dysfunctional" family: "I imploded into a void where a family was supposed to be." Over time, he has learned not to personally blame his parents for their faults but to see that they were merely another link in the chain of a long line of intergenerational trauma, each suffering their own deep emotional wounding. He describes his father as having been severely depressed and "sometimes outright emotionally abusive," and his mother as having severe emotional and cognitive difficulties, having been diagnosed with paranoid schizophrenia.

---

* With the publication of the DSM III (the third version of the DSM) in 1980, the condition previously referred to as *manic-depressive disorder* has been changed to *bipolar disorder*.

Trent believes that the wounding his parents had received was then passed down to him in the form of extreme emotional neglect:

> My parents were very dysfunctional people that totally avoided their childhood hurts. That really stunted them, you know, personally and socially and every other way possible, and they might not be very much to blame because they had very poor role models themselves. . . . It just kept building up generation after generation until it came to me, and I pretty much imploded on it. . . . I had to break the cycle. One therapist in family therapy said, maybe he had a nervous breakdown just so he could save the family, and, you know, there might have been some truth to that. Unfortunately, the family did not see it as an opportunity to save themselves. I'm still dealing with people that..uh..are pretty injured.

Trent recalls that he did sense the impending "collapse" before it happened, but he still was unable to avoid it: "I did do a type of distancing myself from them before I collapsed but it only worked so well. My mind and spirit could not tolerate all the baggage and twisted problems that I inherited through contact within the family and chose to collapse."

Another important factor that Trent feels contributed to his psychosis, one that was directly related to his family situation, was his experience of feeling very isolated prior to the onset (emotionally and psychologically, if not physically). Trent also believes that "the use of tobacco and marijuana also played a role in the collapse," although a much more minor role than that of the difficulties surrounding his family situation.

As Trent's psychosis unfolded, a pattern emerged of feeling overwhelmed alternately by good and evil forces, which included experiences of directly interacting with God and the Devil as well as intense experiences of creativity and destructiveness. One of the most common of these involved the belief that God was sending him messages in the form of actual events taking place within consensus reality: "I really thought like God was trying to give me messages, try to, you know, guide me..uh..a train horn or a flashing light, or a reflection, you know, little reminders." Occasionally, these experiences would increase into "episodes of extreme intensity" in which he felt he could read deeply "into the meaning of things." Though he had other experiences that included a sense of grandiosity, these particular experiences were generally very personal: "[They were] just for my own guidance, for my own wellbeing."

Generally, Trent's experiences of receiving divine messages from God were not particularly distressing; however, these often alternated with extremely distressing experiences in which he struggled with tremendous fear of being persecuted by "the Devil itself." He gives an example of one such time:

> I remember on one occasion, I had the thought that the Devil was gonna come and take me away. I was gonna die and go to Hell or something like that. . . . I was living in my parents' house and we had two cats, and the cats really sensed that there was something wrong. These cats really didn't mingle together, they weren't that type, but they both jumped up on my bed and kept me company. They just added support, you know how they do it, but they just felt that was their place to stay on vigil and be by me, and that's what I needed at the time. But in my own psychotic poor mind, I thought, you know, the Devil might be in these cats, and I was a little fearful of that. So that's a shame, 'cause they were there to help, and here I was thinking that..you know, fearing them, it was silly.

Closely related to his fear of being persecuted by the Devil was the occasional experience of feeling inundated with "evil" thoughts and feelings. He describes one of the most intense of these experiences here:

> *Trent:* One of the worst nights of my life was, you know, I had thoughts, you know, so..so..so negative I was dry-heaving blood. . . . You know, just some evil, horrible, negative, you know, satanic thoughts just filled my head.
>
> . . . .
>
> *Researcher:* Did those thoughts seem like they were your own, or did they seem like they were coming from a source outside of you, or..?
>
> *Trent:* Um..it..it's kind of both, you know. It was something, you know, so deep in me, you know.
>
> . . . .
>
> *Researcher:* Was there a sense that you were kind of evil at that time?
>
> *Trent:* Um..yeah, I was pretty much..uh..part of it, I was enveloped in it, sure. I was immersed. I was definitely immersed [laughs].

Trent also had relatively frequent experiences of profound creativity that stood in direct contrast to these experiences of "evil" and destructiveness: "I really applauded myself later when I got over [my psychosis], you know . . . about how incredibly creative my mind was, and intelligent, because these were some very intelligent insights, wise, wise connections. I couldn't name one right now, but I remember that was very much the case. I really tapped into something very intelligent in my mind."

As Trent fluctuated between experiences of creativity and evil, feeling guided by God and persecuted by the Devil, he also experienced fluctuations between feeling "godlike" himself and feeling utterly powerless and hopeless. He remembers a number of occasions while he was feeling particularly powerful when he believed that by his own actions, he could eliminate all suffering from the world: "I can clearly remember some psychosis episodes that included believing that overnight and in the morning the entire world would be changed. Free from all its miseries because of something I participated in. And from that day on, the world would be free from its sufferings." And, in direct contrast to these times of feeling so much power and hope, he recalls numerous occasions during which he struggled greatly with suicidality, feeling so powerless and hopeless that he felt the only worthwhile action he could perform was to take his own life.

Trent was almost immediately hospitalized at the onset of his first period of psychosis, and he remained committed there involuntarily for about six months. Overall, he believes that the treatment he received in the hospital generally worsened his condition, particularly the lack of compassion and kindness that he experienced from the staff members: "In the mental hospital, . . . I was extremely anxious, and I didn't feel I was getting any type of compassionate treatment there, more of abuse than treatment. . . . They really conveyed the attitude, you know, I could go drop dead for all they cared. That's the attitude I received."

He recalls being stuck with a therapist whom he felt to be "a real, you know, cold, abusive idiot." Trent also felt that they put too much emphasis on medication as the primary form of treatment while putting too little on simple human kindness: "There is no medication that can override a hospital staff including M.D.s [who are] open to laughing in a patient's face at anytime." He also believes he was kept in the hospital longer than necessary to maximize insurance payments, saying "they milked the stay for 6 months."

During Trent's stay in the hospital, he developed overwhelming suicidal feelings and impulses and was very dismayed at the apparent lack of concern the hospital staff had in this regard:

> Before my discharge, I was allowed to go home on weekends, and . . . I'd attempt suicide on a weekly basis. . . . And, you know, before they let me go home on weekends, no one ever asked if I was suicidal. I would have said yes if they asked. They never asked. And I did have weekly attempts. . . . I drank some film cleaner, it said harmful or fatal if swallowed. I had a plastic bag over my head, you know, with rubber bands at night, and I tried cutting my wrist, but I wasn't..yeah, I found that very difficult.

Trent feels that the lack of care "was very much a contributing factor on me, you know, leaving the hospital six months later suicidal. I didn't enter the hospital suicidal. I left the hospital suicidal, you know. It was such an unhealthy environment." He feels that a second and closely related factor in his suicidality was his lack of ever being given a message of hope: "I was very hopeless. I didn't think I'd ever recover. . . . It really, really, really would have helped if they gave it the label of a nervous breakdown because I've heard of that before, I knew people recovered from it."

Trent feels that he experienced both significant harm and benefit from the psychiatric drugs that were given to him. He believes that the main harm that he suffered from these drugs was related to the often haphazard way in which they were used. One particularly harmful event in this regard occurred when he was abruptly taken off of thorazine, an antipsychotic, while in the inpatient unit of the psychiatric hospital. After withdrawal, he recalls behaving in a manner that made it very clear that he was suffering from withdrawal symptoms, but the staff refused to put him back on thorazine and instead put him on a number of other different types of antipsychotics: "I went in on thorazine, and I started pacing when they took me off thorazine. . . . I began pacing, you know, and I couldn't stop. I was pacing like 23 hours a day for months. . . .Then they tried a whole bunch of other stuff, you know, which made me worse. . . . For some reason they really didn't put me back on thorazine for another three months or so, and that's when I stopped pacing." Sometime after leaving the hospital, Trent experienced a second abrupt withdrawal from thorazine, which likely played a significant role in his slipping into another psychotic episode: "I was on, you know, a large amount of thorazine at the time, and I missed some dosages and I think that's what led to a psychotic episode, one of them."

RECOVERY

It took Trent six long and painful years, until he was about 30, to heal to the point where he felt that he was through the worst of his psychosis, and then another 10 years, until the age of 40, to arrive at a level of recovery that satisfies the definition of "full recovery" used in this study.

Throughout his recovery journey, Trent found the support he received from the mental health field to be a mixed blessing. While not entirely sure of the benefits of the psychiatric medications he received over the course of his recovery, and of course relatively clear about the harms, as mentioned above, he feels that two of the drugs in particular—thorazine (a typical antipsychotic) and tofranil (an antidepressant)—"might have helped." He found that while most antipsychotics he tried worsened his condition, thorazine in particular did seem to make him feel somewhat more stable, and he was fortunate in that he experienced very few side effects from it. In spite of his negative experiences with psychiatric drugs, he feels that "overall it was more of a benefit."

Whereas Trent's experience with the inpatient ward of the psychiatric hospital could hardly have been worse, he did find that the two outpatient day hospitals that he attended after being released from the inpatient unit were helpful: "The day hospitals were a much healthier place than the mental hospital. That's where I started to sort things out in individual and group therapy." He feels that the second day hospital was particularly helpful: "[The second day hospital] was a little more goal oriented [than the first] and..uh..for people that were a little less, you know, in the hopeless category, and, you know, I sorted things out there. I had more group therapy and I was there for a couple of months, and then I went out and into the work force."

Trent found that the psychotherapy he received in the day hospital was particularly helpful in that it supported him in connecting with himself and in finding his voice. He later came to recognize that these were particularly important factors in his recovery: "At the time, . . . I didn't even recognize I was doing that. I was just getting things off my chest. I was just speaking my mind. I didn't have a whole lot of self-reflection on the process at the time. I was just doing it, I was just living it. . . . I'd try to get to new levels of honesty and openness. . . . The opportunity to speak my mind [was one of the most helpful resources in my recovery]." In spite of these great benefits when the therapy went well, Trent found his overall experience with therapy and therapists to be "very disappointing": "You know, I don't think I had one therapist try to look into what level of dysfunction my parents had for some reason, and, you know, it really would have saved a lot of time."

Another important factor in Trent's recovery was his ability to take the creativity he tapped into during his psychosis and turn it towards creative forms of expression, especially writing, poetry, and photography: "It adds to stability, and a foundation, self confidence, self worth, . . . self connection, . . . and acceptance, you know, when you get compliments from other people..yeah, recognition. It was definitely helpful."

One more factor that Trent feels supported his recovery was giving up his use of marijuana and cigarettes, which he did "for good within a couple of years of the onset."

◇◇

As of the writing of this book, Trent is 53 years old. He has not experienced any significant psychotic experiences for about 23 years. However, according to the criteria used in this study, he is considered as having been fully recovered for about 12 years since he took antipsychotics for several months at the age of 40 due to some non-psychotic psychological distress that he experienced at that time. In spite of living with significant physical disability from two major physical illnesses for the past twenty plus years, Trent has managed to continue to meet his needs without collecting disability, currently being the sole proprietor of a demanding landscaping business. He has been involved in a mutually rewarding intimate relationship for several years, and he regularly volunteers in the community, finding much enjoyment in supporting the wellbeing of others. He continues to have a strong passion for art and humor, spending a lot of time practicing photography and writing prose and poetry.

As with all of the other participants, Trent feels that his personal paradigm and his general sense of wellbeing and ability to meet his needs have undergone a profound and primarily positive transformation as a result of going through his psychotic process (to be discussed in more detail later).

Chapter 26

The Case of Jeremy

At 21 years of age, within the brief span of a single evening, Jeremy found himself spiraling away from consensus reality and into the grip of a highly persuasive anomalous belief system. Shortly thereafter, he was hospitalized for ten days and diagnosed with Psychotic Disorder NOS (not otherwise specified). This belief system continued to dominate his consciousness for half a year, and then gradually faded away over the next several years.

The Onset and Deepening of Psychosis

Although Jeremy's psychosis came on quite suddenly, he later came to recognize that a number of factors beginning much earlier in his life had culminated synchronistically that evening to create what he calls "the perfect storm."

Jeremy feels that one of the most significant contributing factors was having had a number of highly shameful experiences throughout his childhood:

> My mom was pretty verbally abusive to me growing up, and my dad kind of an emotionally distant person, and I think that what happened was that I had a lot of trauma from that..things that were unresolved about that. . . . I should also add that I was bullied in junior high school terribly, like really, really bad. I ate alone in the cafeteria in seventh grade. I mean, . . . anybody . . . who's had to eat alone in the cafeteria in seventh grade . . . can imagine the kind of..you know, the kind of shame that that brings up.

As is the case with most young people in Western society, as Jeremy reached late adolescence, he began the process of individuating from his family and seeking a meaningful life for himself. By this point, he had established an identity as a well-respected skateboarder with a strong sense of belonging within the skateboarding community and the promise of becoming a professional; yet he found himself beginning to have strong intellectual yearnings: "I pretty much probably could have went professional if I had not gone to college and moved out to California and really applied myself. I was probably one of the top skaters in the Northeast. . . . [But] then I discovered that I really liked thinking. I liked ideas. I liked philosophy."

Being torn between his love for skating and his intellectual aspirations, he found himself at a crucial choice point, and in the end, he chose to let go of his skateboarding dreams and go to college: "That's when I started becoming interested in psychoanalysis, in philosophy, and then I decided I was gonna devote myself to being an intellectual." This choice, however, had a profound impact on his sense of security and belonging, and it ultimately became a very significant factor in the eventual onset of his psychosis:

> So, when I went to college, it was a big decision point to either, you know, go be a professional skateboarder or drop that. And so I chose to go to college. . . . When you're a skater . . . everything revolves around skating. You read the magazines..that's all you do 24 hours a day is think about skateboarding. . . . So, when I got to college, I didn't have that community, I didn't have that identity. I would say that's really the first movement of this episode that culminated on that night, because I was depressed. You know, I wasn't like clinically depressed. I mean I would go to class and do all the stuff, but I was like totally isolated. I was experiencing a lot of anguish, and I had enough insight to know that it was because I had given up skateboarding, and I didn't have anything.

In spite of his difficulties with transitioning from skateboarding to the academic life, Jeremy soon found himself thoroughly enjoying the study of psychoanalysis. An opportunity soon opened up for him to move to London and continue his studies at the Anna Freud Centre, and he jumped on it. Shortly after beginning his studies there, however, he experienced profound disillusionment:

> I idealized analysts, and then when I got to the Anna Freud Center, I think my idealization of them crashed into some kind of..really..ugly reality where they were all sort of very petty and political, and the Kleinians were squabbling with the Anna Freudians, and I was like, wow, these are supposed to be some of the most thoughtful, enlightened people in our culture who look at themselves, and yet here they are behaving like junior high kids.

Jeremy's disillusionment regarding the maturity and wisdom of the psychoanalysts soon expanded to include the entire human race:

> I realized that adults did not really know what they were doing. They were trying very hard on the best information that they had, but there was really nobody kind of like in control. There was no kind of like wise group of people steering us, and I found that to be terrifying. . . . It was terrifying 'cause it was like, wow, you mean this whole..this round ball floating in the universe is just kind of like spinning and people are just kinda like ramming into each other, and it's all one big hurly burly.

After Jeremy's profound disillusionment with the human race, rather than succumbing to fear or despair, he was able to see this insight as an opportunity to make a significant contribution:

> As much as that realization was terrifying, it also was liberating because it was like, wow, you know, I have a voice here. I can make a contribution. I can do something. I can come up with something that might help humanity. . . . I started reading a ton of stuff. . . . I guess I had the realization, I was like, wow, I can like..like I want to make a contribution to the world of ideas and thought, you know.

So, he took the self-discipline and the confident independence that he had cultivated in his skateboarding and directed it towards this new goal:

> I kind of said, fuck college. I mean I still went to my classes, but I was just like, well, the hell with institutionalized learning, I'm gonna learn..you know, it was like the punk-rock DIY [do it yourself] skateboarding thing of like, do it yourself, man, like learn..just do it, you don't need anybody, you know, so I started reading everything.
>
> *Researcher:* Carved your own path.
>
> Exactly.

Jeremy now realizes that this sense of heroic striving, and particularly the striving to be a peacekeeping hero, began much earlier in his life. He realizes that, in many ways, he was trained to take on that role within his family:

> My whole role in my family was to be compassionate [laughs]. I mean I was trained to be a therapist from age zero . . . to care, and be in other people's

> minds, and be trying to read other people..what other people's intentions are. As young as sixteen, I was already..I remember my stepmother saying, what do you want to do, Jeremy, and . . . I said something like, oh, I want to make a major contribution to world peace [laughs]. She was like, holy-.. okay.

As the realization of living in a world without a group of truly wise leaders began to take hold and blossom, and as he began to study this phenomenon more intensely, Jeremy became particularly inspired by the work of the esteemed linguist and political thinker and activist, Noam Chomsky: "I thought, I want to be like Chomsky. . . . I want to think very deeply about the human condition and come up with some kind of response that helps us poor humans down here on this earth, and I really held him up as a model, somebody who was fighting that fight."

With this aspiration in mind, Jeremy began to pay more attention to the behavior and interactions of those around him. First, he began to notice "that people would make other people feel bad a lot of times. As he paid more attention, he came to the realization that this behavior was actually extremely pervasive, even amongst the psychoanalysts. He soon formulated a theory that he felt explained what he was observing:

> I had developed my own theory about how the world was running, and it was based on how people went around constantly scooping up self esteem from everybody else by putting them down subtly all the time. So when I was at the Anna Freud Center, that's what I saw happening. . . . That's what I was zeroing in on. I thought these people would be involved in getting little self esteem nibbles, let's call them that, from their colleagues. And yet, it was even worse, 'cause they were just doing it in an extremely sophisticated language. And that's what really blew my mind.

With this insight, Jeremy felt that he had stumbled onto something really big, perhaps just the insight he needed to contribute to world peace in the profound way for which he had really yearned. It would turn out that this powerful insight combined with his heroic strivings would be another important factor in the eventual onset of his psychosis. Few people would consider this insight particularly bizarre, and it likely even has significant validity within consensus reality; therefore, it would not be appropriate to classify it as a psychotic experience nor even as a particularly anomalous experience. However, Jeremy does feel that he "blew it way out of proportion thus instigating a kind of proto-paranoia."

During this period of deep contemplation of the human condition, Jeremy not only observed the behavior of others, but he also turned his attention inwards and became painfully aware of the battle going on within himself, especially with regard to his social anxiety and associated isolation from others. In time, this penetrating awareness resulted in a profound sense of inner peace:

> I really got into Karen Horney, I think maybe 'cause she wrote about self analysis. That interested me and I was doing a lot of that, and I realized I was a very anxious guy, like I had a lot of social anxiety. . . . [Then] there was a big epiphany, actually. . . . I remember it clear as day. I was in this café . . . [with] my friends . . . and all of the sudden, it dawned on me that it was as if I had been in a war zone, and I had come out of the war zone, but I was still fighting a battle inside of me. And then I realized, what battle am I fighting, and why am I still fighting it within myself if it doesn't exist anymore? And once I had that thought, I experienced this incredible sense of calm and wellbeing at that café, 'cause I was like, I don't need to be afraid [laughs], you know. . . . I [let] go of a lot of anxieties that I had and I started to feel these intense feelings of liberation, like, wow, I don't have to be anxious in a social circle, or maybe everybody else is just as anxious as me, you know, letting all that go, and it was just really liberating.

This moment represented another important turning point in Jeremy's journey towards psychosis. Armed with these powerful insights into the human condition and a renewed sense of liberation and connection with others, Jeremy felt that he was well on his way to making a serious contribution to the human race.

Jeremy returned home for the Christmas holidays soon after his experience of liberation. As enjoyable as these feelings were, however, they continued to strengthen and eventually threatened to become overwhelming:

> When I got back home from London, I had kind of severe jet lag and I kept having these kind of breakthrough experiences, and I remember on the eve of Christmas, I sat up with my mom and we had this really deep discussion and we ended up crying together and hugging, and it was really powerful. But what happened was that I couldn't stop my mind from having these insights, and I just wasn't able to sleep, and I wasn't able to turn it off.

In this highly elated yet vulnerable state, Jeremy then did what he later recognized was a "very silly thing":

> I went over to a friend's house and smoked some pot, and that basically sent me through the roof. I have never experienced..I had a hard time being in my body. The feelings of joy and elation were so intense they were almost painful, and as that was going on, this guy came over and he wanted to form a band with me. I was playing drums at the time, and I got really into that idea, and he found out that I actually had to go back to school, and so he said, no, no, no, let's not do that. And I just thought it was such a great idea that I kept kind of like pushing the issue, and to everybody there who was witnessing the conversation, they were probably like, you know, Jeremy, like stop, he said no. But I was like, no, no, no, let's do this, let's just do it for like a couple weeks, we can make some cool songs. And then one of my friends who was there who now I realize is basically verbally abusive—he had issues with that—he jumped in the conversation and really shamed me in front of a lot of people. And that proved to be the straw that broke the camel's back.

That evening, in a very short period of time, Jeremy plummeted from a state of extreme elation to extreme shame:

> I experienced almost like an existential void open up and I fell through and it was like all these feelings of terror and panic flooded in, and I was so overwhelmed by those feelings, I did not know how to handle them. . . . Psychoanalysts talk about annihilation anxiety, and..uh..I think that's pretty accurate to what I was feeling. . . . I was completely terrified. There was like.. there was no ground..I don't know, it's hard to like..there was no floor to my experience.

After struggling for some time to maintain his sanity in the face of such terrifying groundlessness and overwhelming "annihilation anxiety," Jeremy experienced a sudden and profound shift from terror to grandiosity. In retrospect, he believes that in order to compensate for such extreme terror, his psyche was forced into extreme compensation: "I think what happened was, due to the terror, I became extremely grandiose, and I thought, well, you know, the flip side of that terror is I'm enlightened. I must be enlightened."

Initially, after being overwhelmed by the shaming incident, Jeremy "desperately called a whole bunch of friends" in search of support, but he was not able to connect with any of them. So he returned to his home and to his mother who also happened to be struggling with some anomalous experiences herself during this time. Jeremy has come to believe that it was his subsequent interaction with her while in such a profoundly vulnerable state that "truly kicked [him] over into the psychotic realm":

> Something happened between my mom and I where I told her I was becoming enlightened. I don't know how I even got to that point, and my mom is partly psychotic. She believes all sorts of magical things, and so . . . she said, oh great, you're becoming enlightened. Well, why don't you tell me if my friend is secretly plotting against me. . . . That was the absolute tipping point, 'cause then it was like, oh my god, my mom believes I have special powers and that, you know, . . . and then I just..I went on a whole rant about how her friend was in fact spying on her and there was this whole conspiracy against her.

So Jeremy found himself becoming enmeshed in his own mother's anomalous belief system. As they discussed the supposed plotting of his mother's friend, Jeremy somehow came to believe that his mother's boyfriend was also in on the "plotting," which in turn escalated into the belief that there must be a "black order," and that his mother's boyfriend must be involved with it.

During the *folie à deux** experience between Jeremy and his mother that evening, an entire anomalous belief system unfolded within Jeremy's mind in surprisingly short order:

> Somehow, I got to thinking that [my mother's boyfriend was a member of the black order]. I mean it's all very sort of oedipal†. So . . . for some reason, it became clear that the world was controlled by these very evil people who were very, very smart and actually had it all figured out and were actively manipulating the world for their ends. And I put it together that . . . the mind control weapon that they were using was making people go around, you know, . . . nipping at each other [laughs], essentially, [battling for self esteem]. So they had us all fighting. . . . [I also realized] they were persecuting my family, they were persecuting my mom. . . . [And] this guy, the

* *Folie à deux*, which literally means "folly of two," is a formal diagnosis listed in the DSM-IV to describe the situation in which two people become enmeshed in a psychosis with common anomalous beliefs and/or experiences.

† Jeremy is referring here to the *Oedipus complex*, the concept suggested by Sigmund Freud that in early childhood, we all develop a profound but unconscious conflict involving a deep longing for intimate relations with our opposite-sex parent and jealous, even murderous, feelings toward our same-sex parent. Freud suggested that healthy development involves the relinquishment of these feelings, but that few people are able to eliminate them entirely.

> boyfriend, he was sort of but like a minor functionary in the hierarchy of the black order, but nonetheless, he was part of it. . . . And there was also this good force. They appeared to me as kind of like white beings, and they were kind of like angels, so to speak. They were the people who were also very wise and very smart and compassionate, but who were working to fight the evil people. And, you know, in my thinking, I thought people like Noam Chomsky were part of this white force..and other assorted kind of public figures who, you know, who had shown that kind of intelligence and compassion.

After this belief system regarding the white and black orders came to dominate Jeremy's consciousness, and given his strong desire to make a meaningful contribution to the world, it is not particularly surprising that this heroic striving quickly escalated to the level of messianic striving:

> [Immediately after coming to the realization of the existence of the white and black orders,] I started thinking that I was becoming on that team. I had finally broken through to the realization that that was my future, to be on this force of people and to find them and ally with them, but that the evil people now knew because they were psychic. They knew that somebody had broken through and found them out, and found out their secret plan, and now they were gonna come and get me, and so paranoia set in.

While this anomalous belief system formed in a very short period of time, essentially within just a few hours, Jeremy has since recognized that his earlier insights and experiences had provided all of the essential elements, and that the "magic mix of circumstances" that took place that evening created the ideal set of conditions for the brewing of "the perfect storm": "It's amazing, looking back, how perfect everything felt. It's so incredible."

As Jeremy's paranoia set in that evening, it suddenly occurred to him that the black order was going to attack his brother, who was living at his grandmother's house several hours driving distance away at the time: "Now I was hunted, now I was persecuted, and then I thought they were gonna attack my brother. Oh, well, then I was convinced, actually, that my brother had already been under attack for quite some time, and we had to go rescue him." Surprisingly, Jeremy was able to convince both his mother and his thirteen year old sister of this. So, now in what Jeremy describes as a "folie à trois," the three of them all hopped in the car and headed to his grandmother's house. It turned out that his brother was not there, and they spent the remainder of the night looking for him, but to no avail. This greatly exacerbated Jeremy's paranoia, and by the following morning, he found himself overwhelmed by his paranoid thoughts of persecution and feelings of panic and terror. He began "to say things that were kind of frightening to [his] family" and slept very little, if at all, for at least two more days.

At this point, now several days into his psychosis, Jeremy experienced a temporary "island of clarity" and realized that perhaps he could gain some clarity into his situation if he could manage to get some sleep: "I think I was like, okay, something..something's not right. I think I just need to get to sleep. I think I had a moment of clarity. . . . I think I had one of these islands where my ego kind of recrystallized slightly [laughs]."

Upon having this moment of clarity, Jeremy asked his mother to take him to the hospital so he could get some medication to help him sleep. Once inside the hospital, however, he was not allowed to leave:

> When I got to the hospital, my family had started telling the psychiatrist things I'd been saying and they started to ask me questions about the things I'd been saying, and then I knew that I was caught, so to speak, like the evil forces had actually succeeded in getting me. And so I initially went there just [wanting] to get some medicine to help me go to sleep, but, you know, I spent ten days on the inpatient unit.

The combination of the lack of genuine care, the heavy use of antipsychotics, and the experience of isolation while in the hospital had a particularly devastating effect on Jeremy:

> I was on a very high dose of Risperdal, which is an antipsychotic medication. It made me drool horribly. I looked like a zombie. My friends who came to visit me at that time were just like, wow, where did Jeremy go, and the inpatient unit was just terrible. Nobody talked to me, I was pretty much ignored. I was very frightened, very scared. . . . When I emerged from the inpatient unit fourteen days later, I was a shell of a human being.

After leaving the hospital, Jeremy spent several weeks in their outpatient program: "That was basically another kind of joke. The staff treated us like we were little children and needed to like manage our stress and learn like basic living skills. It was very demeaning." In the end, he emerged from this treatment feeling "totally and utterly broken, empty": "Talk about surviving, I mean that was a war. It was like I'd been through a terrible war, a terrible thing."

After finishing the outpatient program and trying to return to his life, Jeremy found that the treatment he received had instilled in him a terrifying belief, one that he feels greatly hindered his recovery and that took him years to overcome:

"The most terrifying thing that the mental health system did to me was that they made me feel as if I could not trust myself. I could not trust my own mind, because it might happen again..and you have to watch..you can't be too stressed out, and.. That took about 3 years to get over, really, that kind of seed they planted."

As a result of this new belief with which he had been inculcated, he began to have panic attacks:

> I never had panic attacks before in my life. I started to have panic attacks.... The first one I had was when I was still in the outpatient program, or maybe I was just getting out. I had to somehow take a bus up to Boston to like deal with my student housing, and I remember walking into the student union . . . and there are all these students in there, and I felt so incredibly alone and isolated that it was..I pan-..I just..I think I just..it was the terror, you know, I was terrified, utterly terrified. . . . It was like, here are all these people bustling about, seemingly fitting into the social order, having a task, and here I am, I just got out of the mental hospital, and I've been told I have some kind of a brain disorder and made to feel as if there's something extremely, deeply wrong with me, and that it could possibly be there for a long time..forever, and was I gonna..could I ever fit back in.

RECOVERY

When Jeremy reflects upon his path towards recovery, he recognizes that there were numerous factors involved, each of which played an important role. One important factor was the support of his family. Even though there were aspects of his family dynamics that were not particularly supportive, Jeremy acknowledges that their support was crucial at times, especially during his stay in the hospital:

> My family did come through for me in important ways even when I was in the hospital. My grandfather came to visit me every day and brought me good Italian food, while my grandmother took me out to eat when I was allowed out on day passes. My mother brought me clay and art supplies. My father visited or called me every day, simply to check up on me and listen to my concerns and worries. An aunt . . . and step-uncle brought me a CD player with some of my favorite music. It was their simple, repeated, and thoughtful actions that made a difference. In short, they helped me to restore my humanity. I shudder to think how much worse it would have been had they not come through for me. They made me feel like I had worth, at the time in my life when I thought I was worthless.

Another important factor was his encounter with someone who believed in him. While attending the outpatient program after his hospitalization, Jeremy encountered a staff member who really seemed to believe in Jeremy's ability to return to a more normal life:

> I credit [him] with possibly having really unwittingly, or maybe wittingly, turned things around for me because he set me down in his office and he said, you need to go back to school, and I was like, what, school, huh? . . . He said, no, no, no, let's get you enrolled, let's get you some special housing. He really believed in me, and if he hadn't done that, I'm quite sure I would have wound up being a long-term mental health client, or my stay in the mental health system would have been a lot, lot longer.

With the encouragement of this staff member, Jeremy did return to school. He found that returning to school and just going through the motions of being a student again helped him to reconnect with a sense of purpose, in spite of the fact that this purpose initially felt somewhat superficial: "Just being back at [the university] and just walking around and doing my life in that superficial way was really important, like putting, you know, the gears back in motion to a machine that [laughs] needed to..I don't know, you know, keep turning. That was..that was big."

Returning to college not only supported Jeremy by providing him with some meaning and direction, it also provided him with an alternative to living with his family, something he felt would have been a grave mistake: "It was so important for me to go back to college right after my breakdown. If I had taken a leave of absence and gone back to live with one of my parents, I would have unwittingly placed myself right at the center of the cyclone. True to my family role, I would have continued my misguided attempts to heal them at the sacrifice of my own wellbeing."

Jeremy believes that finding a good therapist was another very important factor in his recovery: "I did get into therapy, and that was big. That was big." Jeremy was not open to discussing his psychosis in therapy "because the psychiatry people fucked up so royally"; however, he believes that would not necessarily have been the most helpful thing anyway: "In some way, he just kind of bypassed all that anyway and got to the more root stuff, . . . the real conflicts that were driving a lot of my anxiety in the first place, so we kind of just got down to business, so it was really helpful."

Jeremy believes that among the most helpful qualities of his therapist were his high capacity for tolerance and his choice not to push the medical model:

> He was a good therapist. . . . By that point, I was completely..um..like..you know, schizoid would be the correct psychoanalytic term for what I was. I had to control the whole session. . . . I was extremely critical of him and what he would say. But he knew to handle me with kid gloves and like be gentle and let me control the session, you know, and never be forceful about things, or be forceful in the right way. I mean he just was a very good therapist. I just really lucked out. He never was like, well let's talk about your mental illness, you know, or like, how are you doing with your meds, you know.

Jeremy believes that one of the most important ways that his therapist helped him was in exploring many of the core family dynamics that may have contributed to his development of psychosis. One particularly important dynamic that Jeremy was able to address in therapy was his disconnection with his father, which resulted in a healing that he feels greatly supported his recovery:

> There was a point where I think I had just lost a lot of contact with my dad, probably after going away to college. I mean I never was super close..there was maybe a time in like childhood where like he helped me with sports and took a real interest, but, you know, he's alcoholic, and he has his own family trauma, and..um . . . he's kind of a limited man emotionally, you know. I think he does the best he can, but . . . because of his own damage, we never could talk about our relationship, his and mine, and have any real intimacy, you know. . . . So my therapist put his finger right on that, and helped me understand that, helped me understand that I could do something to change that. Maybe I could begin to talk to my dad about the feelings that I have about him and about his drinking and risk doing that, and I did. The next time I got together with my dad, . . . I started talking to him. It was very uncomfortable. I don't think that he knew what to quite do with that talking, but he hung in there. He heard what I was saying, and it was a real healing that really changed the course of my father and I's relationship profoundly.

Jeremy believes that another very important factor in his recovery was his challenging the belief that he was suffering from a brain disease. He expressed his gratitude for the role his girlfriend at the time played in this process: "And you know, I got lucky again 'cause my girlfriend at the time would challenge me on that. She'd be like, you don't have a brain disorder, what are you talking about. And I'd be like, oh, right, thank you, thank you, and then I go read antipsychiatry, you know [laughs]."

Jeremy still feels a lot of frustration about the harm caused to him and others by the mental health professionals who try to convince those suffering from psychosis that they have a lifelong brain disease:

> *Researcher:* So, it sounds like you went through a period of time where you were kind of buying that framework of having a broken brain—
>
> *Jeremy:* Oh, for sure.
>
> *Researcher:* —and having a lifelong disorder.
>
> *Jeremy:* Yeah, the vulnerability. That's what's fu-..that's what's evil about these fucking people. Their people are so vulnerable at that point that you'll buy just about anything just to get some freakin' coherence.

As a major aspect of his journey towards recovery, Jeremy went through a profound struggle trying to reconnect with meaning and purpose after his experiences, a meaning-making process that has taken place on several different levels. On one level, Jeremy worked hard to make sense of his experiences—"the meaning of what the hell just happened to me." In the fifteen years since the onset of his psychosis, Jeremy has spent considerable time contemplating his psychosis and making some sense out of it, a pursuit that he feels has been very helpful. He continues to feel tremendous frustration, however, at the lack of support offered by the mental health care system in this regard: "So it's 2010. So this happened in 1995. So fifteen years later, I'm finally feeling like all the pieces of my story can be said out loud. I mean, I've said them to myself in various ways, but if we had a truly enlightened system, I would have been encouraged to put all that together as best I could within a..you know, getting on the inpatient unit."

On a second level of meaning-making, Jeremy has struggled with how to make a meaningful life for himself after what he has been through. He found that what has really worked for him in this regard has been to direct his entire life toward his own healing and the healing of others who have had similar experiences:

> I just made my life a reaction and a healing..a deepening, that's what it was. I turned all my energies towards absorbing this process and metabolizing it, and coming up with creative ways to help others. Maybe [laughs] it's kind of an extreme case, but like everything I do is in some way related to that. . . . I mean it just drives me. It drives my waking life . . . I have a purpose. I mean, it does flag, I'm not gonna lie, because I get demoralized, but . . . . at the end of the day, it's what I live for, you know.

◇◇

As of the writing of this book, Jeremy is 37 years old and has not used any psychiatric medications since his hospitalization and outpatient psychiatric treatment over fifteen years ago. According to the definitions used in this study, he considers himself to have been fully recovered for about fifteen years. He is now married, has a three year old daughter, and has been working as a licensed psychotherapist for over ten years, specializing in working with people suffering from psychosis and other extreme states of mind. He also devotes a significant amount of his time to supporting the psychiatric survivors movement in various ways.

As with all of the other participants in this study, Jeremy feels that, as a result of his psychosis and recovery process, he has undergone a profound and primarily positive transformation with regard to his personal paradigm and general wellbeing (more details about this later).

Chapter 27

The Onset and Deepening of Psychosis

Now that we have gone over the details of the DUI model and the stories of all six participants, we're ready to look at the entire psychotic process, exploring what it is that takes place at the most fundamental level of one's experience during the process of psychosis, from onset to full recovery. To do this, we will return to the stories of the six participants' journeys through psychosis and recovery, using the framework of the DUI model as our lens, and dividing their stories into six separate categories of experience: *The onset and deepening of psychosis*, *description of the anomalous experiences*, *recovery*, *lasting personal paradigm shifts*, *lasting benefits*, and *lasting harms*. As we explore the implications that arise by viewing the details of these stories within the framework of this model, we will continue to expand upon the model thus far presented until we arrive at a relatively complete picture of the psychotic process. This will include filling in the details of the psychotic process, exploring the *hows* and *whys* of psychosis and recovery, and discussing the aspects of the psychotic process that are likely to be universal versus those that are likely to be more personal.

Table A.1 at the beginning of Appendix A lists the converging themes for all six categories of experience along with their associated divergences. These came to the surface after thorough cross case analysis of the participants' stories, a laborious process that included extensive coding and recoding of the data and contacting each of the participants numerous times in order to gain additional verification for emerging themes as well as to rule others out. For a more detailed account of the process of cross case analysis and other methodological details of the study, you can find an online link and free download of my published study in the Bibliography section under "Williams, P." For a more detailed account of how each theme manifested for each participant along with any corresponding divergences, see Appendix A. Let's now turn to look more closely at each of these six categories of experience in turn.

When looking at the themes that have emerged for the category of *the onset and deepening of psychosis*, we see that perhaps the most important factor that all participants share is having experienced an overwhelming existential threat to the self just prior to onset (the first theme listed in Table 27.1), and it is likely that the remaining themes in this category all essentially played a role in contributing to this existential threat.

THE ONSET OF PSYCHOSIS

Recall that, in the DUI model, one of the most important roles of the psyche is to maintain the existence of the self, which it does by maintaining our moment to moment experience in alignment with our window of tolerance, which in turn maintains equilibrium on two levels—within the self/other dialectic and between duality and unity. Ordinarily, the psyche has a great capacity to do this, with a variety of strategies in its arsenal (see Figure 23.3 on page 186). However, in spite of all of these powerful strategies, it appears that there are occasions when the balance within the self system is thrown off so radically—when the existential threat to the self is so strong—that the psyche must resort to the extreme

Table 27.1 Converging Themes and Divergences for *The Onset and Deepening of Psychosis*

CONVERGING THEMES	DIVERGENCES
An actual or existential threat to the self just prior to onset	All experienced this.
Childhood isolation	All participants had a significant amount of isolation in their childhood, but to varying degrees.
The significant use of recreational drugs prior to onset	All but Cheryl had significant experiences with recreational drugs prior to onset.
A swing between extreme isolation and extreme connection just prior to onset	All but Sam and Trent had this kind of swing just prior to onset.
A profound shift in one's personal paradigm just prior to onset	All but Trent experienced profound shifts in this regard; Trent, however, did increase his marijuana use significantly just prior to onset, which may be closely related.

strategy of psychosis in a desperate attempt to regain equilibrium within the self system. Perry put this rather eloquently when he said, "when a person finds herself in a state of acute distress, in circumstances that have assailed her most sensitive vulnerabilities, her psyche may be stirred into an imperative need to reorganize the Self"[1]. Before going into the details of how the psyche does this, it will help to first discuss how one comes to find oneself in such a situation in the first place. To better understand this, let's turn now to look at a brief sketch of the onset of psychosis for each participant in this study using the framework of the DUI model, paying close attention to how each of the factors in Table 27.1 is related to their onsets.

Sam. Sam recalls that his parents both struggled with substance abuse and that he was significantly disconnected from them in many ways, especially during his adolescence. We can surmise, then, that by the time of the onset of his psychosis at age 18, his window of tolerance may have been relatively narrow, making him particularly susceptible to being overwhelmed by the existential anxieties, and it may have been skewed somewhat in the direction of the *self* pole (see image #1 of Figure 27.1), while of course acknowledging the limitations of our information in this regard. He had also been using alcohol and marijuana significantly and LSD occasionally since the age of 16, so his cognitive constructs were likely to have already been somewhat unstable.

Upon receiving news that he had been drafted into the Vietnam War, Sam found himself in an untenable existential dilemma. He identified as an anti-war activist at the time—having strong personal values opposed to the war and also having been part of a community that held these values strongly. Now he found himself torn between the severe threat to his actual physical self on the one hand and the threat to the existence of his self within his community (friends, family, country, etc.) on the other. In other words, going into the war would present a severe threat to his physical self and also represent a loss of his community and a contradiction of his personal values; but apparently the only way to avoid this situation was to either got to jail or to hide in some way, which would again likely mean abandoning his community. To use May's language, it is likely that Sam found himself experiencing an apparently insoluble dilemma regarding "the dialectical relation of the individual and his community"[2].

Looking at the succession of strategies listed in Figure 23.3 (on page 186), we can say that this situation was so overwhelming for Sam that his psyche was forced to almost immediately move to the more extreme strategies of destabilizing his cognitive constructs (the third and fourth strategies listed in the diagram in Figure 23.3), which was likely to have been occurring during the initial stages when Sam began to form various anomalous belief systems but he was still somewhat connected to consensus reality. It turned out, however, that even these extreme strategies were not successful in adequately addressing his dilemma and bringing his self system back into balance. Sam's dialectical tension continued to escalate, he began to lose sleep, and soon his window of tolerance was reduced to essentially nothing as he became completely overwhelmed by existential anxiety (see image #2 of Figure 27.1). In the end, his psyche was forced into the final, most desperate strategy—the initiation of psychosis.

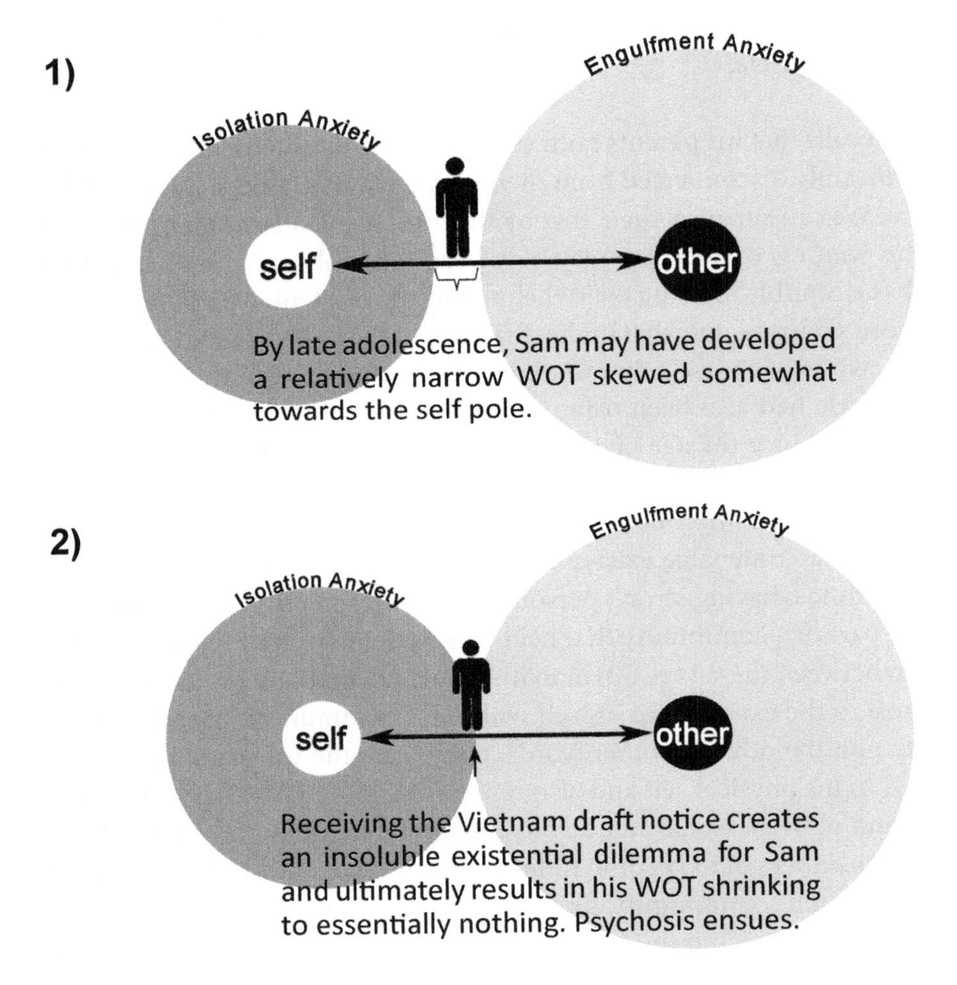

Figure 27.1. Steps leading to the onset of Sam's psychosis

Theresa. Due to a number of circumstances in her childhood, it is likely that Theresa's window of tolerance underwent some significant shifts fairly early in her childhood and ultimately ended up being relatively narrow and skewed by the time of the onset of her psychosis. Theresa believes this development may have first begun with the birth of her younger sister, at which point she experienced the pain of losing a significant degree of connection with her family members as their primary attention suddenly shifted from her to her sister. As a result of this experience of rejection/abandonment, it is likely that her isolation anxiety increased initially. While it is very common for young children to experience the reduction in attention that results from the birth of a sibling as a kind of rejection and/or abandonment, the fact that Theresa seemed unusually sensitive to this suggests that she may have already had a window of tolerance somewhat skewed in the direction of the *other* pole at this stage (see image #1 of Figure 27.2).

Theresa at first attempted to rekindle this lost connection by reaching out in various ways, but her efforts to reach for connection were repeatedly met with still further rejection/abandonment experiences. This occurred during the numerous times her parents separated, came back together, and then separated again. It also occurred to some extent within her peer relationships at school. In her late adolescence, Theresa suffered two more extremely painful abandonment experiences when both of her parents died within the span of just a couple of years. Due to the lack of any real support, she was unable to grieve these tragedies properly. As a result of this repeated reaching for connection followed by devastating disappointment, Theresa lost hope that she would ever be able to satisfy her desire for connection and she lost the capacity to tolerate the pain that connection so often entailed for her. As a result, then, her strategy for dealing with others changed from one of desperately reaching for connection to one of increasing self reliance. In the language of the DUI model, we can say that Theresa's window of tolerance went through a process of shifting from the *other* side of the self/other dialectic to the *self* side.

As Theresa went through the process of integrating this shift of her window of tolerance, she learned to experience herself more as an autonomous being than as one in connection with others. Her personal paradigm shifted to accommodate this new way of being in the world until she was able to establish some semblance of comfort (window of tolerance) on the *self* side of the dialectic, and her intolerance for intimate connection along with her fear of being engulfed by others correspondingly increased. In other words, her isolation anxiety had become quite strong early in life, but as a result of her multiple traumas related to connection and the corresponding development of a self-reliant character strategy, her engulfment anxiety eventually became even stronger. She therefore found herself entering early adulthood with a barely tenable window of tolerance skewed towards the *self* side of the self/other dialectic (see image #2 of Figure 27.2). In simpler terms, she continued to have a strong desire for intimate connection, but her fear of such intimacy was even stronger.

As Theresa entered her 20's, her desire for intimate connection (and her corresponding isolation anxiety) continued to increase until her window of tolerance had become virtually nonexistent. Then, when she arrived at the kibbutz at age 21, she found this deep thirst for connection suddenly and overwhelmingly met. She now found herself in a situation that placed her experience far onto the *other* side of the self/other dialectic, and far outside of her window of tolerance. On one hand, she found herself relishing the connection for which she had been longing for so many years; on the other hand, she was overwhelmed by engulfment anxiety and had to resort to the excessive use of alcohol and hashish as a way to tolerate this: "I was having a great time, you know, but . . . I practically just wasn't sober. . . . It was all there but it was too much. . . . It was overwhelming and I couldn't cope with it, so I had to kind of suppress it the best way that I could" (see image #3 of Figure 27.2).

Looking at the succession of strategies in Figure 23.3 (on page 186), we can safely say that Theresa's existential dilemma was strong enough that the first two more ordinary strategies did not appear to be options for her, so she resorted to the use of psychoactive substances to facilitate the use of the fourth strategy—destabilizing her cognitive constructs in order to bring about a radical/immediate change in her window of tolerance. However, the use of psychoactive substances in this way only provided a temporary coping strategy (as is very often the case with psychoactive substances), and by remaining in a situation that placed her so far outside of her ordinary window of tolerance, she did not have the opportunity to work towards a more gradual and lasting expansion and/or shift of her window of tolerance. In time, the alcohol and hashish could no longer contain her existential anxiety, the tension within her self/other dialectic completely overwhelmed her, and a psychotic reaction ensued. Eventually, "it just kind of bubbled through, you know, and..and just put me over" (see image #4 of Figure 27.2).

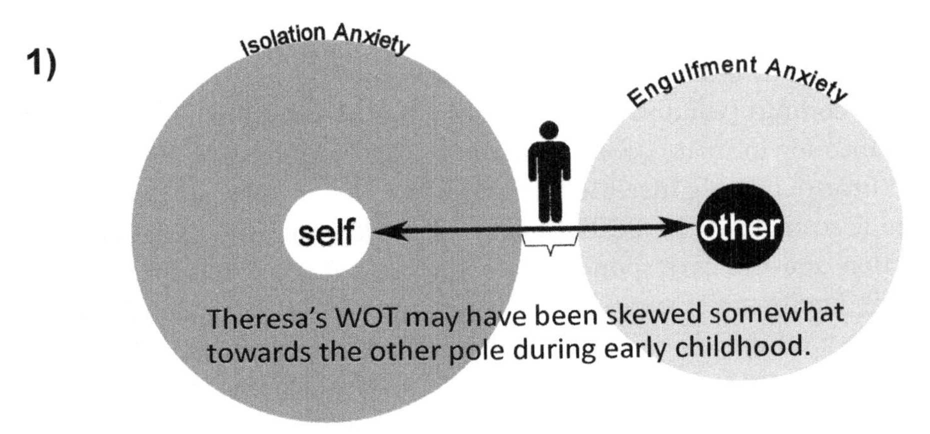

Figure 27.2. Steps leading to the onset of Theresa's psychosis.

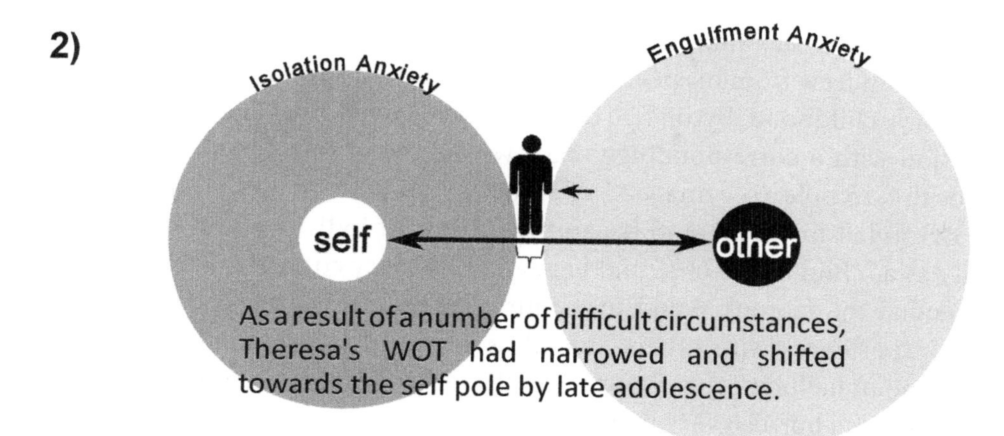

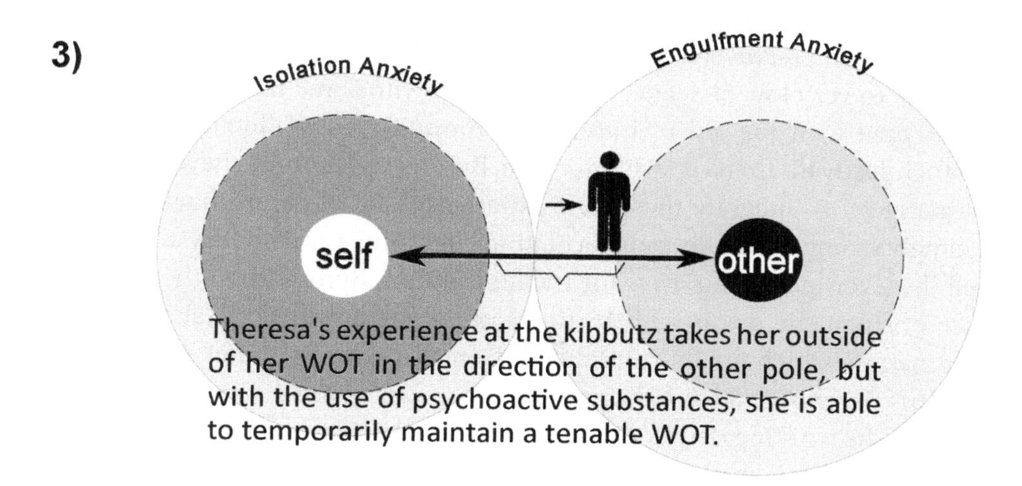

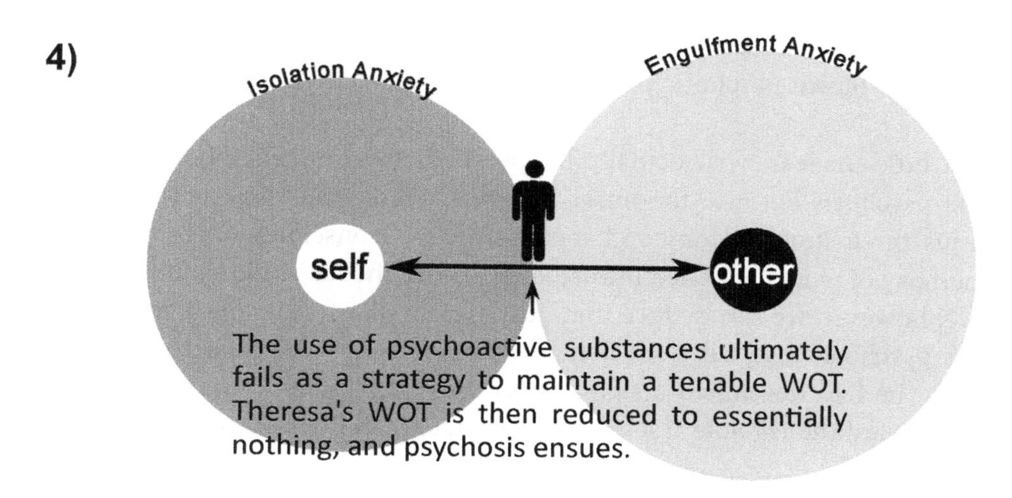

Figure 27.2 (cont.). Steps leading to the onset of Theresa's psychosis.

Byron. Byron recalls that he had a "very troubled adolescence" in which he generally withdrew from his peers and others. It would appear, then, that beginning in early childhood, Byron had developed a relatively high degree of dialectical tension with a correspondingly narrow window of tolerance likely skewed towards the *self* pole (see image #1 of Figure 27.3).

As Byron left his house and began the process of individuation, he came to identify as a "child of the 60's" and began to have very connecting experiences. He attended the famous Woodstock music festival and shared powerful connecting experiences with others as well as powerful transliminal experiences with the aid of hallucinogenic substances. He describes this experience as being incredibly joyful but also very destabilizing, and in fact the onset of his psychosis occurred just six weeks later.

What we can see, then, is that in a relatively short period of time, Byron went from experiencing relatively high dialectical tension with an imbalance towards the *self* pole to very low dialectical tension along with some movement towards the *other* pole with the aid of intimate community and hallucinogens at the Woodstock festival. There is evidence that Byron's psyche may have been trying desperately to integrate these new experiences, as he recalls having felt an increasingly strong desire for transformation beginning immediately afterwards. Feeling this ever increasing drive for transformation, Byron began the intensive practice of shamanic death/rebirth rituals in an attempt to bring about such a transformation, which finally culminated in an all-night ritual that ended in the onset of his psychosis. Such behavior could be seen as the outward manifestation of a psyche working desperately to integrate particularly challenging experiences. Successful integration of these experiences could possibly have led to a genuine growth experience for Byron, with his window of tolerance significantly centered and expanded; but unfortunately, he would have to traverse the chaotic seas of a full-blown psychotic process before experiencing this kind of genuine growth.

It's a little unclear what actually happened at the moment of the onset of Byron's psychosis because the onset apparently took place just before he fell from his three story balcony and was knocked unconscious, interfering with his memory of what happened immediately before the fall*. Taking in the larger picture, however, we can deduce that, just as with all of the other participants, Byron's psychosis resulted from being overwhelmed by the existential anxieties. First, he had developed a significantly narrow window of tolerance likely skewed toward the *self* pole as a result of a difficult childhood. Then, he had very

* He did eventually attain full recovery in regard to both his psychotic experiences and his cognitive faculties, so any brain damage that he may have received from the fall could not have been particularly extensive, if there was any at all.

profound experiences of intimate connection with others along with powerful transliminal experiences as a result of the intense experience of community and psychedelic drug use.

The profound euphoria and sense of liberation that Byron initially experienced as a result of these transliminal experiences indicate that his dialectical tension was initially greatly diminished, allowing strong unitive feelings into his consciousness (see Image #2 of Figure 27.3): "I was feeling incredibly ecstatic, waves of bliss coursing through my body. Every cell in my body exploded in bliss. I felt that I was on a heavenly plane of experience, I felt tuned into multiple dimensions simultaneously. Every desire was satisfied, there was want for nothing." As with any powerful transliminal experience, however, there is the potential for the sense of self to be profoundly threatened (as discussed earlier), leading to a recoil response and escalating dialectical tension. It appears that this is exactly what happened in Byron's case: "I had a transcendent experience. I believe this was a contributing factor to my going off the deep end two months later. Why? Because the experience was so powerful."

So Byron's psyche apparently worked very hard to try to integrate these transliminal experiences into his cognitive constructs, which, if successful, would have prevented the escalating dialectical tension and allowed him to return to balance within a new and perhaps healthier configuration than that which existed previously. Considering the intensity with which Byron was practicing the potentially consciousness altering shamanic rituals, it is likely that Byron was trying desperately to use the third and/or fourth strategies listed in Figure 23.3 (on page 186) to integrate his transliminal experiences (both of which involve the significant destabilization of his cognitive constructs in order to bring his experience and his window of tolerance back into alignment with each other). It is quite clear, however, that he was not ultimately successful in using these strategies to regain equilibrium within his self system, and psychosis ensued. (See image #3 of Figure 27.3).

Cheryl. As a result of a difficult childhood, having been raised by "immature" parents including a verbally abusive mother and an emotionally distant father, Cheryl developed a very painful relationship with herself. She recalls feeling very depressed and harboring strong feelings of self hatred from very early in her childhood until the onset of her psychosis at age 25. It's likely that she developed a relatively narrow window of tolerance skewed somewhat towards the *other* pole of the dialectic, having developed particularly strong isolation anxiety and finding more comfort being with others than being alone with her despised self (see image #1 of Figure 27.4). What this meant was that she identified existentially primarily as a being in connection with others and she was therefore particularly

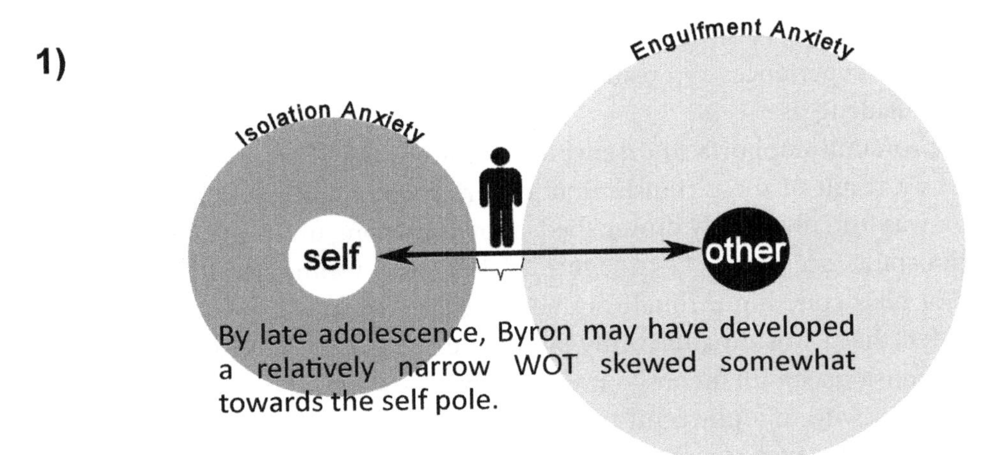

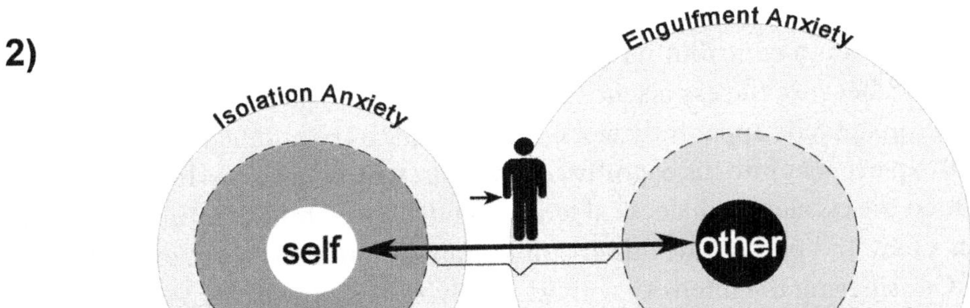

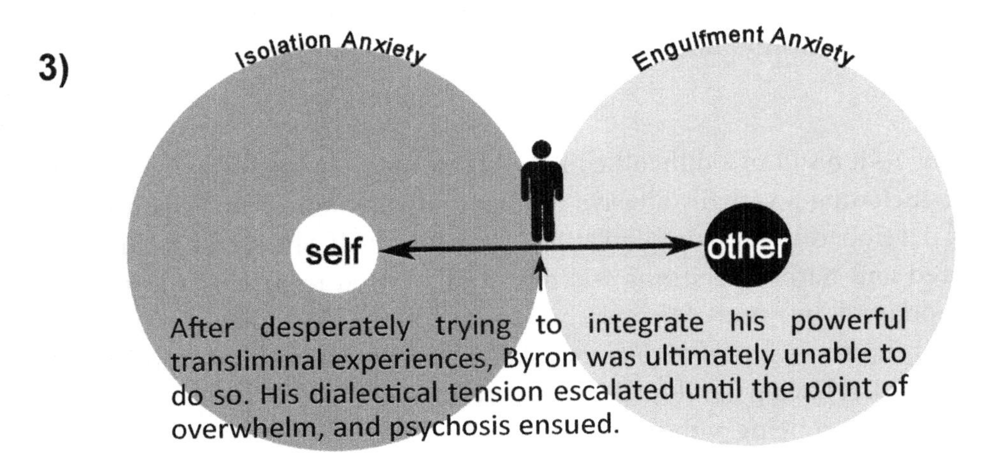

Figure 27.3. Steps leading to the onset of Byron's Psychosis.

vulnerable to experiencing isolation and/or abandonment as a severe existential threat to her self.

At the age of 25, Cheryl experienced a series of incidents that carried her ever further into isolation and ever further outside of her window of tolerance in the direction of the *self* pole. As her psyche attempted to adjust her window of tolerance to accommodate for these new life circumstances, it is likely that it destabilized her cognitive constructs to some extent (see image #2 of Figure 27.4). At the same time, she had been practicing a new kind of meditation and was also eating very little (as a result of a difficult job situation), both of which led to disorientation and perhaps even further instability of her cognitive constructs. She was clearly in a very fragile state, then, when she attended a spiritual counseling course that was designed to teach her how to contact spirits and other benevolent entities. The course seemed to go remarkably well for her at first, as she began to hear the voices of disembodied entities whom she initially mistook to be her spirit guides. Initially, they were very benevolent and supportive, telling her that they "really cared about [her]," and Cheryl initially experienced genuine relief from her painful loneliness as a result of her encounter with them.

Whether or not these "spirit guides" were actually living entities existing in a legitimate realm outside of consensus reality or whether they were merely the manifestations of a desperate strategy initiated by her psyche, it is clear that they provided Cheryl with some genuine relief from her overwhelming isolation anxiety and shifted her experience back to within her window of tolerance. We can also recognize this experience as a transliminal experience, since it is clear that it initially corresponded with feelings of "mild euphoria and liberation" and a significant reduction in her dialectical tension; however, it is also clear that the integration of this experience was very shaky and that this experience was ultimately only a temporary solution to her existential crisis (see image #3 of Figure 27.4). In some ways, we can say that Cheryl's experience of the "spirit guides" served her in a manner similar to the way that psychoactive substances served Theresa and Trent—providing some temporary relief from her nearly overwhelming existential anxiety but with the costs of further instability to her cognitive constructs and allowing her to remain far too long in an unsustainable situation.

Ultimately, Cheryl was not able to integrate her experience of the "spirit guides" in a way that led to a lasting expansion of her window of tolerance, and she soon found herself overwhelmed by both isolation anxiety and engulfment anxiety with highly unstable cognitive constructs. Within a very short period of time, the voices changed from representing the love and support she had always longed for to representing the deep-seated self hatred that she had struggled against for so long. She now found herself in the impossible dilemma of feeling unbearably alone while also feeling on the verge of being completely engulfed by these voices as they relentlessly assaulted her with unimaginably vicious remarks.

Looking more closely at the self/other dialectic, we can say that Cheryl first experienced a fairly radical shift from isolation and high dialectical tension to intimate connection and diminished dialectical tension when the spirit guides first appeared in their benevolent form. I don't feel it's appropriate to consider this first experience of the voices psychosis, since Cheryl was not initially distressed by them and she was still relatively connected to consensus reality. Looking at the diagram in Figure 23.3 (on page 186), I believe that these voices were associated with the psyche's use of the strategy of intentionally destabilizing her cognitive constructs in an attempt to bring her experience back into alignment with her window of tolerance. I suspect that this was primarily the third strategy in the sequence listed in Figure 23.3—destabilizing her cognitive constructs to alter her perception of her situation—though the fourth strategy may also have been used to some extent—destabilizing her cognitive constructs to radically change her window of tolerance.

It is clear that these strategies did provide some temporary relief; however, it is also clear that they did not provide a lasting solution, perhaps partly due to the fact that Cheryl was already in such a fragile state at this point. Since she was not ultimately able to integrate this fairly radical restructuring of her cognitive constructs and window of tolerance, her dialectical tension continued to escalate, and a full-blown psychotic process ensued; see image #4 of Figure 27.4).

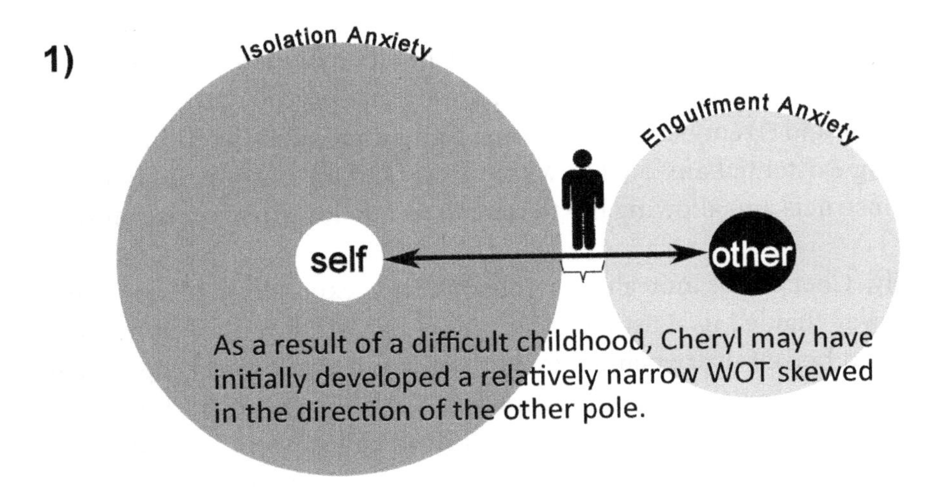

Figure 27.4. Steps leading to the onset of Cheryl's psychosis.

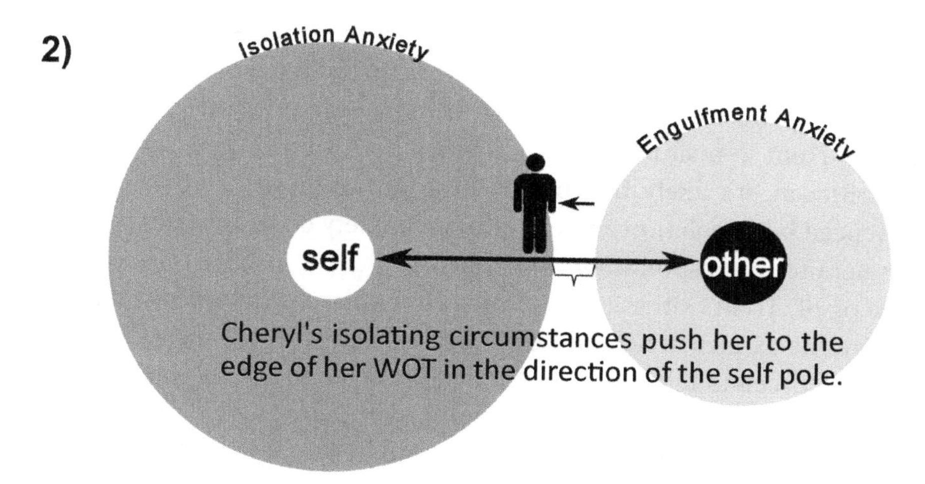

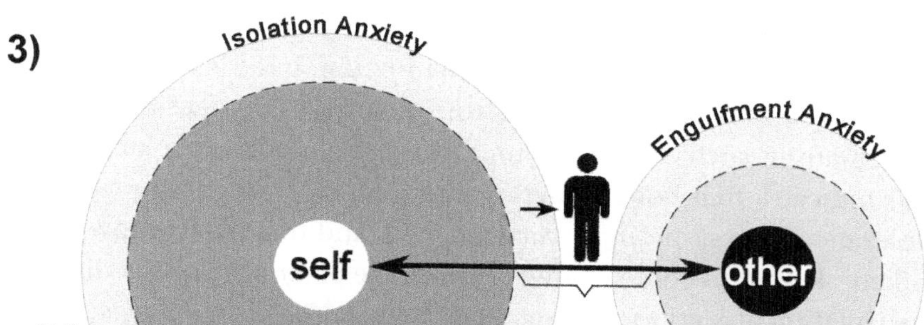

When Cheryl's "spirit guides" first arrive, their benevolent presence puts her back within her WOT and even expands her WOT somewhat, reducing her dialectical tension and allowing some unitive feelings to rise into her consciousness. Her cognitive constructs are unstable, however, as she attempts to integrate this experience.

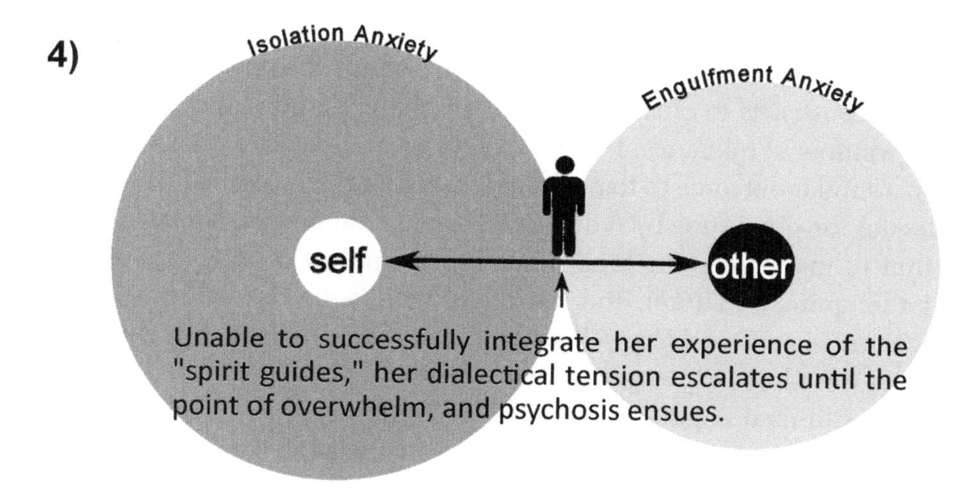

Figure 27.4 (cont.). Steps leading to the onset of Cheryl's psychosis

Trent. Trent reports having been raised in a severely troubled family system with a mother who was diagnosed with paranoid schizophrenia and an emotionally abusive father who suffered from chronic depression. With this unhealthy system placing Trent in a situation that included both unhealthy enmeshment and severe criticism, it's likely that by the time of the onset of his psychosis, Trent experienced both isolation and engulfment anxiety quite strongly, with his window of tolerance correspondingly very narrow.

By the age of 24, Trent's situation became truly intolerable. He was still living with his family, his ex-girlfriend and her 2 year old son had just moved in with him, and he was forced to take on the role of caretaker in many ways within this unhealthy and chaotic family system. He had a strong desire to move towards autonomy and self development, as is ordinary and healthy for someone at that age, yet it is clear that his fear of isolation was even stronger and was therefore holding him back. In other words, it appears that his engulfment anxiety was very strong, leading to a strong desire for individuation, but his isolation anxiety was even stronger and prevented him from doing so. The fact that he had continued to remain in such a situation until the age of 24 (the age of onset of his psychosis) indicates that both anxieties were indeed quite strong (with his window of tolerance correspondingly very narrow), and that his window of tolerance was likely skewed to some degree towards the *other* pole of the dialectic (having more isolation anxiety; see image #1 of Figure 27.5).

Finding himself in this very painful dilemma, Trent began to significantly increase his use of marijuana and tobacco as a way to mitigate his anxiety. Looking at the succession of strategies listed in Figure 23.3 (on page 186), we can say that the intensity of his dilemma did not allow him to make use of the first two more ordinary strategies (changing the current aggravating conditions or allowing his window of tolerance to gradually adjust to his situation), so he resorted to the use of psychoactive substances to jump straight to the fourth strategy—destabilizing his cognitive constructs to create an immediate though highly unstable expansion of his window of tolerance. His use of psychoactive substances in this way resulted in a similar outcome to that of Theresa's use of them. While this strategy was successful for some time by reducing his dialectical tension somewhat and helping him to maintain some kind of a tenable window of tolerance, it only provided a temporary solution, and it brought with it the risk associated with unstable cognitive constructs as well as the risk of allowing him to remain in a situation that was ultimately unsustainable (see image #2 of Figure 27.5).

In time, this limited coping strategy failed, Trent's window of tolerance essentially disappeared, and he became completely overwhelmed by the existential anxieties (see image #3 of Figure 27.5). A psychotic reaction ensued and, to use his own words, he "imploded into a void where a family was supposed to be."

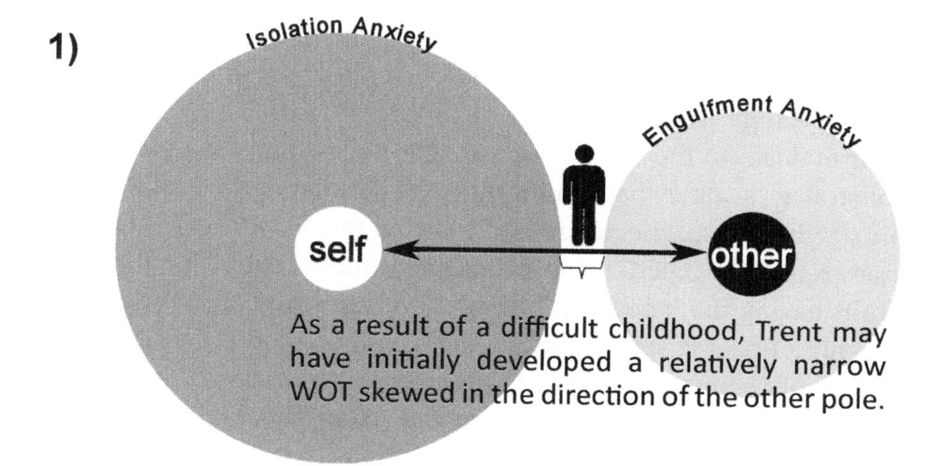

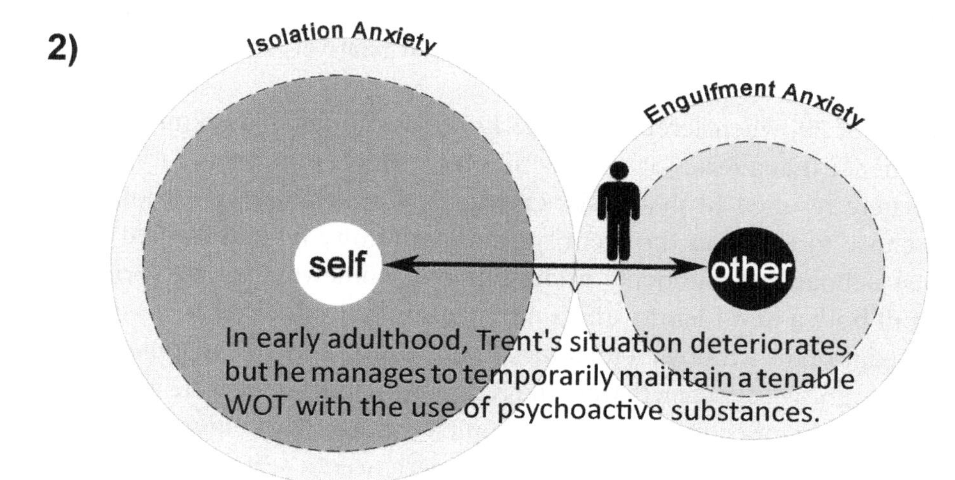

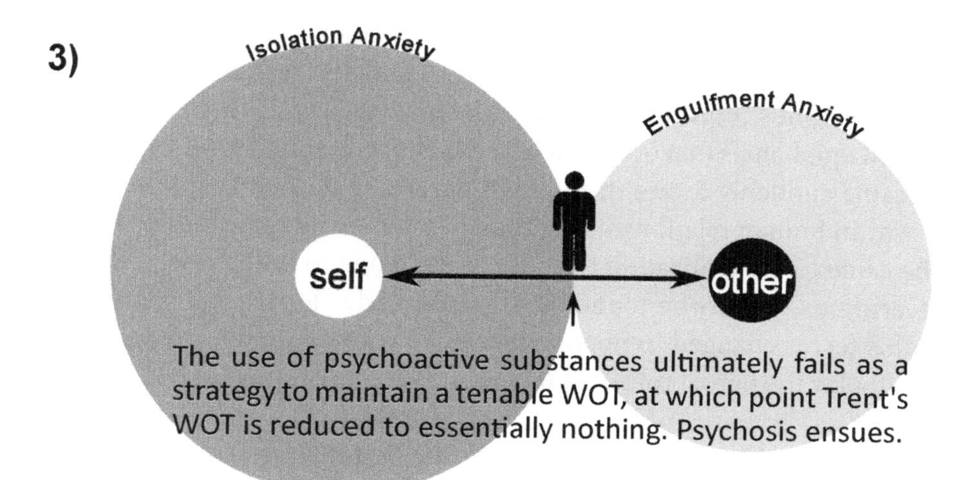

Figure 27.5. Steps leading to the onset of Trent's psychosis.

Jeremy. It appears that Jeremy may have developed a relatively narrow window of tolerance during his childhood as the result of significant trauma in his relationships with his family and peers. Referring to Yalom's two strategies for fending off the fear of death to the self[3], it appears that Jeremy may have developed both types of strategies somewhat significantly. On one hand, he recalls having strong heroic strivings (in particular, related to being the "savior" of his family), which is closely related to experiencing strong engulfment anxiety; on the other hand, he recalls having the belief that he and all of humanity are protected by some "wise group of people steering us," which could be seen to be associated with the strategy of "being protected . . . by an ultimate rescuer"[4], and which would therefore be closely related to experiencing strong isolation anxiety. The fact that both of these strategies were relatively prominent suggests that Jeremy experienced both types of anxieties fairly acutely, which corresponds with a window of tolerance that is relatively well centered but relatively narrow (see image #1 of Figure 27.6).

At the age of 20, when Jeremy went to London to pursue academic studies, there is evidence that a series of shifts within his window of tolerance took place that ultimately resulted in psychosis. Perhaps the first major shift took place when he chose to abandon the identity and community he had formed in relation to skateboarding. Suddenly finding himself somewhat isolated with the need to craft both a new identity and a new community was likely to have taken him to the edge of his window of tolerance in the direction of the *self* pole, and as a result, his cognitive constructs were likely somewhat unstable as he attempted to integrate this major change in his situation (see image #2 of Figure 27.6).

Shortly after arriving in London, Jeremy experienced a series of epiphanies, which suggest that he was in the process of successfully integrating his new situation with corresponding shifts in his window of tolerance and his cognitive constructs. An epiphany, in fact, can be seen as simply the subjective experience of a shift in one's cognitive constructs. Jeremy recalls initially having three closely related epiphanies that occurred within a short period of time. In the first, Jeremy became suddenly aware of the high degree of conflict that takes place regularly within human relationships, although often being quite hidden. In the second, he came to the realization that "there was no kind of like wise group of people steering us," which he found to be particularly "terrifying." Very soon, however, this terror changed to opportunity as he realized, in a third epiphany, that he could play the heroic role of making a major contribution to our lost and troubled race.

In the context of this model, then, we can say that Jeremy first experienced an increase in his engulfment anxiety when his sense of security within human relationships was somewhat challenged by his recognition of the pervasiveness of conflict within them combined with the loss of his belief in a "wise group of people steering us." Very close to the same time that his engulfment anxiety increased, however, his isolation anxiety *de*creased as is evidenced by the strengthening of his hero strategy and his sense of self reliance and self confidence. Corresponding to both of these changes (the increase of his engulfment anxiety and the decrease of his isolation anxiety) was an overall shift in his window of tolerance in the direction of the *self* pole (see image #3 of Figure 27.6). So, in short, after finding himself in a much more isolating situation than he was used to, Jeremy's self system responded with a successful shift in his window of tolerance to match this new situation.

Shortly after this shift, Jeremy had another powerful epiphany when he realized that the interpersonal "battles" that he was observing in others were also taking place within himself to some extent. Upon having this realization, he almost immediately experienced a significantly increased sense of peace along with a corresponding increase in his self confidence, his connection with others, and his connection with himself. By recognizing that these battles were also taking place within himself, he probably felt more empathy towards others and less threat from them as well; and he also probably experienced more empathy for himself along with a generally increased sense of self connection. We can say, then, that this was a genuine growth experience for him—that his window of tolerance had now expanded in both directions and perhaps also became more centered, and that he therefore experienced some movement in the direction of the optimal personality (see image #4 of Figure 27.6).

Jeremy had now undergone a series of very significant shifts within his being at the existential level in a very short period of time, and even though it appears that he experienced some genuine growth during these transformations, some danger remained in that his cognitive constructs probably remained somewhat unstable for some time as they continued to integrate these new feelings, experiences, and understandings (see image #4 of Figure 27.6).

When Jeremy returned home for the holidays shortly after all of this had occurred, he was relishing in the experience of much more enjoyable relationships that had resulted from his recent growth, and he was likely also feeling a general increase in unitive feelings since his dialectical tension would have been significantly diminished. Unfortunately, while in this relatively fragile state, Jeremy attended a party and smoked some marijuana, which likely further destabilized his cognitive constructs ("[it] sent me through the roof"). Then, while in what is now an extremely precarious condition with regard to his cognitive constructs, he experienced a profound "shaming," and a psychotic reaction ensued.

Shame is possibly one of the most painful feelings that we can experience, and it is probably directly associated with the experience of very high dialectical tension. With shame, one experiences both types of anxieties very acutely. On one hand, there is the fear that one will be destroyed and/or consumed by others, which is clearly related to engulfment anxiety; on the other hand, there is the fear of being ostracized and isolated, which is clearly associated with isolation anxiety. In Jeremy's case, we can say that just prior to his experience of profound shame, he was attempting to integrate being in the world with a significantly reduced dialectical tension (a stronger experience of unity) than what he had been used to. When the shaming occurred, he suddenly experienced a dramatic swing from relatively strong unity to relatively strong duality, or to put it in different terms, from relatively low dialectical tension to very high dialectical tension (see image #5 of Figure 27.6). This radical swing represented a severe threat to his self and overwhelmed the precarious balance of his system, throwing him into profound dysregulation at the root existential level and psychosis as his psyche tried desperately to restore balance. What is particularly unfortunate about this situation is that it appears that Jeremy had experienced profound growth immediately prior to this and was in the process of integrating it. Now, after the initiation of psychosis, he would be required to continue the process of integrating this growth in a particularly chaotic and haphazard manner.

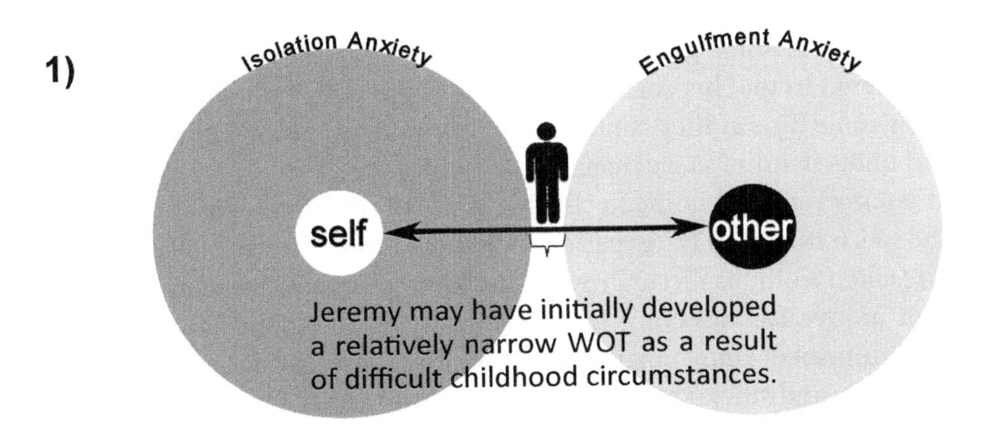

Figure 27.6. Steps leading to the onset of Jeremy's psychosis.

2)

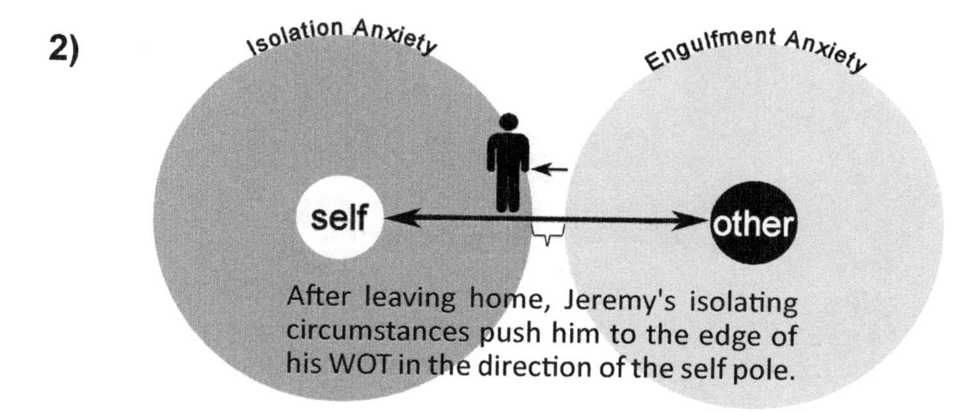

3)

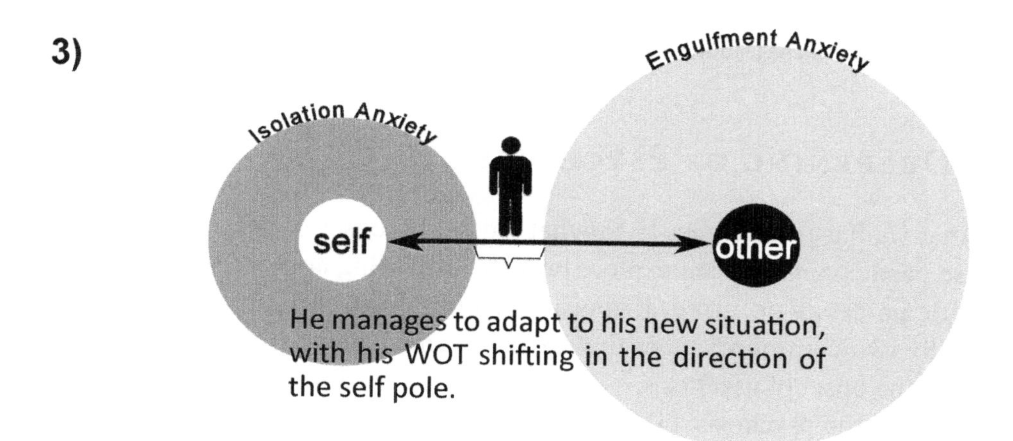

4)

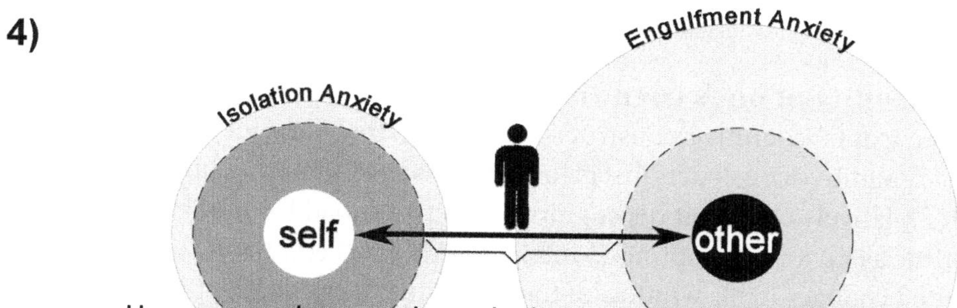

He apparently goes through the process of attempting to integrate a genuine growth experience, which would entail a significantly expanded and centered WOT if completed successfully. His cognitive constructs remain unstable at this point, however, as integration is not yet complete.

Figure 27.6 (cont.). Steps leading to the onset of Jeremy's psychosis.

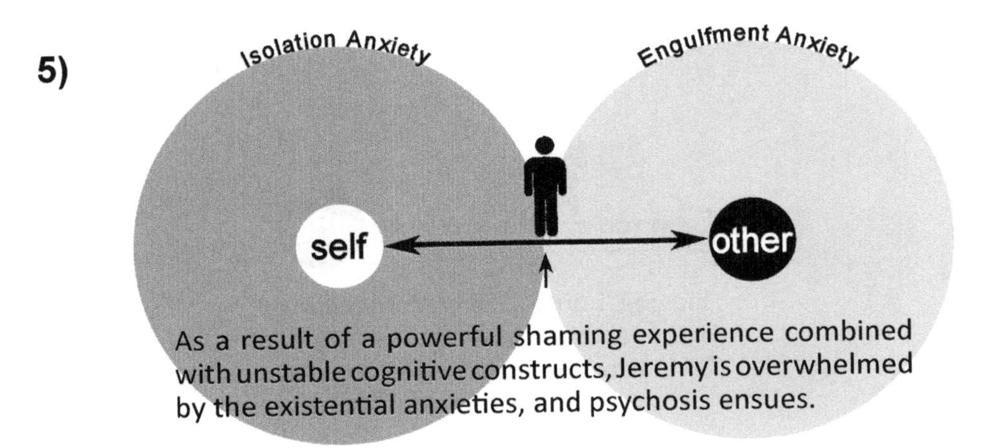

Figure 27.6 (cont.). Steps leading to the onset of Jeremy's psychosis.

THE DEEPENING OF PSYCHOSIS

Now that we have some idea as to what led to the onset of psychosis for each of these participants, we can explore how the unfolding and deepening of their psychotic processes occurred. It appears that for all six participants, there were essentially two common factors that existed immediately prior to and concurrent with the onset of psychosis, as well as during the subsequent unfolding and deepening of the psychosis: (a) the instability of their cognitive constructs, and (b) an overwhelming degree of dialectical tension (which is closely related to a severe existential threat to the self, as discussed earlier). Let's look more closely at the role that each of these factors plays in generating the psychotic process.

The instability of one's cognitive constructs. Regarding the instability of the participants' cognitive constructs, it is likely that they all had significantly unstable cognitive constructs just prior to the onset of their psychosis. In Jeremy's case, it's relatively clear that his cognitive constructs had been quite unstable for some time as he was attempting to integrate profound epiphanies along with the associated changes in his window of tolerance. In Cheryl's case, it's also clear that her cognitive constructs were relatively unstable as she began to experience the voices of the "spirit guides" prior to having a full psychotic reaction. In Byron's case, it's also evident that he was attempting to integrate profound transliminal experiences prior to onset, which would certainly have entailed unstable cognitive constructs. In the cases of Sam, Trent, and Theresa, they all had been using significant amounts of psychoactive substances during that period of time, which implies that some instability within their cognitive constructs was likely present.

As mentioned previously, the evidence suggests that, in all of their cases, it is likely that their psyches had been attempting to use the more extreme strategies of significantly destabilizing their cognitive constructs in order to bring their experience and their windows of tolerance back into alignment with each other after experiencing significant existential threats to their selves (the third and fourth strategies listed in the diagram in Figure 23.3 on page 186). For some, this destabilization had occurred relatively organically; and for others, it was facilitated by the use of psychoactive substances.

As discussed earlier, such strategies entail significant risk, acting somewhat like a double-edged sword. On one hand, the destabilization of one's cognitive constructs offers powerful leverage for making profound changes to one's perceived experience and/or window of tolerance. On the other hand, it appears that unstable cognitive constructs are likely to increase one's vulnerability to being completely overwhelmed by existential anxiety. This may occur either as the result of failure to arrive at a lasting solution to the aggravating conditions, which leads to continuously escalating dialectical tension until the point of overwhelm, or as the result of opening the door to overwhelming transliminal experiences that cannot be integrated. If such a strategy fails to lead to a lasting realignment of one's experience with one's window of tolerance, then it appears that psychosis is likely to ensue.

An overwhelming degree of dialectical tension. All of these participants experienced an overwhelming degree of dialectical tension immediately prior to and concurrent with the onset of their psychosis. As discussed earlier, it appears that there are primarily two ways that this can occur—either one experiences overwhelming duality (too much dialectical tension) or one experiences overwhelming unity (via an overwhelming transliminal experience). These two different ways are actually very closely related in that once one experiences one or the other, there appears to be a recoil response in the opposite direction, resulting in an oscillation between the two states as the psyche attempts to regain equilibrium (see the top of Figure 27.7 below).

The main distinction, then, is simply with regard to which occurred first, and regardless, it appears that the result is very similar—a highly unstable fluctuation between these two states. While it is quite clear that all participants experienced an overwhelming degree of dialectical tension immediately prior to onset, the manner in which this came about was somewhat different for each of them. For Sam, Trent, and Theresa, it seems that there was simply a steady increase in the dialectical tension until the point of overwhelm and subsequent psychotic reaction. It appears that Trent and Theresa, and maybe Sam to some extent, were able to use psychoactive substances to prolong the point of overwhelm, but these strategies ultimately failed and may have even increased the likelihood of a psychotic reaction by allowing them to remain so long in an unsustainable situation and by further destabilizing their cognitive constructs. The other three participants (Cheryl, Byron, and Jeremy), on the other hand, seem to have experienced a fairly radical swing from high dialectical tension to relatively low dialectical tension and then back again to an overwhelmingly high degree of dialectical tension, at which point a full-blown psychotic reaction ensued. One could even argue that this preliminary swing could be considered a mild form of psychosis, in the sense that some significant dysregulation had clearly begun to take place, although all three of them were still generally in contact with consensus reality during this period of time. In any event, it is likely that such a swing would make one particularly vulnerable to experiencing a more genuine psychotic process.

Seeing psychosis as a desperate attempt to regain equilibrium of the self. So, regardless of whether the psychosis was initiated after a gradual increase in the tension until the point of overwhelm or was initiated after a swing between relatively high and low degrees of dialectical tension, what occurred afterward appears to be quite similar for all of them. We see evidence of a profound dysregulation of the self system involving dramatic fluctuations within their experience on two levels—between unity and duality (between very high and very low overall dialectical tension), and between the *self* and *other* poles of the self/other dialectic—resulting in the myriad of experiences we associate with psychosis (see Figure 27.7). It's important to point out that these fluctuations are generally not nearly as linear and ordered as these illustrations might suggest.

These findings are in very close alignment with the research of Batson and Ventis[5], and Jackson[6], in which their findings suggest that both mystical and psychotic experiences begin with the building of emotional and cognitive tension in relation to an existential problem (see Chapter Thirteen). Then, upon reaching a particular threshold of tolerance, the psyche attempts to solve this problem by initiating a profound personal paradigm shift. In the context here, this attempted personal paradigm shift consists of increasing (and possibly centering) one's window of tolerance. Jackson suggested that there is essentially just one key difference between mystical experiences and psychotic experiences. Mystical experiences lead to the successful completion of a personal paradigm shift that allows for a solution to the problem and the subsequent diminishment of tension (dialectical tension within the context of the DUI model); whereas in psychosis, a personal paradigm shift takes place, but for whatever reason, this does not resolve the tension and may even generate more tension followed by further paradigm shifts. The result can be a relatively rapid alternation and fluctuation

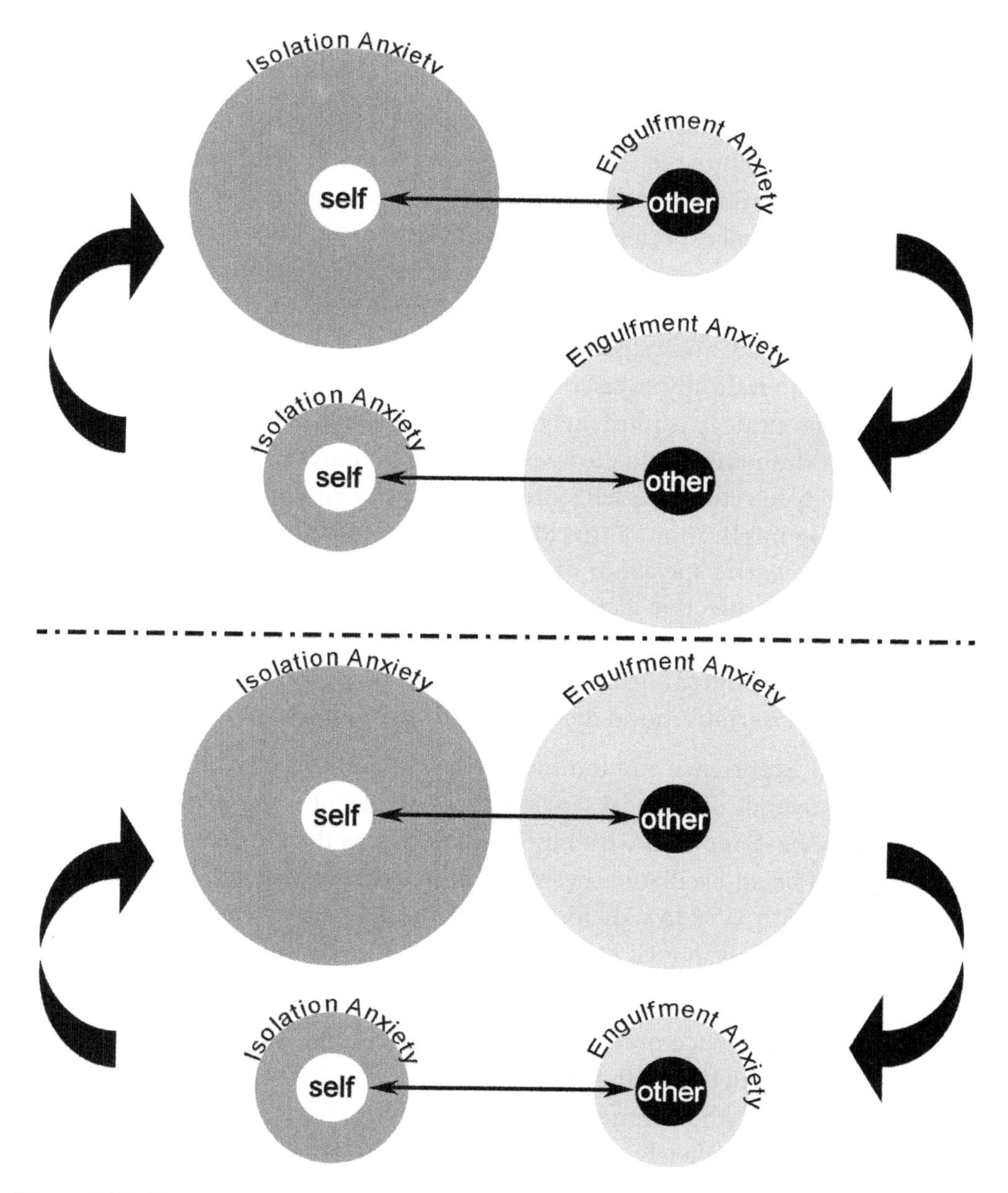

Figure 27.7. During psychosis, the self system undergoes dramatic fluctuations on two different levels: between the strengths of the isolation anxiety and the engulfment anxiety (top), and between high and low dialectical tension (bottom). The type of fluctuation most predominant within an individual's experience can vary significantly from one individual to another and even from one time period to another within the same individual. The fluctuation represented by the top figure is likely to be more predominant in so called schizophrenia, and the fluctuation represented by the bottom figure is likely to be more predominant in so called bipolar disorder. It's also likely that in the majority of cases, a significant blend of both types of fluctuations will be experienced to a greater or lesser degree.

between different belief systems and experiences, a condition we often equate with florid psychosis.

To place these ideas within the context of the DUI model, we can say that upon facing an existential threat to the self, if the psyche is successful in bringing about a radical transformation of one's window of tolerance as a result of "cracking open the lid" to transliminal experiences, then a mystical experience has occurred. This corresponds with a successful resolution to the severe existential threat by using the fourth strategy (and also possibly the third strategy to some extent) in the diagram in Figure 23.3 on page 186—the destabilization of one's cognitive constructs in order to radically adjust one's window of tolerance. In psychosis, on the other hand, this strategy is ultimately unsuccessful, and upon reaching complete overwhelm, a dramatic fluctuation within the self system occurs as the psyche tries desperately to return the self system to a state of equilibrium (Figure 27.7).

A fascinating implication of this idea is the surprisingly cogent explanation it provides for the entire spectrum of patterns we observe within psychosis, from the predominantly affective disturbances we see in so called bipolar disorder to the predominantly cognitive disturbances we see in so called schizophrenia, as well as the more "blended" disorders we see in between (so called schizoaffective disorder and the various mood disorders with psychotic features):

- Those who experience predominantly a fluctuation between very low and very high overall dialectical tension (see the bottom of Figure 27.7) would most likely be diagnosed with bipolar disorder, as their predominant experience would be an oscillation between the very expansive and euphoric states that occur with very low dialectical tension (so called mania) and the very constrictive and dysphoric states that occur with very high dialectical tension (severe emotional distress, or so called depression).
- Those who experience predominantly a fluctuation between the strengths of the isolation anxiety and the engulfment anxiety (see the top of Figure 27.7) would most likely be diagnosed with schizophrenia, as these heavily skewed configurations are closely associated with anomalous beliefs and perceptions, as discussed previously.
- Considering that these are actually just different manifestations of a common phenomenon, we would expect the vast majority of cases to fall somewhere between these two extremes, exhibiting a significant blend of both types of fluctuations. And in fact the evidence does support this—that when thorough statistical analysis has been used, and when people are diagnosed at different time points, we find that most people's experiences (a) do fall somewhere between the extremes of so called schizophrenia and so called bipolar disorder[7], and (b) often satisfy the criteria for more than one disorder[8].

Chapter 28

The Anomalous Experiences

After the psychotic process was initiated, the scope and intensity of the anomalous and psychotic experiences that these participants went through is somewhat breathtaking. In spite of this diversity, however, the cross case analysis revealed that there are actually only a small number of different core themes within this vast array of experiences, of which there was strikingly low divergence among the participants (See Table 28.1). Both the nature of the converging themes and the low degree of divergence between them suggest that all participants were experiencing a very similar process, which the evidence suggests is the attempted reestablishment of the equilibrium of the self, as discussed above. And as we look more closely at these core themes using the framework of the DUI model, we find a remarkably cogent explanation for these experiences that continues to offer strong support for this idea.

Good/Evil and Creation/Destruction

All participants reported having polarized experiences of good and evil and of creative and destructive forces. I suspect that these are essentially different manifestations of the experience of unity and duality at the root experiential level. According to the DUI model, we can say that all participants experienced powerful fluctuations, to a greater or lesser degree, between very high dialectical tension (a strong experience of duality) and very low dialectical tension (a strong experience of unity).

Recall that with very high dialectical tension and the corresponding experience of an extreme dualistic split, the dualistic feelings (the fears and desires associated with the dialectic poles along with their many derivatives) are

Table 28.1 Converging Themes and Divergences for *Description of the Anomalous Experiences*

CONVERGING THEMES	DIVERGENCES
Polarized experiences of good and evil	All experienced this.
Creative and destructive forces	All experienced this.
Fluctuating between omnipotence and powerlessness	All experienced this.
Heroic striving (fighting evil and/or ignorance)	Trent and Cheryl experienced striving against evil forces within themselves.
Being watched over by malevolent and/or benevolent entities	All experienced this. Jeremy experienced being watched over by primarily malevolent entities, though sensed a powerful presence of both types.
Groundlessness	All except Sam mentioned experiencing profound groundlessness.
Parallel dimensions	All but Trent experienced different realms of experience occurring simultaneously to some degree.
Feelings of euphoria, liberation, and/or interconnectedness	All but Sam recalled having these kinds of experiences, to a significantly greater or lesser degree. Jeremy had these just prior to his psychosis, but not so much after onset.

particularly strong, and when taken to their extreme, the dualistic feelings correspond closely with feelings often associated with evil (hatred, animosity, envy, greed, selfish lust, etc.).

With very low dialectical tension, the dualistic feelings are diminished, and so the unitive feelings (unconditional love, compassion, sympathetic joy, equanimity, etc.) are able to rise into consciousness; and as discussed earlier, unitive feelings are those that are most closely associated with the concept of good. Therefore, when someone is experiencing extreme dysregulation of the self system along with the corresponding fluctuation between very low and very high dialectical tension, we would expect them to experience a fluctuation between strong dualistic feelings and strong unitive feelings—in other words, between so-called "evil" feelings and so-called "good" feelings (See Figure 28.1).

A person in psychosis is also likely to experience significant exposure to the transliminal realm, the realm between unity and duality, and this realm is seething with creative and destructive forces. Recall that, according to this model, all dualistic manifestations arise from a condition of unity, the ultimate creative act, and once in dualistic form, all manifestations are doomed to undergo decay and destruction as they return once again to a condition of unity (See Figure 28.1).

Perhaps one way to distinguish experiences of good and evil from experiences of creation and destruction is to consider experiences of good and evil as representing *feelings* associated with the extremes of unity and duality, whereas experiences of creation and destruction represent *volition* associated with these two extremes. So, as someone in psychosis essentially swings between these unitive and dualistic extremes, they experience a much wider array of feelings, volitions, and experiences than what one would ordinarily experience, and they are also likely to experience these in a much rawer, more archetypal form—for example, having visions or other perceptions of God and the Devil, saints and sinners, and/or other forms representing the archetypes of good and evil that correspond with the individual's society or personal experience, and also experiencing creative and destructive impulses that may be associated with these.

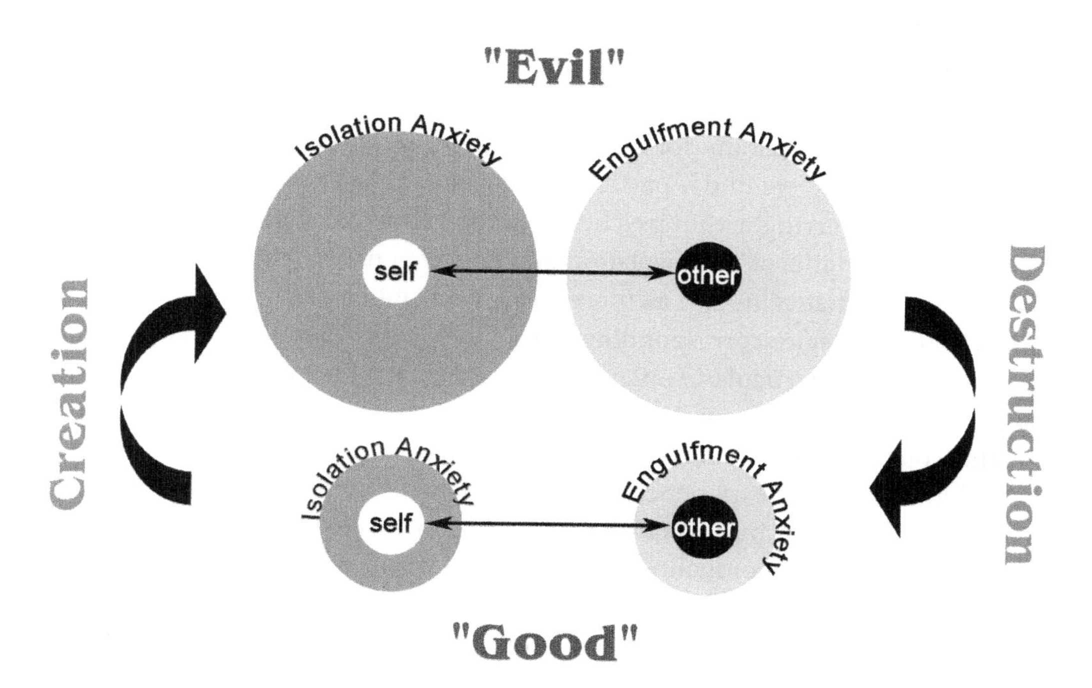

Figure 28.1. Profound experiences of "good," "evil," creation, and destruction associated with the fluctuation between relatively high and relatively low dialectical tension.

HEROIC STRIVING AND BEING WATCHED OVER

All participants had experiences of heroic striving and experiences of being watched over by benevolent and/or malevolent forces. All participants also experienced fluctuations, to a greater or lesser degree, between feelings of omnipotence and/or omniscience at one extreme and feelings of powerlessness at the other. I suspect that these are all closely related, and that they all result from dramatic fluctuations between the strengths of the two anxieties of the self/other dialectic.

As discussed earlier, Yalom contended that all of us use essentially two types of strategies to fend off our existential fear of death—(a) the belief that we are "personally inviolable," which in the DUI model corresponds with the *self* pole, and (b) the belief that we are "protected eternally by an ultimate rescuer," which corresponds with the *other* pole[1]. For the relatively healthy individual, it is likely that both of these strategies will be used to some extent and that they both will take relatively subtle forms. For an individual experiencing a high degree of imbalance between the isolation and engulfment anxieties, however, it is likely that they will rely more heavily on just one or the other of these strategies, and that it will manifest in a particularly extreme form. Those with particularly strong engulfment anxiety will likely experience themselves as being particularly inviolable, and they will likely experience strong heroic strivings. Those with particularly strong isolation anxiety will likely experience a strong sense of being watched over by a very powerful other.

When we place Yalom's formulation of these strategies within the context of the DUI model, we find a particularly cogent explanation for these kinds of experiences occurring within psychosis. In the highly dysregulated state of psychosis, the strengths of one's isolation and engulfment anxieties are likely to be fluctuating dramatically, and as this occurs, we would expect dramatic fluctuations in the strategies corresponding to each type of anxiety. As the engulfment anxiety becomes particularly intense, we would expect the individual to resort to extreme heroic strivings and even feelings of omniscience and/or omnipotence. As the isolation anxiety becomes particularly intense, we would expect the individual to have strong experiences of being watched over by a powerful other(s).

When we keep in mind that these individuals are also experiencing strong fluctuations within the duality/unity continuum, we can understand why experiences of archetypal good and evil and also creation and destruction tend to become so bound up within these experiences. One may become identified with "good" and experience strong heroic striving against evil and/or ignorance in the world (as they all experienced to some extent); or one may become identified with "evil" and experience strong heroic striving against the evil within oneself (as was the case especially for Trent and Cheryl). Also, one may experience being watched over by benevolent and/or malevolent entities, and possibly even some fluctuation between these. All participants with the exception of Jeremy experienced being watched over by both types of entities at different times. Jeremy did experience a strong sense of the existence of both malevolent and benevolent entities; however, he experienced being watched over by primarily the malevolent ones.

It's important to note that while all participants seem to have experienced both types of strategies to some extent, there appears to be a high correlation between the type of strategy that was most prevalent for each participant during their psychosis and the direction toward which their window of tolerance was likely skewed prior to onset. In the cases of Sam, Byron, and Theresa, it appears that their window of tolerance was likely skewed toward the *self* pole prior to onset, and as would be expected, they all experienced significantly more heroic striving than being watched over. In Cheryl's and Trent's cases, it appears that their windows of tolerance were likely skewed toward the *other* pole prior to onset, and as would be expected, they both experienced significantly more being watched over than heroic striving. In Jeremy's case, it is not so clear whether or not his window of tolerance was significantly skewed prior to onset—a rough assessment suggests that it may have been skewed either slightly toward the *self* pole or relatively well centered. His experiences appear to fit quite well with this assessment, however, in that he seems to have had relatively equal amounts of experiences related to heroic striving and being watched over.

Parallel dimensions

All participants except for Trent recalled having had distinct experiences of different realms of experience occurring simultaneously, although the intensity of this varied significantly among the participants. I suspect that this is closely related to the experience of unstable cognitive constructs. Perhaps these participants were aware of multiple sets of cognitive constructs or at least multiple aspects of their cognitive constructs occurring simultaneously. It is interesting to note that all participants reported that one of their lasting personal paradigm shifts has been an increased acknowledgment of the limitations of their construction of reality.

Groundlessness

All except for Sam mentioned having had experiences of profound groundlessness. According to the descriptions given by the participants (see Appendix A), it is likely that their feelings of groundlessness correspond closely with a direct experience of the transliminal realm. It is clear that all participants including Sam experienced the profound instability of their cognitive constructs, so it is somewhat unclear why Sam does not recall having had experiences of groundlessness. I suspect that each time Sam's cognitive constructs shifted, they shifted quickly enough so that he was never consciously aware of the transliminal realm in its raw unconditioned form.

Feelings of Euphoria, Liberation, and/or Interconnectedness

All participants except for Jeremy reported having experiences of euphoria, liberation, and/or interconnectedness to a greater or lesser degree during their psychosis, and Byron had such experiences prior to his psychosis as well. Jeremy was an exception in that he experienced these kinds of experiences just prior to the onset of his psychosis, but not so much during his psychosis. According to the DUI model, these kinds of experiences likely correspond to the unitive experiences that are free to arise when one's dialectical tension has been significantly reduced. It appears, then, that Byron and Jeremy experienced significantly low dialectical tension at times prior to onset and that everyone except Jeremy experienced significantly low dialectical tension at times during their psychosis. According to other aspects of Jeremy's experience as discussed above, however, it's likely that he did experience significant fluctuation between relatively high and low dialectical tension during his psychosis, but that it simply did not reach levels as low as the others during this stage.

The Diverse array of Other Anomalous and Extreme Experiences

Regarding the myriad other more idiosyncratic details found within these participants' anomalous experiences, Perry suggested that individuals in psychosis tend to *identify with* (to associate with the self) and/or to *project* onto the world (to associate with others) the experiences that they are having at this root existential level[2] (see Chapter Ten), an idea that fits very well with the data I've collected here. It's clear that all participants experienced significant identification and projection of numerous experiences such as different archetypes and other manifestations of good and evil, creation and destruction, and many others. One example is Jeremy's projection of his experiences of good and evil in his perception of the black and white orders. Another is Trent's projection of evil in his perception of a persecuting Devil at times and his identification with evil and destruction at other times: "I had thoughts, you know, so..so..so negative I was dry-heaving blood. . . . You know, just some horrible, negative, you know, satanic thoughts just filled my head." They all identified with profound experiences of "good" at different times when they experienced themselves as messianic heroes to varying degrees.

A second compatible explanation for the myriad more personal anomalous experiences reported by these participants involves the idea that our experience can be divided into a number of different yet highly interrelated realms. While the exact division of these realms of experience may be somewhat arbitrary, for the sake of simplicity, I'll consider them as being into four realms: the *environmental*, the *interpersonal*, the *intrapersonal*, and the *existential*. We can see each realm as being progressively closer to core experience, with the environmental being the furthest removed from core experience, and the existential being the closest to our core sense of self (see Figure 28.2).

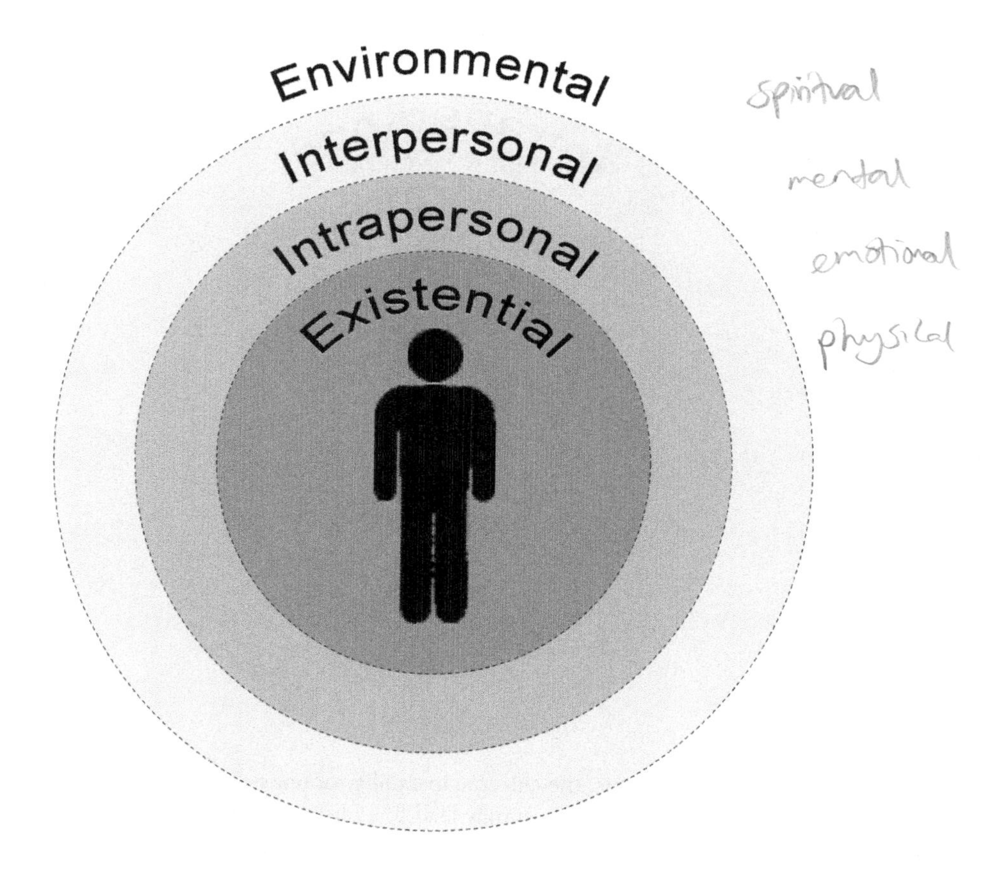

Figure 28.2. A graphical depiction of the different layers of our experience, the boundaries of which are ordinarily maintained to a greater or lesser degree by our cognitive constructs.

If we consider that one very important role of our cognitive constructs is to maintain some coherence and balance between these different realms of experience, then we can see that having unstable cognitive constructs combined with a highly dysregulated self system can open the door for past and present experiences related to all of these different realms to flood into our consciousness, blending with each other in haphazard and confusing ways (see Figure 28.3). It's apparent that this occurred to a greater or lesser degree at different times for all of these participants. Since each individual has unique experiences and memories determined by their personal environment and personal history, we would expect the anomalous experiences related to such mixing to be highly idiosyncratic, which was certainly the case with these participants.

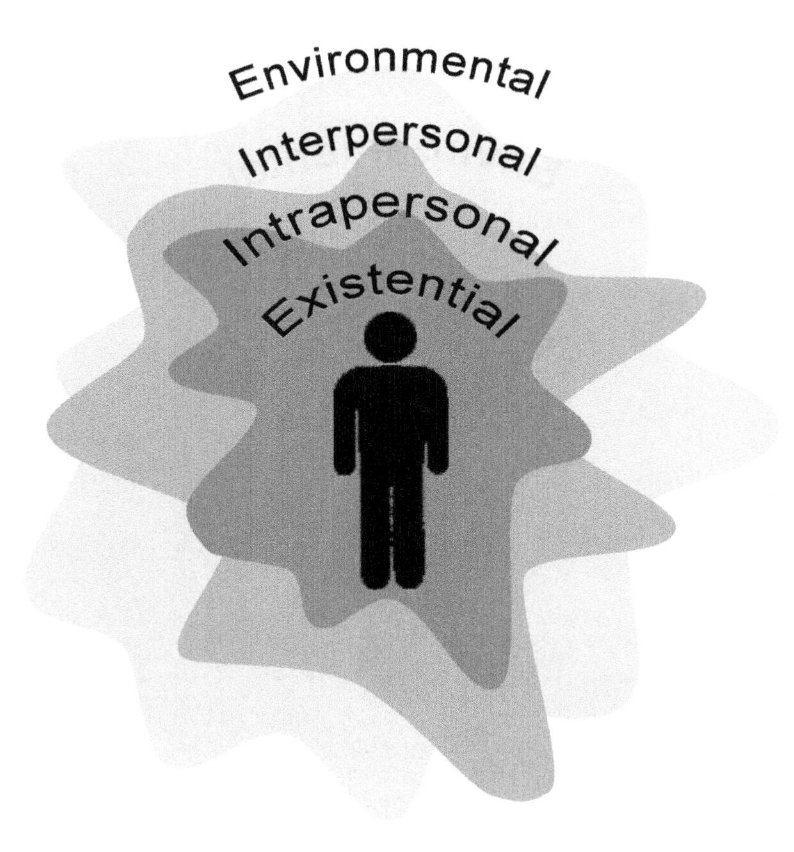

Figure 28.3. During psychosis, the extreme instability of one's cognitive constructs combined with a highly dysregulated self system may lead to a blurring of the boundaries between one's different realms of experience. This can lead to a confusing blend of anomalous experiences unique to each individual.

Chapter 29

Recovery

If we consider psychosis to be essentially a process initiated by the psyche in an attempt to regain equilibrium within the self system after experiencing an overwhelming existential threat to the self, then we can see recovery as movement in the direction of regaining that equilibrium. Full recovery, then, would refer to having attained a degree of equilibrium that is equal to or greater than that which existed prior to the onset of psychosis.

In the context of the DUI model, such an overwhelming existential threat to the self is associated with very high dialectical tension and the corresponding reduction of one's window of tolerance to essentially nothing. Therefore, in order to have any hope of regaining equilibrium, the psyche must find some way to increase the window of tolerance, and possibly center it to some degree as well, in order to bring it to a more stable, sustainable, and balanced position. We can say, then, that at the onset of psychosis, the self has come to be in a configuration that is diametrically opposite to that of the optimal personality; that during recovery, the psyche is in the process of attempting to reorganize the self at the root-most level of our experience; and upon full recovery, the psyche has succeeded in bringing the self into a much healthier and more balanced configuration that is significantly closer to the optimal personality than that which existed immediately prior to onset.

The Benefit of Unstable Cognitive Constructs

In order to carry out such a profound reorganization of the self, it seems that one of the most important resources available to the psyche is the ability to destabilize the cognitive constructs, as has already been discussed. There is some evidence for this within the field of psychotherapy research. Hallucinogenic substances (such as LSD, peyote, and psilocybin mushrooms) are well known for their ability to destabilize one's cognitive constructs, and there is evidence that, when used carefully and with appropriate guidance, they offer the potential for healing at very profound levels[1]. All of the participants in this study have also apparently experienced profound healing as a result of the successful resolution of their psychotic processes, so the evidence does suggest that the destabilization of their cognitive constructs may very well have played an important role in allowing this healing to take place at such a deep level. Cheryl articulated this well when she said, "the psychotic experience really just like . . . ripped off all the layers and got right to the core . . . and then when I was able to look at my core of self hatred and heal that, then I just..I built my mental health on that from there."

The implication, then, is that the psyche first dramatically destabilizes the cognitive constructs in order to facilitate this healing and growth (the modification of the window of tolerance), and then the cognitive constructs must go through a process of integrating this new way of being and understanding and eventually regain relative stability in this new configuration. The data presented so far demonstrates that this explanation matches the experiences of these participants very well. As we look more closely at the themes regarding what best supported these participants' recovery (see Table 29.1), we see continuing evidence that such a natural healing process was indeed taking place.

THE IMPORTANCE OF SUPPORTING THE PROCESS

The major themes that emerged for the category of *recovery* show striking convergence for all six factors (see Table 29.1), with virtually all participants having expressed the importance of all of these factors in their recovery process. The only exception was with regard to Sam and the fourth factor listed in Table 29.1—Sam was the only participant who did not explicitly express that arriving at a more hopeful understanding of his psychosis was helpful in his recovery, although he did develop an understanding of psychosis that is clearly much more hopeful than the brain disease model.

Note that the first five factors in the list all supported recovery, whereas the sixth factor represents what they all felt was the most significant hindrance to their recovery—*harm from the psychiatric system hindering recovery*. Returning again to the idea that psychosis may be a natural healing process, it stands to reason that the most helpful support we can offer people in recovery is to directly support this process rather than to interfere with it; and as we look more closely at the details for each of these six factors, we can see that they are all in direct alignment with this reasoning.

Table 29.1 Converging Themes and Divergences for *Recovery*

CONVERGING THEMES	DIVERGENCES
Finding meaning in life	All participants expressed that finding meaning and/ or a purpose in life that made life worth living was important in their recovery.
Connecting with one's aliveness	All expressed the importance of fostering a deep connection with their aliveness—particularly their feelings, needs, and sense of agency.
Finding hope	All expressed the importance of maintaining hope in the possibility of full recovery.
Arriving at a more hopeful understanding of their psychosis	All expressed having arrived at an understanding of psychosis that is more hopeful than the brain disease model, which all but Sam expressed was important in their recovery.
Healthy vs. unhealthy relationships	All expressed the importance of cultivating healthy relationships. Trent and Jeremy also expressed the importance of distancing themselves from unhealthy relationships.
Harm from the psychiatric system hindering recovery	All experienced this.

The Triad of Hope, Meaning, and Connecting with One's Aliveness

The first three factors listed in Table 29.1—*hope, meaning*, and *connecting with one's aliveness*—seem to have worked together in a particularly symbiotic manner for all participants. In the face of the intense pain, confusion, and other challenges that all of these participants faced during their process, hope provided a sense of possibility—that working towards full recovery was even an option—meaning provided the guidance, and connecting with their aliveness provided the energy (Figure 29.1 offers a helpful metaphor for the symbiotic relationship between these). These three factors were essential for all six participants in their path towards recovery, and yet all participants found an enormous hindrance to all of these, ironically, from the mental health care system, and particularly from psychiatry.

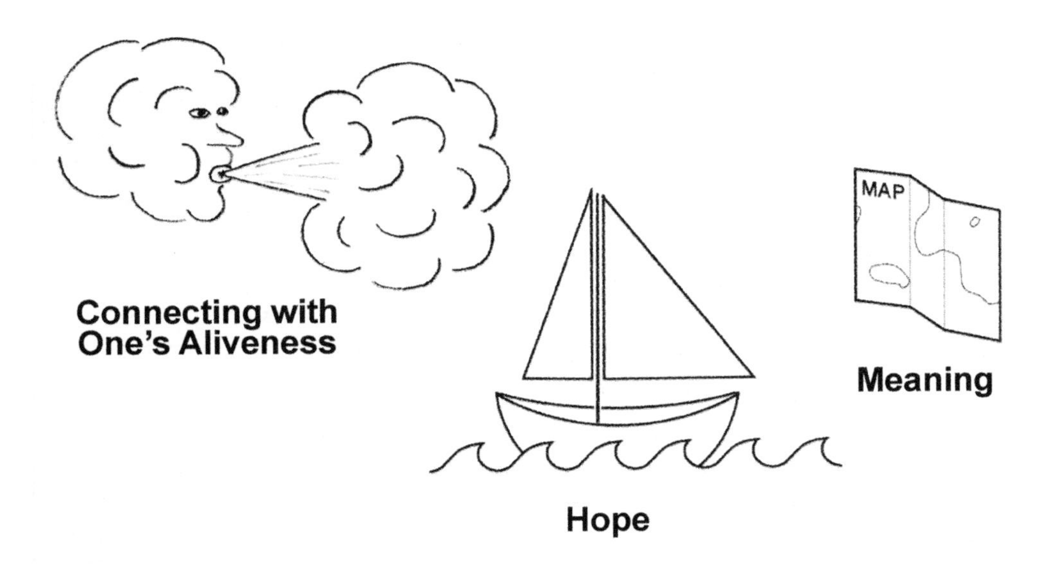

Figure 29.1. The three factors of connecting with one's aliveness, hope, and meaning acted together symbiotically to support all of these participants in their recovery. Above is a graphical depiction of a helpful metaphor in this regard. Hope is represented by a sailboat, in that it provided the vehicle—the actual sense that it was even possible to embark upon the journey towards full recovery. Connecting with one's aliveness is represented by the energy of the wind—by connecting with their feelings, needs, and sense of agency, they were able to find the energy to go on. Finally, meaning is represented by a map—the cultivation of meaningful pursuits provided the guidance necessary to continually move towards an increasingly rich and meaningful life.

Within Western society, psychiatry is the field of mental health generally considered to hold the position of highest authority with regard to treating those suffering from psychosis. Unfortunately, as discussed in Part One, psychiatry has a paradigm of care that is in direct contradiction to the philosophy of supporting the process of psychosis, instead doing virtually everything in its power to try to stop this process in its tracks. Based on the model presented here, one would expect that such a paradigm of care would be more of a hindrance than a benefit to recovery, and in fact, this is exactly what the research shows (as discussed in detail in Chapter Five), and this is also what every participant in this study expressed to a greater or lesser degree. As we look more closely at how the factors of hope, meaning, and connecting with one's aliveness supported these participants in their recovery, we will also look at how the psychiatric system seriously interfered with all of these factors.

Hope. It seems that the most harmful aspect of psychiatric care that these participants received was the undermining of their hope as the psychiatric system relentlessly inculcated them with the message that they were suffering from a lifelong degenerative brain disease and that they would have to remain on debilitating and highly toxic drugs for the rest of their lives (in spite of significant evidence to the contrary—see Part One). For all of these participants, maintaining hope in the possibility of full recovery and moving on to a rich and meaningful life in the face of this message of hopelessness was very difficult yet extremely important for their recovery.

One aspect of being so heavily inculcated with the brain disease model that was clearly a hindrance for most if not all of these participants was that they began to fear and mistrust their own minds. If we see psychosis as a natural though precarious process that is initiated by the mind (the psyche), there is no doubt that the mistrust and fear of one's mind can lead to greatly increased internal conflicts as one becomes involved in an intrapsychic struggle against the very process of healing that is attempting to take place. While it is of course helpful to have a healthy sense of mistrust for one's cognitive constructs (our understanding and interpretation of the world), especially while going through psychosis, it is an altogether different and potentially much more harmful matter to mistrust the nature of one's mind itself.

In spite of the significant obstacle of being told by so many "experts" in the field that genuine recovery is not possible, it is truly amazing that all of these participants managed to hold onto some hope, each in their own way (see Appendix A for details), and there can be no doubt that this was crucial for all of them in their making a full recovery.

Meaning. For all of these participants, being able to connect with a sense of meaning and purpose that made their life worth living was an essential factor in their recovery; but in order to do so, they found themselves face to face with the myth of hopelessness that is so heavily proselytized within the psychiatric system. This myth generally includes the message that someone experiencing long-term psychosis should abandon serious ambitions and devote their life to minimizing stress, remaining compliant with a lifelong regimen of heavy psychiatric drugs, and strictly monitoring their "illness" to reduce the risk of relapse. In spite of being bombarded by such a potentially devastating message that essentially promotes the abandonment of attempts to pursue a truly rich and meaningful life, all of these participants somehow managed to resist this message and to cultivate meaningful pursuits that allowed them to channel their energy and passion. They each expressed in their own way that their ability to do so was crucial in their recovery, and all of them have since gone on to develop rich and meaningful lives.

Connecting with one's aliveness. For all participants, having the opportunity to foster a deeper connection with themselves—particularly with their feelings, needs, and sense of agency—was another crucial factor in their recovery. And yet we discover that, once again, the psychiatric system interfered significantly in this regard. All participants reported receiving antipsychotics (and other psychiatric drugs) in a manner that was excessive, haphazard, and/or lasting significantly longer than necessary. Another term for antipsychotics, one that much more accurately reflects their actual effects, is *major tranquilizers*, and it is clear that there are few methods more detrimental to one's connection with one's aliveness than the excessive and/or haphazard use of major tranquilizers. All participants except for Trent expressed that the use of antipsychotics was more harmful than beneficial in their recovery. Even Trent reported having experienced significant harm from them, especially including one or more relapses of psychosis upon sudden withdrawal, but he was fortunate in that, after the particularly intensive use of them during his hospitalization, he was able to use them judiciously for the most part (typically using them for just a few months at a time on an as-needed basis).

In the end, in spite of the repeated message that they would need to remain on these drugs for the rest of their lives, and in spite of the often great difficulty in coming off them, all participants did manage to come off of them and they all expressed having much more success connecting with their feelings, needs, and sense of agency as a result of having done so. Along with this renewed connection came increased energy and motivation, which in turn reinforced their hope and ability to pursue a meaningful life.

The symbiotic relationship between hope, meaning, and connecting with one's aliveness. As mentioned above, these three factors—hope, meaning, and connection with one's aliveness—apparently formed an important symbiotic triad for all of these participants, with each factor reinforcing the other, and with the weakening of one being detrimental to the others. One metaphor that may be appropriate for this relationship is that of a tripod, with all three legs being equally important for the tripod's ability to remain upright. When we look more closely at the recovery process of these participants, we can see how, in spite of the hindrance to all three of these factors coming from the psychiatric system, all participants managed to find a way to at first barely hang on to each of these factors and then to eventually strengthen all of them until they were standing firmly on their own feet once again.

We see that all participants managed in their own way to maintain some connection to hope throughout their process, even though this connection may have felt hardly stronger than a strand of thread at times. Yet, by maintaining even this weak strand of hope, they were able to find the strength and the courage to resist the message of hopelessness and to come off the drugs, each at their own pace. Then, with the reduction of the drugs and/or other means, they each were able to connect more fully with themselves and their aliveness in various ways and to find a sense of trust in their basic sanity that lay beneath the often bewildering experiences of the psychotic process and the toxic messages of the psychiatric system. As their connection with their aliveness strengthened, each participant was then able to develop some sense of mastery over their psychotic experiences and to harness their agency, which in turn provided the fuel for them to strengthen their hope and pursue a meaningful life. Finally, by putting their passion and their energy into meaningful pursuits, their sense of hope and their connection with their aliveness were further reinforced. Like this, the factors of hope, meaning, and connecting with one's aliveness mutually supported each other and provided all of these participants with the means to maintain the faith that they could work through their psychosis and overcome the obstacles they faced from the psychiatric system and elsewhere.

While the medical model paradigm of care that these participants received was clearly much more harmful than beneficial in their recovery, it is important to note that four of the participants (Sam, Trent, Theresa, and Jeremy) found psychotherapy to be particularly helpful*. It's not difficult to see that a good psychotherapist, especially one who does not subscribe to the medical model, can support all three of these factors—maintaining hope, working towards a meaningful life, and connecting with oneself in a deep and healthy way.

Arriving at a More Hopeful Understanding of their Psychosis

Five of the six participants expressed that arriving at a more hopeful understanding of their psychosis was an important factor in their recovery. Sam was the only participant who did not explicitly report this as being an important factor in his recovery; however, he did develop an understanding of his psychosis that is clearly more hopeful than the degenerative brain disease model. We can see this factor as being very closely related to all three of the factors listed above. Its relationship to hope is obvious, in that a more hopeful understanding of psychosis allowed them to move beyond the message of hopelessness associated with the brain disease model. It is closely related to meaning in that the development of a deeper understanding of their psychosis offered a tangible framework and corresponding guidance in their path towards recovery. Finally, it is closely

* Trent was the only one who received what he felt was helpful psychotherapy within the psychiatric system—the others had to seek helpful psychotherapy from private practitioners outside the system.

related to connecting with their aliveness in that working to develop a deeper understanding of their psychosis required stronger self-connection which in turn reinforced their connection with their aliveness and self-agency.

HEALTHY VS. UNHEALTHY RELATIONSHIPS

All participants expressed that cultivating healthy, supportive relationships with others and healing unhealthy relationships (as was especially the case for Cheryl and Jeremy) was very important in their recovery. Returning to the DUI model, we can say that a healthy relationship involves the paradoxical incorporation of both self and other—the capacity to be relatively comfortable and connected with one's self and also relatively comfortable and connected with others—which corresponds to a relatively wide window of tolerance. Since, in psychosis, it appears that this is exactly the configuration that the psyche is working towards, it stands to reason that having relationships that support this process would be very helpful.

In addition to the importance of cultivating healthy relationships, Jeremy and Trent also expressed the importance of distancing themselves, at least to some degree, from unhealthy relationships within their families. This finding also fits very well with the DUI model. It is, after all, unhealthy relationships that typically create a skewed and/or narrow window of tolerance in the first place. If the psyche is attempting to reorganize this window of tolerance, then it stands to reason that distancing oneself from relationships that contributed to this dysfunctional configuration is important.

We can see that for someone suffering from psychosis who does not currently have any particularly healthy relationships in their lives, a well-trained psychotherapist can play an especially supportive role. They can support the individual in distinguishing healthy relationships from unhealthy ones, provide support in healing unhealthy relationships and in creating appropriate distance from relationships that appear to be currently beyond the possibility of healing. They can also provide a role model for a healthy relationship, ultimately acting as a bridge as the person learns how to cultivate healthy relationships outside of the psychotherapeutic relationship.

Chapter 30

Lasting Personal Paradigm Shifts

We find that all six participants in this study experienced a profound and lasting transformation of their personal paradigms as a result of their journeys through psychosis and recovery, and the convergence in this regard is striking. All six participants experienced every major theme within this category, to a greater or lesser degree (see Table 30.1).

An Increased Window of Tolerance and Reduced Dialectical Tension

All of the personal paradigm shifts that the participants reported experiencing fit well within the DUI model. I suspect that the first four factors listed in Table 30.1—*a significantly changed spectrum of feelings with more depth and more unitive feelings*, *an increased experience of interconnectedness*, *a strong desire to contribute to the wellbeing of others*, and *an integration of good and evil*—are all very closely related in that they all suggest a generally reduced dialectical tension and an expanded and more centered window of tolerance.

As previously discussed, when the dialectical tension is reduced, more unitive feelings are able to arise in one's consciousness. Also, while the extreme forms of the dualistic feelings associated with high dialectical tension (hatred, animosity, greed, envy, etc.) are most likely reduced along with the reduction in the dialectical tension, the feelings that do remain are likely to be imbued with an increasing sense of depth and richness as a result of being intertwined with more unitive feelings. All participants expressed having experienced a lasting shift in their spectrum of feelings that is in very close accord with this understanding. While the specific details of this shift in the spectrum of feelings varied somewhat

Table 30.1 Converging Themes and Divergences for *Lasting Personal Paradigm Shifts*

CONVERGING THEMES	DIVERGENCES
A significantly changed spectrum of feelings with more depth and unitive feelings	All experience this.
An increased experience of interconnectedness	All experience this.
A strong desire to contribute to the wellbeing of others	All experience this.
An integration of good and evil	All experience this.
Appreciating the limits of consensus reality	All experience this.
A greater understanding of psychosis	All experience this.

among the participants, they all described experiencing more unitive feelings and less of the more extreme negative feelings (the dualistic feelings associated with a very high degree of dialectical tension).

All participants expressed feeling a sense of interconnectedness, which refers to their experiences of perceiving all manifestations of the universe as fundamentally interconnected. This is clearly a unitive experience and is in contrast to feeling *connected* with others, which may or may not include a sense of interconnectedness.

All participants expressed feeling a strong desire to contribute to the wellbeing of others and they are all currently doing this in some way. Jeremy was somewhat an exception in that he expressed that the overall strength of his desire to contribute to others had not changed much, whereas all of the others expressed that this desire had increased significantly. Jeremy said that the reason for this, however, was that, prior to his psychosis, he had a strong sense that it was his duty to save others. Therefore, while the strength of his desire to contribute to others has not changed significantly, what has changed is his desire to contribute in a much more balanced way, remaining much more connected to himself and his own needs now as he supports others, and acting more from a sense of choice and genuine compassion rather than primarily a sense of duty, as had been the case previously. This desire to contribute to the wellbeing of others out of a sense of compassion and sympathetic joy, as reported by all of them, is clearly related to unitive feelings, and again, it suggests that the dialectical tension has diminished significantly for all of them.

One would also expect an integration of the experiences of good and evil as the dialectical tension decreases and more unitive feelings and experiences are available. While each participant described their new understanding of good and evil somewhat differently, every one of them moved towards a personal paradigm that is more inclusive, compassionate, and understanding of all, generally seeing evil as being associated with ignorance and/or woundedness, rather than something that is an innate quality within anyone.

More Flexible Cognitive Constructs

All participants expressed that they are now much more aware of the limits of consensus reality. We might say that their cognitive constructs have become more flexible, which I suspect is closely related to having gone through an experience in which their former cognitive constructs were so profoundly reorganized. I also suspect that having more flexible cognitive constructs may offer more capacity for growth and resilience (more on this in the next chapter). There is likely an important balance in this regard, however, since excessive flexibility may make one susceptible to having further overwhelming transliminal experiences, which may then lead to further psychotic episodes.

A Greater Understanding of Psychosis

It's hardly surprising that these participants emerged from their psychosis with a greater understanding of this process, or at least of their own particular process. It's particularly interesting, however, that we find significant parallels between their different understandings, which suggests that they may very well have developed significant insights about psychosis in general: (a) all participants have emerged with the strong sense that their psychosis was a natural response to finding themselves in an untenable situation (and clearly was not a degenerative brain disease); (b) all participants have come to believe that if antipsychotics are to be used at all, they should be used minimally and in a very judicious manner; and (c) all participants have come to believe that their psychosis has resulted in profound healing and far more lasting benefits than harms. These understandings clearly have important implications for developing beneficial systems of support for people going through psychosis, something we will look at more closely in the Conclusion.

Chapter 31

Lasting Benefits

All participants reported experiencing far more lasting benefit than lasting harm as a result of their psychotic process. Many readers may find this claim surprising and possibly even difficult to believe. Yet, as discussed in the introduction to Part Two, there is significant evidence that such positive transformation is relatively common. In the case of these participants, we find significant evidence of their having experienced such positive growth from two separate lines of inquiry: from their individual narratives that took shape as a result of our multiple interviews (their stories that have been presented throughout the book), and also from the results of a quantitative assessment tool that they all completed.

The Interview Data

Table 31.1 lists the themes for the category of *lasting benefits* that emerged from the participants' interview data. As with most of the other categories, the parallels in their experiences in this regard are very strong, with no significant divergences with regard to any of these major themes. All participants expressed that they now experience a greatly increased overall sense of wellbeing and ability to meet their needs, greater equanimity, greater resilience, and healthier relationships with themselves and with others. Again, as with all of the other categories of experience, The DUI model provides a strikingly cogent explanation for all of these changes.

As discussed in detail earlier, these participants likely experienced psychosis in the first place because they found themselves with a window of tolerance that was insufficient to cope with their circumstances at the time; and this limited

Table 31.1 Converging Themes and Divergences for *Lasting Benefits*

CONVERGING THEMES	DIVERGENCES
Greatly increased wellbeing	All experience this.
Greater equanimity	All experience this.
Greater resilience	All experience this.
Healthier, more rewarding relationships with others	All experience this.
Healthier relationship with oneself	All experience this.

window of tolerance with a correspondingly high dialectical tension would naturally be associated with a poor sense of wellbeing. Since the successful resolution of psychosis, necessarily entails a significant widening of one's window of tolerance and possibly some centering of it, along with the associated reduction of dialectical tension, we would expect to see an overall increased sense of wellbeing, which is what all of the participants reported.

Regarding the increase in equanimity* reported by all participants, this can also be explained by their having a wider window of tolerance. With a wider window of tolerance, by definition, one can tolerate and experience relative peace with a wider range of feelings and experiences.

Regarding the increase in resilience† reported by all participants, I believe that this may also be a result, to some extent, of having a wider window of tolerance. However, I believe that the increased flexibility of their cognitive constructs is perhaps an even stronger contributing factor. After having gone through such a profound reorganization of one's cognitive constructs, it stands to reason that one is likely to have more awareness of the limitations of them (as indeed they all expressed) and that one would not cling to them so firmly. The natural result of this, it seems, would be a greater capacity to honor multiple perspectives and to shift one's own perspective as necessary. To put this in the context of the DUI model, we can say that those who have relatively flexible cognitive constructs can more easily make modifications to their window of tolerance when finding

* *Equanimity* is defined here as the capacity to maintain a balanced state of mind in the midst of challenging circumstances.

† *Resilience* is defined here as the capacity to regain the balance of one's mind after having lost it.

themselves in a situation that places them outside of it, which I believe is the hallmark of resilience. In other words, I suspect that there is a direct correlation between flexible cognitive constructs and the capacity for resilience. Of course, as mentioned above, there is a balance here, since if someone's cognitive constructs are overly flexible, they may be susceptible to overwhelming transliminal experiences and potentially further episodes of psychosis.

Regarding the participants' movement towards healthier relationships with themselves and with others, we can also see this as a direct result of their increased window of tolerance. As discussed earlier, the widening of one's window of tolerance corresponds with growth in the direction of the optimal personality and a greater capacity to incorporate both a healthy connection with the self and a healthy connection with others, making for much more fulfilling and rewarding relationships.

The Posttraumatic Growth Inventory

As an important adjunct to collecting data from the participants during the interviews (which resulted in the stories presented in Part Three), the participants were also required to complete the Posttraumatic Growth Inventory (PTGI), which is a standardized instrument that measures the lasting benefits of a traumatic crisis[1]. The traumatic crisis in this case refers to the psychotic process itself. The PTGI consists of a total of 21 questions that are divided among five factors: *relating to others, new possibilities, personal strength, spiritual change,* and *appreciation of life.*

All participants' scores for the five factors plus their overall mean scores are listed in Table 31.2. Note that all six participants experienced a significant degree of beneficial change (a score of 3 or greater) in all five factors as a direct result of their psychotic process. When the mean scores for all factors are taken into consideration (the bottommost row in Table 31.2), we arrive at a score that gives us a sense of each participant's overall change as a result of their psychosis. When rounding the mean score to one significant digit, we find that two of the participants experienced an overall "moderate degree of beneficial change" as a result of their psychosis, two participants experienced an overall "great degree of beneficial change," and two participants experienced an overall "very great degree of beneficial change." These results match the interview data very well in that all participants reported having experienced much more lasting benefit than harm as a result of having gone through their psychosis.

While these results do match the interview data very well, it's important to note that two of the participants (Theresa and Byron) expressed the challenge

Table 31.2 Comparison of the Participants' PTGI Results

1 = I *did not experience any* beneficial change as a result of my crisis.
2 = I experienced a *very small degree* of beneficial change as a result of my crisis.
3 = I experienced a *small degree* of beneficial change as a result of my crisis.
4 = I experienced a *moderate degree* of beneficial change as a result of my crisis.
5 = I experienced a *great degree* of beneficial change as a result of my crisis.
6 = I experienced a *very great degree* of beneficial change as a result of my crisis.

Factor	Sam	Theresa	Byron	Cheryl	Trent	Jeremy	All Part.s
Relating to Others	4	5	4	5	4	4	5
New Possibilities	5	5	4	4	5	6	5
Personal Strength	6	5	6	6	6	4	5
Spiritual Change	6	4	6	6	6	3	5
Appreciation of Life	6	3	6	6	5	5	5
Mean Score	*6*	*4*	*5*	*6*	*5*	*4*	*5*

Note. Each figure is rounded to one significant digit.

of trying to capture their experiences within such a constrained quantitative measurement. Byron also expressed that he has continued to experience significant growth since the point at which he would be considered fully recovered according to the definitions used in this study, so he found it challenging to draw the line between the growth he had experienced as a result of the psychotic process itself and the growth that he experienced afterward. Still, having accounted for these limitations, the degree of growth that we find for all six participants as a result of their psychotic process is striking, especially when we consider the widely prevalent myth within Western society that very few people return to one's pre-psychotic level of wellbeing and functioning, let alone experience such tremendous growth beyond it.

Chapter 32

Lasting Harms

Regarding lasting harms, one participant (Byron) named two significant lasting harms, four of the participants (Sam, Trent, Cheryl, and Jeremy) could name only one significant lasting harm, and one participant (Theresa) said she could not think of any lasting harm at all. All participants reported that any lasting harms are either relatively minor or are a necessary cost of the lasting benefits.

Cheryl and Sam still occasionally experience some relatively minor anxiety that they feel is in some way related to their psychosis; Trent expressed his challenge with the social stigma of being labeled mentally ill; and Byron and Jeremy both expressed some regrets for the closing of other possible paths in their lives, but they both acknowledged that this was a necessary trade-off for being led to more meaningful paths as a result of their psychosis (see Appendix A for more specifics of each participant's lasting harms).

Table 32.1 Converging Themes and Divergences for *Lasting Harms*

CONVERGING THEMES	DIVERGENCES
All except Theresa expressed some lasting harm, although they all expressed experiencing much more benefit than harm overall.	Each participant expressed a significantly different harm.

Conclusion

Towards a New Paradigm

After having taken this long journey down the "rabbit hole" and through the challenging topics of fundamental human experience and the phenomena of anomalous experiences, we find ourselves at a unique vantage point where we can take a step back from it all and take in the broader picture. So the time has come to consider the very important questions of how this understanding can guide us in offering real support to those struggling with distressing anomalous experiences, and just as importantly, how it can offer guidance to all of us as we each struggle to find peace in our own lives and in the world in general.

Not Mistaking the Map for the Territory

Before looking at the implications of the model and research so far presented, it's important to first acknowledge some of the limitations we're dealing with in doing so. Perhaps the largest such limitation is the constraints that automatically come into play anytime we try to squeeze human experience into a theoretical model or framework. In other words, the map is simply not the same as the territory, and we have to be careful not to confuse the two. Similar to the way that no geographical map could ever come close to capturing the entirety of the territory that it represents, I don't believe that any theoretical model, regardless of how well conceived or supported it may be, can ever come close to capturing the enormous depth, complexity, and ultimate mystery of human experience.

Looking at this limitation as it applies to the model presented here, then, even though I have gone to great lengths to try to shape this model so that it fits all of the relevant research as well as the wide variety of experiences reported by those suffering from psychosis, I doubt that it will ever be possible to entirely capture the sheer "messiness" of human experience within a perfectly orderly and coherent model. As long as we keep this limitation in mind, however, there is no doubt that a sound theoretical framework can offer an invaluable source of guidance. Just as a good geographical map can be indispensable in aiding a traveler across difficult terrain, a good model of psychosis has the potential to guide someone lost in the chaotic seas of psychosis move in the direction of successful resolution.

Holding the Terminology Lightly

A second important limitation that is important to acknowledge when trying to make the leap from a theoretical model to supporting actual human beings is the limitation that is inherent in the use of language. The simple fact is that words are very slippery. They're prone to misinterpretation, distortion, and personal bias, and this is especially true when we talk about subjective human experience, which of course includes anomalous experiences. Therefore, it's of utmost importance that we make every effort not to hold the concepts that we use to talk about these kinds of experiences or any aspect of subjective human experience too tightly. Words such as "schizophrenia," "psychosis," "madness," and "recovery" can all too easily take on lives of their own, developing connotations that are often very far from the actual experiences that these words are attempting to point to.

A perfect example of this is the concept of "schizophrenia," which, as we have seen, has so far proven to be little more than a mirage. The term "psychosis" has become similarly problematic. In our exploration here, we have seen that experiences often labeled "psychotic" can be seen as manifesting from a very profound organismic intelligence devoted to the survival and/or growth of the organism/self; and yet the implications that most commonly arise during the mainstream use of the words "psychotic" and "psychosis" are nearly diametrically opposed to this understanding. Currently, the term "madness" has surprisingly become one of the most acceptable terms to those who have been challenged by anomalous experiences, perhaps because it is the term least associated with the medical model, but it too clearly has its problems.

So, how do we choose the best terms to use when discussing anomalous experiences? Some terms, such as "schizophrenia," appear to be so corrupted that perhaps it is best to discard them altogether (in fact, there is currently a major push to do exactly this within a number of Western nations). However, it seems that some kind of compromise needs to be made between the need to steer clear of problematic terms and the need to maximize clear communication, and so we often find little choice but to begin with the terms with which everyone is familiar. In short, then, I think there are no simple answers for how best to choose the terms we use, but we can certainly make the effort to hold all such terms very lightly, allowing their meanings to flow and adapt as our understanding of a particular phenomenon evolves, and to be willing to make the transition to altogether different terms when this seems most appropriate.

The Metamorphosis of Madness

It is clear that all six of the participants presented in this book have been on incredible journeys to the very depths of their beings and back, having integrated what they experienced and finally rejoining the rest of us within consensus reality. They have all experienced to a greater or lesser degree the extremes of human suffering and of human joy; they have all spent time mired in utter chaos and confusion and have somehow emerged with a renewed sense of equilibrium and lucidity. What is perhaps even more impressive is that they have all experienced profound healing from their journeys, having emerged with greater equanimity and resilience, a richer feeling realm that includes less negativities and more unitive feelings, more rewarding and enjoyable relationships with themselves and with others, and a greater overall sense of wellbeing. What we find in the stories of these participants is further validity to the idea that psychosis is a natural process of the psyche—there is no doubt that it is a radical and very risky process that has the potential to greatly exacerbate one's suffering, but there is also no doubt that it offers the potential to result in profound healing at the deepest levels of one's being when successfully resolved.

When we reflect upon the profound and ultimately beneficial transformations that took place within the most fundamental structures of these participants' beings, we find remarkable parallels with the process of metamorphosis that takes place within the development of butterflies. In order for a larva to transform into a butterfly, it must first disintegrate at a very profound level, its entire physical structure becoming little more than amorphous fluid, before it can reintegrate into the fully developed and much more resourced butterfly. In a similar way, when someone enters a state of psychosis, we can say that their very self, right down to the most fundamental levels of their being, undergoes a process of profound disintegration; and with the proper conditions and support, there is every possibility of their continuing on to profound reintegration and eventual reemergence as a renewed self in a significantly changed and more resourced state than that which existed prior to the psychosis.

IMPLICATIONS FOR SUPPORTING THOSE STRUGGLING WITH PSYCHOSIS

The importance of supporting the psychotic process. When we consider the metaphor of metamorphosis for the process of psychosis, and bring in the findings of the recovery research, we arrive at a particularly important implication for how best to support people going through psychosis. Just as a larva requires an environment free from predators and the extra protection and sustenance provided by a cocoon in order to go through the extremely vulnerable process of metamorphosis, someone experiencing psychosis requires a similarly dependable sense of protection and sustenance. The research we have studied demonstrates quite clearly that those most likely to make a full recovery are those whose psychotic process is allowed to carry through to a natural resolution with minimal interference.

We see this firsthand from the reports of the very high recovery rates experienced at residential facilities such as Diabasis house and the Soteria houses. In such facilities, an environment of maximal freedom contained within a structure of maximal safety is maintained in several ways: the residents are allowed the freedom to follow their experiences and maintain full choice regarding the use of psychiatric drugs while firm limitations are placed on activities that may cause harm to themselves, others, or property; they receive dependable support in the form of having their basic needs met—healthy food, water, shelter, clothing, and relative comfort; and they receive continuous nourishment in the form of 24-hour care by staff who are trained to hold them within an atmosphere of empathy, unconditional positive regard, and authenticity. In other words, we can say that these kinds of residential facilities attempt to create a safe and supportive cocoon that allows the metamorphosis of the psychotic process to resolve with minimal hindrance.

We can also see this same principle at work within the societies that have shown a particularly high natural rate of recovery. These societies—such as are found in India, Nigeria, and Colombia—while very poor materially, tend to hold the values of family and community very highly, rarely abandoning a family member regardless of their degree of disability, and generally holding the assumption that family members going through psychosis will eventually recover. In addition, coercive psychiatry and the use of psychiatric drugs are rare within these societies. As a result, individuals experiencing psychosis within these societies often find a "cocoon" of support, security, and nourishment naturally established within their very own communities without the need to resort to special residential facilities; and as would be expected, a high percentage of these individuals go on to make full recoveries.

We can see a similar "cocoon" being spun within the very successful Open Dialogue Approach, which was developed in Lapland, Finland, and is beginning to spread to other Western countries. In Lapland, they do not naturally have quite as high a degree of community/family support as that found in many of the so-called developing countries, so the mental health care system has come up with an effective strategy for building this kind of support within the families and communities that surround individuals suffering from psychosis. While the details of the Open Dialogue Approach are too complex to go into here, the essence of it is simply healing and strengthening the social web surrounding the individual by facilitating and encouraging open, authentic, and intimate communication and connection between the various members of this web. Also, as with the other methods mentioned above, the individuals receiving this kind of support are allowed to maintain maximal freedom and agency, with psychiatric drugs used very judiciously and only with full consent if they are used at all.

Another therapeutic system worthy of mention here is *Windhorse therapy*, a system of treatment developed in Boulder, Colorado in the early 1980s and inspired by the teachings of Tibetan Buddhist master Chogyam Trungpa Rinpoche. Similar to the other approaches mentioned above, the general philosophy of this approach is to trust and support the profound wisdom and powerful movement towards health and wholeness that exists within all organisms. This innate wisdom is referred to as *basic sanity* and this innate movement towards health is referred to as *windhorse energy*. The essence of this approach is similar to those mentioned above—by placing the primary emphasis on creating a healthy, harmonious, and nurturing environment for the individual in distress, there is trust that movement towards recovery will naturally occur. There is yet to be formal research on the recovery outcomes of this approach, but there are numerous accounts of clients of this approach who have experienced profound recovery[1].

One thing we find in common with these different methods of support is that they all have the capacity to provide *all* of the factors of support for recovery listed in Table 29.1. By not subscribing to the brain disease model and instead expecting that these individuals will recover and eventually move on to rich and meaningful lives, the factors of *hope*, *meaning*, and *the development of a hopeful understanding of their psychosis* are supported. By not losing sight of the humanity of these individuals and maximizing their freedom and sense of agency, they are supported in *connecting with their aliveness*. In being surrounded by an empathic, caring, supportive community, they are supported in *cultivating healthy relationships and distancing from and/or healing unhealthy relationships*.

When there is simply not the availability of a highly supportive "cocoon" such as what is offered within the systems mentioned above, traditional psychotherapy can play an important role in creating a significant degree of nourishment and safety, and in supporting individuals in developing other important resources. The factors of recovery mentioned above suggest that the most helpful types of psychotherapy are likely to be those methods that support the individual in: (1) creating a coherent understanding of their psychotic process, particularly one that is more hopeful than the brain disease model; (2) connecting with their feelings, needs, and sense of agency (i.e., their aliveness); (3) cultivating healthy relationships and/or healing/distancing from unhealthy ones; and (4) developing methods of coping with the distressing anomalous experiences themselves. There exists a wide array of psychotherapeutic modalities and theoretical orientations, but the research suggests that those modalities likely to be particularly beneficial to individuals undergoing this kind of process are: existential/humanistic; relational/attachment-based/family systems oriented; somatic (mind/body) and trauma focused; mindfulness based; psychodynamic/depth oriented; and cognitive behavioral*. Fortunately, research on the efficacy of these kinds of approaches has become increasingly common, and the results so far have been quite promising[2].

And last but certainly not least is *peer support*. The term *peer support* simply means receiving support directly from others who have "been there." It can be used either as an adjunct to any of the above methods, or even stand entirely alone as the primary source of support in areas with a strong peer support network. Many of the harms caused by mainstream treatment can be avoided when peers are involved—peers are generally much more understanding and validating, are less likely to push the brain disease model and forced "compliance" with the use of drugs, and of course they have access to the wisdom they have personally gained from their own recovery process. The peer support movement is currently growing by leaps and bounds, bringing with it a strong emphasis on the importance of human rights for all and a genuine democratic process within the mental health care system. It also offers a number of excellent viable alternatives to the mainstream paradigm of care. Some of the largest components of this movement are peer-run crisis homes, 24-hour a day crisis hotlines, support groups and classes (such as those offered within the *Hearing Voices* movement), and overarching peer-run organizations that are not influenced by the pharmaceutical industry†

* It's important to point out that many, and perhaps most, psychotherapists in the West have themselves been heavily inculcated into the degenerative brain disease model of psychosis. Having bought into this belief system themselves, there's a high likelihood that they will try to push it onto their clients (even though they may have the most benevolent of intentions as they do so). When this occurs, even otherwise highly skilled psychotherapists may unwittingly cause more harm than benefit in their attempts to offer support.

† Be cautious of groups who claim to be "peer support" or "grassroots" but are actually covert arms of the pharmaceutical industry. The largest such group is (continued on the next page)

and act as hubs for these other groups (see the *Resources* section in the back of the book for more information on these groups).

Mainstream mental health care interfering with the process. When we turn our attention to look closely at the primary method of support for those suffering from psychosis within the Western mental health care system today—the mainstream psychiatric system—we see that it stands in stark contrast to the methods mentioned above. Whereas all of the above methods can be seen as simply various methods of providing a safe and nurturing cocoon that allows a person the possibility of moving through their psychotic process with support and minimal interference, the psychiatric system can be seen as making every effort to prevent such a cocoon from ever being built, instead trying to stop the psychotic process dead in its tracks.

We cannot say that this is necessarily out of any kind of malicious intention—certainly there are many people working within the mainstream psychiatric system who have tremendous care and compassion for those that they care for. Rather, as was discussed in Part One, the mainstream psychiatric system operates under a radically different paradigm—seeing psychosis as the manifestation of a diseased brain—and therefore operates under the belief that the most compassionate thing to do is to make every effort to minimize the symptoms of the psychosis with the hope of averting any further damage and/or suffering that this "brain disease" might otherwise cause (which is understandable given this paradigm). As the recovery research continues to accumulate, however, we see ever increasing evidence that this paradigm is profoundly misguided and that the treatment model arising from it is likely causing much more harm than benefit, as we have been discussing throughout this book.

Returning, then, to the metaphor of metamorphosis and the importance of providing a safe and nurturing cocoon that allows the psychotic process to resolve unhindered, we can see clearly that the mainstream psychiatric treatment model interferes with this process profoundly. In this system, as we find in the stories of the participants of this study and within so many other similar accounts,

---

(continued from the previous page) the National Alliance on Mental Illness (NAMI), which claims to be "nation's largest grassroots mental health organization." After decades of refusing to disclose the names of its contributors, a recent US Senate probe revealed that NAMI had been receiving well over half of its funding directly from pharmaceutical corporations, and so is clearly not "grassroots." While NAMI (and other similar groups) may offer some useful resources, they are heavily steeped in the brain disease model and have serious obligations to their largest contributors. Therefore, any involvement with them is likely to result in being pressured to subscribe to the paradigm of care that is most profitable to the pharmaceutical industry (i.e., the medical model).

people suffering from psychosis are often institutionalized against their will in very unpleasant environments. Again, while the staff of such facilities often includes very well-intentioned people, the reality is that they are very often heavily overworked and undertrained, and so the task becomes more about "managing" the patients rather than creating a particularly warm and nurturing environment. Also, being trained primarily with the medical model understanding of "mental illness," it is all too easy for the staff to interpret the unusual behavior of the patients as being merely the manifestation of a diseased brain and to lose sight of the human being suffering underneath. This all too often results in the staff treating the patients in a way that is easily perceived by the patients as cold, dehumanizing, and even downright hostile. Adding to the often profound sense of confusion and insecurity created by such treatment, the patients' free will and sense of agency are generally stripped away from them, making it virtually impossible for them to feel any sense at all of genuine safety and comfort.

Furthermore, as these patients are told that the unusual experiences they are having are caused by a lifelong degenerative brain disease, it is very likely that they will develop profound intrapsychic conflicts (in addition to any conflicts already existing within the psychosis itself) as they lose faith in the innate wisdom of their own psyches and struggle to fight against their very own healing process. They now find themselves in the terrifying predicament of finding no sense of security either *outside or inside*. In what is yet further interference to the natural healing process of psychosis, these patients are typically forced to take heavily tranquilizing drugs or even undergo electroconvulsive shock therapy, severely impairing their most important resources—hope, meaning, and connection with their aliveness.

How could we ever expect anyone to establish a secure cocoon and move towards successful transformation under such debilitating conditions? Yet, incredibly, many people still do, as we have seen with the participants here. I believe that the fact that such genuine recovery and transformation continues to take place in spite of these incredible odds is a testament to the power of the organismic wisdom within our beings—that innate wisdom within all organisms that relentlessly pushes for survival, healing, and growth. Just as the vulnerable earthbound larva contains within its being the profound wisdom to transform itself into a beautiful, mature butterfly with the capacity to fly thousands of miles in some cases, so we see evidence that a profoundly wounded individual has within her or his being the wisdom to transform into a much healthier, more mature individual with the capacity to live a rich and meaningful life and contribute greatly to society.

Where Do We Go From Here?

So, when looking at the recovery research that has accumulated over the past century, we find that there are two messages that come across quite clear: (1) full recovery from long-term psychosis is not only possible, but can be the most common outcome given the right conditions; and (2) our mainstream mental health care system is seriously failing to create the conditions that maximize this possibility. We have explored some of the reasons this system remains so broken and seriously misguided, and it is essential that we continue this exploration until we can make the society-wide paradigm shift necessary to move towards a system that is much more beneficial.

Fortunately, as we have seen with the alternative methods mentioned above, we already have some excellent foundations off which to build in transforming our system in this way. But in order to move more seriously in this direction, we still have before us the hard work of pulling the deep-seated myths of hopelessness out by their roots, a task that seems especially daunting when we consider that we are up against enormously powerful players who rake in obscene profits from the current system—many members of the psychiatric-pharmaceutical complex, in particular.

In spite of this daunting task, the good news is that a grass roots movement dedicated to utilizing the very hopeful findings of the recovery research and exposing the corruption within the psychiatric-pharmaceutical complex is gaining considerable momentum (see the *Resources* section in the back of this book for more information). Hopefully, it is only a matter of time before enough dust is wiped from our collective eyes and a tipping point is reached that will break the stranglehold of the psychiatric medical model, and we can make a society-wide shift towards a system of support that is much more in line with the research, much more beneficial to those struggling with psychosis, and much more beneficial to society as a whole.

Implications for Future Research

As mentioned above, what we find within the research presented here, as well as within a number of other similar studies and first person accounts, is still further evidence for the possibility of full recovery and further evidence that our mental health care system is seriously failing to support this possibility. Yet we also find substantial hope in the possibility of transforming our mental health care system to one that is much more beneficial in this regard. There is no doubt that ongoing research can play a very important role in this transformation, but simply continuing along the same research track that we've been pursuing for the past century is clearly not the answer.

The so called developed countries in the West have conducted much more research on psychosis and schizophrenia in the past century than the so called developing countries, and yet, as discussed earlier, there is overwhelming evidence that members of these developed countries (the United States, Europe, etc.) have much worse rates of recovery than members of the poorest countries of the world, suggesting that our mainstream method for treating psychosis is generally worse than no treatment at all. Since our mainstream treatment methods are presumably guided by our research, it's clear that our general research focus has been seriously missing the mark. With this in mind, then, we can see that we do not necessarily need to do more research, but rather, we need to radically change the kind of research that we do.

The vast majority of psychosis research in the West has paid almost no attention to the subjective experiences of the participants, instead operating under the misguided notion that psychosis is the manifestation of a diseased brain and therefore that it is much more important to seek out biological correlates than to pay attention to the "mad ramblings" of psychotic individuals. But, after many thousands of such research studies and many billions of dollars spent, we still have no clear evidence of a biological cause of schizophrenia and the other psychotic disorders, and our recovery rates remain extremely poor, and are possibly even declining. Isn't it about time that we acknowledge that we've been barking up the wrong tree and turn our attention to the actual lived experiences of the individuals who have had direct personal experience with psychosis?

Returning to look at the research study presented here, we find six individuals who have undergone profound journeys through psychosis and recovery, and it is clear that each one of them has much to offer us in the form of lessons and guidance in shaping a much more supportive mental health care system; and as one solo and relatively novice researcher, there is no doubt that my work here has only touched the surface of the wisdom that these participants and others like them have to offer. Just imagine what can be done if even a small percentage of the researchers currently looking for psychosis in the brain were to shift their focus to the countless individuals who have recovered from psychosis or are currently in the process of recovery and inquire into their actual lived experiences.

There is no doubt that the potential for a profoundly beneficial transformation of our mental health care system exists, and in fact we have all of the necessary components to do so right here and now. The wisdom is here, the means are here, and as researchers, we have the potential to contribute greatly to this transformation by providing the skills necessary to bring this wisdom and these means into conscious awareness; but in order to do so, we must learn to treat the mental health care consumers as partners and even as teachers. With this in mind, I hope that my work here inspires many other researchers to continue the journey down this very important path.

Madness and Beyond . . . Appreciating the Benefits for Society

When we contemplate the current conditions in our society and in the world, there is no doubt that we find ourselves at an extremely crucial juncture in the trajectory of the human species. And as difficult as it might be for some to believe, the research strongly suggests that those who have experienced and are experiencing so called psychosis may find themselves in a mutually beneficial relationship with their societies: On one hand, it's clear that many of these people need significant support, sometimes much more support than the average person; but on the other hand, it's also clear that these individuals have the potential to attain profound insights into the human condition, perhaps the very insights that our species so desperately needs in order to survive.

The key to understanding this is in the ever increasing evidence that the person we think of as "psychotic" is simply entangled in a profound wrestling match with the very same core existential dilemmas with which we all must struggle. One implication of this is that the boundary between madness and sanity is surprisingly thin, an idea that is likely to be deeply unsettling to some. There is, however, another implication that offers us some very real hope, not only in our pursuit to offer genuine support to those who are the most caught up within these struggles, but also in our pursuit to find real peace on all levels—individually, socially, and globally. It appears that those who have these kinds of experiences often find themselves dipping beneath the layers of their cognitive constructs and catching glimpses of the more fundamental qualities of the world and dilemmas that shape all of human experience.

While some may consider this idea to be a "romanticization of psychosis," this actually couldn't be further from the truth—many of these people become utterly lost and confused for significant portions of their lives as they essentially drown in these deeper waters. In fact, it's all too clear just why it is that the typically "healthy" psyche is so effective at preventing one from falling into these chaotic seas. But the reality is that many people do fall in, and thankfully, many people do eventually learn how to swim and find their way back to the "shores of consensus reality" (to use the participant Byron's expression). And as we find in the stories of those who have been able to successfully integrate these experiences, as presented in this book and elsewhere, we discover that one real gift that often emerges from this journey is the ability to share some important truths with the rest of us, truths that may very well be exactly what our species needs to hear if we are to make it through these trying times:

With the recognition that the suffering with which each of us struggles is fundamentally universal, we are likely to find it a little easier to develop equanimity and self compassion for our own difficulties, and also more tolerance and compassion for others. With the recognition that we each understand and experience the world through our own individually constructed lenses, we are likely to find it easier to hold our own perspectives more lightly while being more open to the different perspectives held by other individuals and other societies. And by appreciating the profoundly impermanent and interconnected sea of life to which we all belong, we are likely to find it easier to act from a place of love and compassion for all of our fellow living beings, great and small.

APPENDIX A

Details for the Categories of the Psychotic Process

In performing cross case analysis of the interview data from these participants, I arrived at a set of themes for each category of experience—*description of the anomalous experiences*, *onset and deepening of psychosis*, *recovery*, *lasting personal paradigm shifts*, *lasting benefits*, and *lasting harms*. Table A.1 lists the converging themes for each category along with their associated divergences. Following is a detailed outline of all of these themes, including the most relevant data pertaining to each theme for each participant:

THE ONSET AND DEEPENING OF PSYCHOSIS

A physical and/or existential threat to the self just prior to onset. All participants experienced an existential threat to their self, and Sam and Byron also experienced the threat of actual physical death. The existential threat resulting either from an overwhelming sense of connection with others or from an overwhelming experience of isolation.

- Sam had received a draft notice for the Vietnam War, which represented the threat of physical death as well as "death" from his sense of identity as an anti-war activist.
- Theresa was overwhelmed by such a strong sense of connection at the kibbutz that she felt as though she were losing a sense of herself.
- Byron was actively seeking a death/rebirth process just prior to onset, and then just after experiencing some profoundly liberating anomalous experiences, he fell three stories onto his head, barely surviving, and awoke in the hospital to more distinctly psychotic experiences.

Table A.1 Converging Themes and their Associated Divergences for the Six Categories of Experience

	Converging Themes	Divergences
Onset and Deepening of Psychosis	(1) An actual or existential threat to the self just prior to onset (2) Childhood isolation (3) The significant use of recreational drugs prior to onset (4) A swing between extreme isolation and extreme connection just prior to onset (5) A profound shift in one's personal paradigm just prior to onset	(1) All experienced this. (2) All participants had a significant amount of isolation in their childhood, but to varying degrees. (3) All but Cheryl had significant experiences with recreational drugs prior to onset. (4) All but Sam and Trent had this kind of swing just prior to onset. (5) All but Trent experienced profound shifts in this regard; Trent, however, did increase his marijuana use significantly just prior to onset, which may be closely related.
Description of the Anomalous Experiences	(1) Polarized experiences of good and evil (2) Creative and destructive forces (3) Fluctuating between omnipotence and powerlessness (4) Heroic striving (fighting evil and/or ignorance) (5) Being watched over by malevolent and/or benevolent entities (6) Groundlessness (7) Parallel dimensions (8) Feelings of euphoria, liberation, and/or interconnectedness	(1) All experienced this. (2) All experienced this. (3) All experienced this. (4) Trent and Cheryl experienced striving against evil forces within themselves; and all except for Cheryl experienced striving against evil and/or suffering "out in the world." (5) All experienced this. Jeremy experienced being watched over by primarily malevolent entities, though sensed a powerful presence of both types. (6) All except Sam mentioned experiencing profound groundlessness. (7) All but Trent experienced different realms of experience occurring simultaneously to some degree. (8) All but Sam recalled having these kinds of experiences, to a significantly greater or lesser degree. Jeremy had these just prior to his psychosis, but not so much after onset.
	(1) Finding meaning in life (2) Connecting with one's aliveness	(1) All expressed the importance of this in their recovery. (2) All expressed the importance of fostering a deep connection with their aliveness—particularly with their feelings, needs, and sense of agency.

Recovery	(3) Finding hope (4) Arriving at a more hopeful understanding of their psychosis (5) Healthy vs. unhealthy relationships (6) Harm from the psychiatric system hindering recovery	(3) All expressed the importance of hope in their recovery. (4) All expressed having arrived at an understanding of psychosis that is more hopeful than the brain disease model, and all but Sam expressed that this was important in their recovery. (5) All expressed the importance of cultivating healthy relationships. Trent and Jeremy also expressed the importance of distancing themselves from unhealthy relationships. (6) All experienced this.
Lasting Personal Paradigm Shifts	(1) A significantly changed spectrum of feelings with more depth and unitive feelings (2) An increased experience of interconnectedness (3) A strong desire to contribute to the wellbeing of others (4) An integration of good and evil (5) Appreciating the limits of consensus reality (6) A greater understanding of psychosis	(1) All experience this. (2) All experience this. (3) All experience this. (4) All experience this. (5) All experience this. (6) All experience this.
Lasting Benefits	(1) Greatly increased wellbeing (2) Greater equanimity (3) Greater resilience (4) Healthier relationship with oneself (5) Healthier, more rewarding relationships with others	(1) All experienced this. (2) All experienced this. (3) All experienced this. (4) All experienced this. (5) All experienced this.
Lasting Harms	(1) All except Theresa expressed some harm, though they all expressed experiencing much more benefit than harm overall.	(1) Each participant expressed a significantly different harm.

- Cheryl experienced profound and overwhelming isolation and despair just prior to onset.
- Trent's dysfunctional family system left him feeling paradoxically both profoundly isolated and also profoundly enmeshed with the others with whom he was living, ultimately resulting in the sense of "imploding into a void where a family was supposed to be."
- Jeremy experienced a profound and overwhelming shaming and associated isolation immediately prior to onset.

Childhood isolation. All participants expressed having experienced significant isolation in their childhoods, an experience that several of them expressed was associated with feelings of profound shame.

- Sam felt significantly disconnected from his parents, especially in adolescence.
- Theresa became very withdrawn and isolated at a very young age, and she expressed having layers of isolation stacked on top of each other culminating with the death of both of her parents in late adolescence.
- Byron's relationship with his idiosyncratic mother made it difficult for him to fit in elsewhere, and he remained very isolated at school and with peers throughout the majority of his adolescence.
- Cheryl experienced emotional abuse and neglect from "immature" parents. As a result of this and perhaps other factors, she experienced significant shame, self hatred, and depression from a very early age.
- Trent describes his family as "highly dysfunctional," having a father who was chronically depressed and was often verbally and emotionally abusive, and a mother who had been diagnosed with paranoid schizophrenia.
- Jeremy was raised by an emotionally neglectful father and a verbally abusive mother who struggled with psychosis at times. He also experienced severe shaming and bullying in his adolescence.

The significant use of recreational drugs prior to onset. All participants except Cheryl expressed having consumed a significant amount of alcohol and/or other recreational drugs prior to the onset of their psychosis.

- Sam had used marijuana and alcohol relatively frequently for the three years leading up to the onset of his psychosis (from age 16 to age 19). He also occasionally experimented with LSD during this time, though he did not use it at all in the six months prior to onset.

- Theresa consumed a significant amount of alcohol and hashish at the kibbutz immediately prior to the onset of her psychosis—"I practically just wasn't sober."
- Byron had been using LSD and other hallucinogens for some time prior to onset. He had a particularly profound experience while under the influence of LSD at the well known Woodstock music festival just six weeks prior to onset.
- Trent believes he was using marijuana "too often" prior to onset. Immediately prior to onset, a number of stressful incidents occurred in his life which led to him increasing his marijuana intake significantly as a means to cope with these stressors.
- Jeremy had not been consuming a significant amount of drugs or alcohol prior to his onset; however, he was quite high on marijuana at the time of onset (during the shaming incident at the party that led to his experience of annihilation anxiety).

A swing between extreme isolation and extreme connection just prior to onset. A particularly interesting pattern shared by four of the participants is having experienced a swing from an extreme experience of isolation to an extreme experience of connection just prior to onset.

- Sam and Trent did not express having experienced such a swing prior to their psychosis.
- After living a very isolated childhood and adolescence, which was greatly exacerbated by the death of both of her parents, Theresa quite suddenly became overwhelmed with an extreme experience of connection and community at the kibbutz.
- After an isolated childhood, Byron experienced profound connection—both with other people and also a more universal connection—at Woodstock via the aid of hallucinogens and the intense level of community in that environment.
- After experiencing a lot of shame and isolation in her childhood, Cheryl experienced a series of highly shaming, isolating incidents that further exacerbated her sense of isolation. This was then followed by a sense of deep

connection and support from her apparent contact with benevolent and supportive "spirit guides" as a result of her spiritual counseling class.

- After a childhood filled with much shame and isolation, Jeremy experienced profound connection with himself and others after his insights into the inner war and the war between others. This continued until the night at the party when he had a profoundly shaming and isolating experience, leading very quickly to psychosis.

A profound shift in one's personal paradigm just prior to onset. All participants except for Trent expressed having experienced a profound shift in their personal paradigm just prior to the onset of their psychosis.

- Sam received his draft notice for the Vietnam War just prior to onset. Considering the intensity of the war at that time and his involvement with anti-war activism, he found it impossible to integrate this news.
- The experience at the kibbutz profoundly changed Theresa's experience of herself and the world, and she found herself overwhelmed with the task of trying to integrate the possibility of such connection into her personal paradigm.
- After a very isolated youth, Byron was at first somewhat able to integrate the degree of connection he experienced with others as he came to identify as "a child of the 60's." However, he found it nearly impossible to integrate the profound experience of universal interconnectedness that he experienced while at Woodstock.
- The spirit guides that contacted Cheryl during her spiritual counseling class were at first a pleasant anomalous experience (not causing significant distress and/or limiting her functioning) and are therefore not defined as psychotic according to the definitions of this study. However, this experience entailed a profound shift in her personal paradigm that she was unable to integrate, and it was followed by a relatively rapid descent into psychosis soon afterwards.
- It doesn't appear that Trent experienced this as much as the others; however, a series of particularly stressful incidents occurred immediately prior to onset, including his having to take care of his ex-girlfriend and her 2-year-old son while still living with his very troubled family, which led to a significant increase in his use of marijuana as a means to cope with this, which in turn can be seen to be associated with a personal paradigm shift.
- Jeremy went through a series of personal paradigm shifts prior to onset:

1. Coming to realize that the human race is not guided by wise elders.
2. Coming to perceive that interpersonal relationships are plagued by self-esteem battles.
3. Coming to realize he had been struggling with an intrapsychic war and that such awareness provided significant relief from it.

Jeremy found it somewhat challenging to integrate each of these into his personal paradigm to a greater or lesser degree, the sum total of which apparently made him vulnerable to the psychosis that eventually occurred.

Description of the Anomalous Experiences

Polarized experiences of good and evil. All participants experienced a polarization of good and evil within their anomalous experiences, though the details vary significantly. The use of the word "good" here is best defined as "consisting of predominantly creative and/or beneficial actions or forces," and "evil" being best defined as "consisting of predominantly destructive and/or harmful actions or forces."

- Sam experienced the good and evil of war—the good of stopping evil forces in the world but the evil of causing destruction and harm in the process, as well as the evil of being harmed. He also experienced the good and evil of work conducted by the "initials agencies" (such as the FBI and CIA)—the good of catching evil criminals, but the evil of being hunted by these agencies.
- Theresa had profound experiences of Heaven and Hell during her first episode, and experiences of being watched over by both good/benevolent and evil/malevolent entities.
- Byron had numerous visions of the battle between good and evil forces, including being guided by the benevolent dakinis into the depths of the underworld, and going from the heavens to the hells to rescue prisoners and hospital patients from the psychiatric hospital. He also played an important role in bringing about the transition of the ages—from the Stone Age (the age of ignorance, and therefore evil in many contexts) to the Age of Aquarius (the age of wisdom, and therefore good in many contexts).
- Cheryl experienced the good of benevolent and supportive spirit guides which then became profoundly evil and malevolent, followed eventually by the lasting sense that perhaps there is some benevolent and protective force in her life. She also had profound experiences of a God who despised the evil

within her, threatening to send the entire world to Hell because of her utter evilness.

- Trent experienced messages of God offering him guidance, but was also persecuted by the Devil, the "satanic" energies of whom completely "immersed" him at one point.
- Jeremy had an elaborate nonconsensus belief system comprised of the battle between the good forces of the "white order" and the evil forces of the "black order."

Creative and destructive forces. All participants experienced both powerful creative and destructive forces, which were often closely related to good and evil.

- Sam experienced the creative role of actively directing major movies, but he sometimes took on destructive roles in these scenarios (such as being the demolitions expert at one point). As an "initials" agent, he took on the very creative role of directing a complex assault operation in Iraq and in capturing D. B. Cooper, yet these roles often entailed severe destruction, such as bringing bombs down on the people of Iraq.
- Theresa experienced profound creativity associated with the process of channeling a child and giving birth, and she spent significant time experiencing the destructive forces of Hell and fire.
- Byron had numerous visions of the death/rebirth process of his self and of the world, which entailed qualities of both profound creativity and profound destruction.
- Cheryl experienced a period of profound creative intelligence and insights in which she felt she was on the verge of formulating a paradigm-shifting theory related to the illusion of time; and she also spent much time filled with the destructive energy of hatred, both directed towards herself and experienced as being directed from others (e.g., God, family members, her future husband).
- Trent experienced periods of profound insights and creative intelligence and imagination, but he was also inundated with very violent and destructive "satanic thoughts" that resulted in him having a severe bout of "dry-heaving blood."
- Jeremy experienced profound creative intelligence when formulating a theory regarding the destructive forces in the world. His insights took him

from realizing the destructive "war" within himself to seeing this enacted interpersonally in others and finally to seeing this war on a grand scale directed by the forces of the black order.

Fluctuation between omnipotence and powerlessness. All participants expressed having fluctuations to a greater or lesser degree between feelings of omnipotence and feelings of powerlessness. (Ironically, they all express having felt extreme powerlessness as a direct result of their psychiatric treatment, though this was of course taking place within consensus reality.)

- Sam experienced the tremendous power of being a leading "initials" agent, a famous movie star, and movie director. He also experienced the powerlessness of being caught in a war and being hunted by the "initials agencies."
- Theresa went from the omnipotent sense that she could save others from suffering merely by overcoming certain challenges to feeling too powerless to even save herself while experiencing Hell in the psychiatric hospital. She also experienced significant powerlessness along with feelings of persecution during her second episode of psychosis.
- Byron had experiences of carrying the entire weight of the world on his shoulders at times (both literally and figuratively), and at other times, feeling at the complete mercy of malevolent doctors.
- Cheryl felt that she was on the verge of a paradigm-shifting truth that could alter the course of mankind to feeling at the mercy of her spirit guides and a God seething with hatred for her.
- Trent experienced numerous fluctuations between feeling "godlike" on one hand and profound despair and/or terrifying persecution by the Devil on the other.
- Jeremy went from the profound terror of utter self annihilation to a sense of omniscience and enlightenment, then to paranoia and persecution.

Heroic striving (fighting evil). All participants experienced a kind of heroic striving, especially in the sense of a mighty striving against evil. Trent and Cheryl experienced a somewhat different form of heroic striving than the others—whereas the others primarily experienced a striving against evil out in the world, Trent and Cheryl experienced a striving against the evil within themselves (and in Cheryl's case, this was *how* she would save the world). Cheryl experienced primarily this kind of internally-directed striving, whereas Trent experienced a significant amount of both internally-directed and externally-directed striving.

- Sam experienced the striving of having to do "certain things" to avoid global catastrophe. He also strove very heroically in his work for the "initials agencies" while pursuing criminals and also while directing strategic military operations against the evil forces within Iraq.
- Theresa found herself striving very heroically in overcoming various challenges as the means to prevent the suffering of others.
- Byron found himself performing rituals for world peace (e.g., "turning poison into nectar") during a number of his visionary periods. He felt at one point in the visionary realm that he had literally taken the world on his shoulders (giving it to Heracles and taking it back again). At other times, he strove mightily to help transition the world into the Age of Aquarius, and he made daring rescue attempts to rescue all beings from Hell (which, in consensus reality, involved attempts to rescue the hospital patients from the psychiatric hospital).
- Cheryl strove to save the entire world by fighting the evil that was within herself. She also went through a period during which she was conducting various rescue missions directed by her spirit guides, including attempting to rescue a juvenile delinquent from the rescue facility at which she worked.
- Trent had experiences in which he believed that "the world would be free from its sufferings . . . because of something that [he] participated in." He also experienced intense striving against the evil and "satanic" forces within himself.
- Jeremy strove to rescue the world from the evil forces of the black order, at one point embarking on a heroic quest to save his brother from these forces.

Being watched over by malevolent and/or benevolent entities. All participants seem to have experienced some degree of being watched over by benevolent and/or malevolent entities (thought not necessarily at the same time). Jeremy was the only participant who does not recall feeling watched over by both types of forces, having only experienced being watched over by a malevolent force, though he had a strong awareness of the existence of a powerful benevolent force. Sam had a slightly different experience than the others in that the benevolent forces he felt were watching him were somewhat less archetypal, taking the form of fellow "initials" agents and an enrapt audience when he was performing stunts for a movie.

- Sam experienced being watched over by supportive non-malevolent entities as he starred and performed stunts in various movies, and as he communicated with intelligence agencies while carrying out various missions. He experienced a malevolent watching over in the form of being hunted by the initials agencies and assaulted by enemies in a war.

- Theresa experienced the benevolent forces of a large black man in her closet and a guard dog protecting her, and she also experienced being persecuted by unknown malevolent entities.
- Byron experienced the guidance of the benevolent dakinis, and he experienced being imprisoned in Hell and tormented by the doctors and the psychiatric establishment.
- Cheryl experienced a nearly constant watching over from spirit guides who began as benevolent, then became extremely malevolent, and finally left her with a residual sense of a being protected by some benevolent force. She also experienced being persecuted by her father, her future husband, and others.
- Trent experienced being watched over and personally guided by a benevolent God, and he was also persecuted by the Devil, both from within and without.
- Jeremy experienced being watched over and hunted by the black order. Although he sensed the presence of the white order, he did not experience their watching over him quite so acutely.

Groundlessness. All participants except for Sam recall having profound experiences of groundlessness, typically accompanied by terror, though occasionally accompanied by feelings of liberation and euphoria, as in the cases of Byron and Cheryl, or deep despair, as in Trent's case.

- Theresa recalled being "pretty much 'groundless' the whole time during both psychotic experiences": "I didn't even feel 'physically' connected to the ground most of the time, or 'psychically' connected to the planet – and quite importantly (I think) didn't feel connected to another human being thru pretty much the whole of the time."
- Byron had powerful experiences of groundlessness throughout much of his psychosis, often accompanied with the sense of spiritual or bodily disintegration, terror, and at times rapture.
- Cheryl experienced extreme terror and groundlessness for the majority of the first two months of her psychosis.
- Trent experienced his onset as "imploding into a void where a family was supposed to be," resulting in a sense of profound despair and disconnection with himself and the world.

- Jeremy experienced profound annihilation anxiety at the onset of his psychosis. It was relatively short-lived and resulted in a very quick flipping over into feelings of omniscience and enlightenment, what he believes was a compensatory reaction.

Parallel dimensions. All participants except for Trent expressed having experiences of different realms or dimensions of experience coexisting. The degree to which the participants experienced this varied significantly, with Sam on one extreme experiencing this regularly and intensely, and Trent on the other extreme not recalling anything like this happening at all—"It was clear boundaries between one experience to the next."

- Sam had the strong sense most of the time that his various missions were taking place concurrently, like "parallel worlds." He also occasionally had the experience of this happening with consensus reality, as when the hospital staff jumped on top of him, forming a "dog pile" after he had panicked at the psychiatric hospital. He recalls having the distinct awareness of what was taking place within consensus reality while also having the experience of being in a war "buried in people and rubble" after a bomb had gone off.
- Theresa described occasions in which "several different realities [were] happening at the same time [and] the boundaries weren't clear at all."
- Byron had a vision of bardo, in which he was descending through many concurrent layers of the world. He also had many visions in which he had some awareness of what was happening in consensus reality but was also profoundly involved with what was happening within his visions.
- Cheryl had the experience at one point of seeing time as an illusion and perceiving all of our past lives to be existing concurrently with our current life. She also had times when she experienced the different layers of her experience simultaneously—the "spirit guides," her own sense of self and agency, the presence of Hell, and consensus reality—and being profoundly confused by this.
- Jeremy did not recall having "layers" of experience in quite the same way as those mentioned above; however, he does recall "tracking two levels of phenomena at once," in the sense of being aware to some extent of what was taking place within consensus reality, while simultaneously being aware of applying a significantly nonconsensus interpretation onto this.

Feelings of euphoria, liberation, and interconnectedness. Trent, Theresa, Byron, and Cheryl and Sam to a lesser extent, expressed having experienced feelings of euphoria, liberation and/or interconnectedness during their psychosis. Jeremy had such feelings prior to his psychosis, but not so much after the onset. *Interconnectedness*, as discussed here, refers to the experience that all manifestations of the universe, including oneself, are fundamentally interconnected.

- Sam did not recall having particularly strong experiences of liberation or euphoria during his psychosis, but he did recall having some profound experiences of interconnectedness: "I think spiritually I had always thought that all experiences and manifestations of the universe are interconnected. These experiences helped me to solidify these ideas in my mind."
- Theresa recalled having a number of different experiences of profound universal expansiveness and interconnectedness.
- Byron had profound experiences of liberation and interconnectedness just prior to his psychosis, and he recalled that the initial stages were "full [and] enrapturing." He recalled, however, that as his psychosis progressed, he primarily experienced a fairly strong dualistic (self/other) split, although he "was intensely seeking to attain 'unification.'"
- Cheryl did not recall strong feelings of interconnectedness per se, but she felt "mild euphoria and liberation" when first contacted by the "spirit guides."
- Trent recalled having experiences of liberation, euphoria, and interconnectedness during his psychosis, even though these kinds of experiences were generally quite rare.
- Jeremy did not recall experiencing a strong sense of interconnectedness per se, but during the few weeks leading up to his first psychotic experiences, he did experience fairly strong feelings of liberation, deep peace within himself, and intimate connection with others.

Recovery

Finding meaning in life. All participants reported that finding meaning and/or a purpose in life that made their life worth living was very important in their recovery. It is significant that five of the participants (all except Trent) have become professionals within the mental health field and are now working directly with people suffering from extreme states, finding that contributing in this way has provided important meaning.

- After coming off of antipsychotics about ten years after his first psychotic episode, Sam was able to return to college, obtain a bachelor's degree, and find a meaningful career within the mental health field.
- Theresa has found meaning in her life by creating a healthy family and also by becoming a professional within the mental health field.
- Byron first began to connect with some spiritual meaning in his life when he read *Be Here Now* by Ram Dass. He continued to find meaning within spiritual traditions, eventually arriving at Tibetan Buddhism and a supportive teacher. He has also found some important meaning working within the mental health field.
- Cheryl worked within the mental health field both before and after her psychosis, and reconnecting with this work in a more meaningful way afterwards has been particularly helpful to her.
- Trent has found meaning in his life by taking the creativity he had tapped into within his psychosis and channeling it towards expressive arts, especially writing, poetry, and photography.
- Jeremy has found important meaning by devoting his life to the work of developing more helpful ways to support people who suffer from psychosis.

Connecting with one's aliveness. All participants expressed that fostering a deeper connection with their authentic selves—particularly their feelings, needs, and their sense of agency—was very important in their recovery.

- Sam was first able to reconnect with his "spark" after coming off of antipsychotics. He also used self-hypnosis tapes, which helped him cultivate self esteem and confidence, he learned the value of self care, and he began to turn to a number of different creative outlets such as writing and playing music as a way to connect with his own aliveness.
- For Theresa, becoming pregnant was an important factor in connecting with herself and finding her sense of agency. Although many might see the manner that she became pregnant as highly irresponsible (having done so with a short-term partner and while in severe crisis), this represented the beginning of a genuine sense of connection with herself and her dream of a family that ultimately initiated the motivation and agency necessary to do the hard work of recovery.

- With the help of meditation and other contemplative practices, Byron was able to connect more deeply with himself, his spirituality, and his sense of agency, as well as to find some grounding for his anomalous experiences.
- Cheryl found her agency when she stood up to the voices, insisting that she would not let them rule her life anymore. Shortly thereafter, she began to intentionally practice self-love and self-forgiveness, which further contributed to her self connection and sense of agency.
- Trent found his voice and authenticity during psychotherapy and then in his relationships with others. He also gave up marijuana and cigarettes and learned how to be with himself without the need of chemical aids.
- Jeremy found that returning to school was a very important turning point in his recovery, making the effort to "put . . . the gears back in motion" even while in the midst of such a profound struggle. His sense of self connection and agency continued to grow as he focused his energy on the healing of himself and others.

Finding hope. All participants expressed that connecting with hope was crucial in their recovery. Also, meaning and hope seem to be closely related for most of these participants—all but Sam expressed that making meaning of their experiences gave them hope.

- Sam found significant hope when he realized that he did not have to remain on antipsychotics forever. He realized he could use them on as as-needed basis, which he did for several years, and return to a full life.
- Theresa's first moment of hope came when she realized that her experiences were very similar to what other people had have experienced—that she was not a "freak." She had another important experience of hope when she met the psychiatric survivors, another after "hitting bottom" and realizing that there was nowhere to go but up, and yet another when finding a supportive partner. She also expressed feeling hope when she realized that her psychosis had a purpose, which she came to believe was to heal her so that she could have a family.
- Byron found both hope and meaning when he connected to a spiritual path and then eventually with Tibetan Buddhism, which provided him with a particularly good framework for making sense of his experiences.

- Cheryl was fortunate to have read the WHO recovery studies prior to her psychosis, and so she had some hope all along that it was possible to fully recover. Later, when she felt immediate relief after a healer performed an "exorcism" on her, she experienced some hope that she could actually improve. Then, she experienced further hope when a psychic was able to explain her experiences in a way that made sense to her. Another particularly hopeful moment occurred when her cat demonstrated that she really was lovable.
- Trent expressed that what gave him hope was being able to recognize the healing that was taking place for him with the passage of time: "Knowing I felt better than I did a month or two previously gave hope."
- Jeremy felt the first spark of hope when he encountered a therapist at the day hospital who "believed in [him]" and encouraged him to return to school. Then, his hope increased as he challenged the medical model and began to believe that full recovery really was possible.

Arriving at a more hopeful understanding of their psychosis. Everyone but Sam expressed the importance of making sense of their psychosis for their recovery, and in particular the importance of arriving at a more hopeful understanding than the degenerative brain disease model.

- Sam did come to an understanding of his psychosis that is clearly much more hopeful than the brain disease model, but unlike the others, he did not explicitly mention this as a particularly important factor in his recovery.
- Theresa believes that being able to make some sense of her psychosis and related experiences with the help of an individual psychotherapist has been helpful in her recovery.
- Byron found that the teachings within Tibetan Buddhism and other similar spiritual traditions resonated very strongly with his experiences, and these provided him with a very helpful framework for making sense of his experiences and ultimately making peace with them.
- Cheryl found some important meaning for her experiences when she met a psychic who described her experiences in a way that resonated with her, which became a very important turning point in her recovery.
- Trent found that making sense of the roots of his psychosis via group and individual psychotherapy, reading, introspection, and contemplation was helpful in his recovery.

- Jeremy found a "very good therapist" who was able to help him explore the roots of his psychosis, which in turn offered him important guidance in his path towards recovery.

Healthy vs. unhealthy relationships. All participants mentioned the importance of developing healthy relationships as an important part of their recovery. Trent and Jeremy also expressed the importance of creating distance from unhealthy relationships.

- Sam expressed how important it was to have loved ones willing to stand by his side when things got difficult.
- Theresa was fortunate to have come across a group of psychiatric survivors while she was homeless on the street and still struggling with intense psychotic experiences. Shortly afterward, she met the man who would become her lifelong partner, and his support and care for her played a very important role in her recovery.
- After Byron had recovered significantly, he met the woman who would become his lifelong partner, and her love and support was very important in his ongoing recovery.
- While Cheryl had significant problems with her parents when younger, she feels that they came through for her during her psychosis, supporting her in finding alternatives to psychiatry and being steadfast in their love and support for her even while she was vocally insisting that they must despise her and want to see her dead.
- Trent was able to cultivate several relatively healthy and supportive long-term romantic relationships during his recovery. He also created some distance between himself and his family, wanting to avoid returning to the dysfunctional relationships within it that he feels played a significant role in the onset of his psychosis.
- Jeremy received significant support from his friends and family while in the hospital, which he believes greatly reduced the harm he otherwise might have experienced there. He also realized that creating distance between himself and his family after getting out of the hospital was important for his recovery, since he came to recognize particularly unhealthy dynamics in his family system that likely played a role in the onset of his psychosis. He also developed a supportive relationship with a romantic partner, and he was

able to create a much healthier relationship with his father with the help of a therapist.

Harm from the psychiatric system hindering recovery. All participants expressed having suffered severe harm and even trauma from their interactions with the psychiatric system, and they generally found the psychiatric "treatment" they received to be much more of a hindrance than a benefit in their recovery. Only Sam and Trent found the psychiatric system to be of any significant benefit at all.

- Sam experienced harm from the social system on several different levels—physical abuse from the police officers when he was arrested for his psychotic behavior, increased fear and terror as a result of treatment within the hospital (feelings that he experienced very little outside of the hospital), and harm from the use of antipsychotics. He believes that the sudden withdrawal of antipsychotics at times led to psychotic experiences, and that the long-term use of them greatly hindered his recovery; he does, however, believe that they were somewhat helpful when he was able to use them judiciously and for short periods of time during the final stages of his recovery.
- Theresa described her experience in the hospital "as bad as if not worse than the trauma of anything that had actually caused [the psychosis]." She believes that receiving antipsychotics during her first episode interrupted a natural process, thereby necessitating a second psychotic episode later.
- Byron suffered significantly as a result of the generally poor care and strict confinement he received while in the hospital. He describes the antipsychotics as being a "poison" to him, believing they interfered significantly with his recovery process.
- Cheryl feels that the psychiatrists were "not helpful at all" and that the antipsychotics were "irrelevant to recovery," perhaps helping with sleep, but exacerbating her psychosis at times and causing physical problems. She also believes that the "myth of hopelessness" perpetuated by the psychiatric establishment was very detrimental to her recovery, and she expressed gratitude for having come across hopeful recovery research prior to her psychosis.
- Trent expressed having received neglectful and abusive treatment while an inpatient resident at a psychiatric hospital, although he did find the therapy he received in outpatient treatment to be helpful. He also likely experienced *withdrawal psychosis* and other severe withdrawal symptoms as a result of

antipsychotic use; however, like Sam, he believes that the judicious short-term use of antipsychotics "might have helped."

- Jeremy feels that he emerged from his hospitalization "a shell of a human being." He struggled tremendously for several years afterwards with having been told by the doctors that he could never fully recover. "[It] made me feel as if I could not trust myself. I could not trust my own mind."

Lasting Personal Paradigm Shifts

An integration of good and evil. All participants expressed having arrived at a very different understanding of good and evil as a result of having come through their psychosis. While all of them have come to somewhat different understandings in this regard, it appears that they all share several common themes: They have all come to see good and evil in a more integrated and less polarized way, and they have all come to distinguish evil acts from evil people, believing that evil is not something that is innate within anyone, but that evil acts result from those who are particularly ignorant or wounded.

- Sam finds that he is able to more easily see the humanity in others now, even those who have committed very harmful acts. He sees evil as acts that are caused by people who "need as much of healing of their soul as they can get," and he finds that he really enjoys supporting such wounded people.
- Theresa has come to believe that there is valid meaning behind any act, whether harmful or not, and that validating and understanding the person's experience provides more guidance for how to deal with any problems than judging whether or not something is good or evil. Her choice of profession, which is within the field of forensics psychology, is closely related to her beliefs in this regard.
- Byron has come to share the Buddhist perspective that evil (i.e., harmful) acts are done out of ignorance of the true nature of the world, and that everyone struggles with this ignorance to a greater or lesser degree.
- Cheryl has come to see the world as "not so black and white anymore," seeing evil actions as caused by people who are not inherently evil but are "wounded or ignorant in some other way."
- Trent has come to see good as that which arises when one is connected with the "true heart" and evil as that which arises when the true heart is obstructed.

He sees the true heart as the heart that we all share; therefore, he now sees that everyone has the capacity to connect with the true heart.

- Jeremy has come to see good and evil as different degrees of tolerance to the anxiety of mystery. He has come to believe that people who have particularly rigid belief structures, such as those who subscribe to positivism and/or the medical model of mental illness, are likely to have a particularly low tolerance for the anxiety of mystery.

A significantly changed spectrum of feelings with more depth and unitive feelings. All participants have expressed that the spectrum of feelings that are available to them now has changed significantly since prior to their psychosis, especially having shifted in the direction of having more unitive feelings and experiencing other feelings with more depth and richness.

- Sam feels he experiences more "bliss" and general contentment in his life now.
- Theresa finds that she has developed increased awareness of her experience beyond her "thinking mind," including her body and creativity.
- Byron feels that his overall experience is "richer and broader" with a greater sense of presence. He also has a sense that his self "is not so solid," which goes along with feeling more unitive feelings in general.
- Cheryl finds that she now experiences a wider range of feelings, which includes generally more joy, love, and compassion, and less hatred and judgment.
- Trent describes his "vision" as being much less limited now. He experiences much more fulfillment, beauty, and possibilities in the world now. Along with this, he has a greatly increased appreciation for creativity—of himself, others, and the world in general—and also an increased appreciation for humor.
- Jeremy has discovered that the range and depth of feelings available to him has increased significantly.

An increased sense of interconnectedness. All participants expressed experiencing a significantly greater sense of interconnectedness now than they did prior to the onset of the psychosis, generally speaking, and to a greater or lesser degree. As mentioned earlier, *interconnectedness* as used here refers to the concept that all manifestations of the universe are fundamentally interconnected.

A strong desire to contribute to the wellbeing of others. All participants expressed a significantly stronger desire to contribute to the wellbeing of others. Trent has been seriously involved in the *Big Brothers, Big Sisters* project, in which adults mentor/befriend troubled youths. All five of the others have become professionally involved within the mental health care field, each working in some way or another with people suffering from psychosis and other extreme states, and all of them apparently going well beyond the call of duty, striving to bring about genuine beneficial change within the field.

Appreciating the limits of consensus reality. All participants expressed experiencing the limitations of consensus reality, though each in somewhat different ways.

- Sam had profound experiences of "parallel worlds" during his psychosis, and though he has not experienced any psychosis in many years, he continues to maintain some sense of this. Closely related to this is an increased capacity to be aware of and to tolerate paradox.
- Theresa has expressed that she has loosened her attachment to her thinking mind. Along with this has come the recognition that there are other equally valid realms of experience, and the importance of recognizing that there is always some validity to anyone's experience of the world.
- Byron has come to see that we all see the world with vision that is distorted significantly by our own conditioning and ignorance regarding the true nature of the world—"everybody's crazy."
- Cheryl has come to appreciate that there are different valid realms of experience, including those that are not experienced within consensus reality.
- Trent has come to see the world as "limitless" and full of possibilities beyond the constraints of consensus reality.
- Jeremy has come to recognize the frailty of our cognitive constructs—how we all "cobble together little rafts and ride them around" in the face of the "awesomeness of the universe."

A greater understanding of psychosis. All participants expressed having come through their process with a much greater understanding of their psychosis, and even a better understanding of psychosis in general. All participants have come to see their own psychosis as a natural process that resulted from finding themselves in an intolerable situation. All except for Trent now work professionally with people suffering from extreme states and so have the unusual perspective of having been on both sides of the mental health care worker/ consumer relationship.

- Sam understands his psychosis as having possibly two different causes which are not necessarily mutually exclusive: He believes his psychosis may have come on as a result of the severe stress of being drafted combined with not sleeping, the lack of self care, and possibly the use of recreational drugs; and he also believes there may be some validity to the idea that his "consciousness and corporeal self were jumping between dimensions of the multiverse." He has also come to appreciate the harms and benefits of the use of antipsychotics in the recovery process, particularly the harm of using them as "long term prophylactics" and the potential benefit of using them in a very judicious manner.
- Theresa has come to see her psychosis as a healing process of her psyche that allowed her to heal at a very deep level so that she could transcend her isolation and inability to meet her needs and eventually raise a healthy family.
- Byron has come to see that "everybody's crazy" in the sense of being ignorant to the true nature of the world. He sees his own psychosis as the expedited settling of his "karmic debt," in the sense that it allowed him to work through a lot of his own ignorance and suffering. He has also come to believe in the importance of minimizing psychiatric drugs and forced restraint, and to learn to trust the process, as was done in J. W. Perry's Diabasis house. He has also come to see that all psychoses have a spiritual component and that attempting to separate spiritual problems from so called genuine psychosis is ultimately an exercise founded on a false distinction.
- Cheryl has come to see that her psychosis has allowed her to heal at a very deep level—"[it] ripped off all the layers and got right to the core." She has also come to see antipsychotics as generally more of a hindrance than a benefit, and she is a strong advocate for the "message of hope" for full recovery within the mental health field.
- Trent has come to see his psychosis as an "awakening and a new direction from an ill path from a very young age."
- Jeremy has come to see psychosis in general as "an activity that the psyche has taken up in order to achieve some kind of balance" in the face of overwhelming feelings and experiences. He sees a strong link between isolation and psychosis, and also between the individuation process of a young adult and psychosis. He also has come to greatly appreciate the primacy of healthy relationships in supporting the recovery process.

Lasting Benefits

Greatly increased wellbeing. All participants expressed experiencing a significantly greater sense of wellbeing now than they did prior to the onset of the psychosis, generally speaking, and to a greater or lesser degree.

Greater equanimity. All participants expressed experiencing significantly greater equanimity now than they did prior to the onset of their psychosis, generally speaking, and to a greater or lesser degree. Equanimity, as I am defining it here, refers to the capacity to maintain a relatively calm and balanced state of mind, even while under stress.

Greater resilience. All participants expressed experiencing significantly greater resilience now than they did prior to the onset of the psychosis, generally speaking, and to a greater or lesser degree. Resilience, in contrast to equanimity, refers to one's ability to return to a relatively calm and balanced state of mind after having lost one's balance (which they all acknowledged does still occur at times).

Healthier relationship with self. All participants expressed having a healthier relationship with themselves now as compared to prior to the onset of their psychosis, though they each emphasized different aspects of this.

- Sam feels he has developed an increased capacity for self awareness, which includes the ability to identify when to ask for support. He has also gained a greater appreciation of the value of self care, and he feels that his self esteem and confidence have increased significantly.
- Theresa finds that she has a greater sense of self worth and self acceptance, which is closely related to her decreased need for approval from others.
- Byron experiences his self as less "solid" now, which entails being much more at ease with himself.
- Cheryl feels that she has a more innate sense of self worth, based on who she is, not on what she does. Also, her sense of self love and self forgiveness has greatly improved.
- Trent feels that his self confidence and his courage to be authentic with himself and with others has increased significantly. He also expressed having a significantly greater awareness of his needs and the capacity to meet them.
- Jeremy finds that he has "far more confidence and assertiveness" now.

Healthier, more rewarding relationships with others. All participants expressed that their relationships with others have become far healthier and more rewarding.

- Sam feels that he has an increased ability to be supportive of others and an increased capacity to see the humanity of others, both of which contribute to significantly improved relationships.
- Theresa finds that, since she experiences much more self worth and less need for approval, she is able be more at ease in relationships now, and she therefore finds them much more rewarding.
- Byron feels that he has more presence and empathy with others, both of which have improved his ability to listen to and connect with others more deeply.
- Cheryl feels that, prior to her psychosis, she had harbored "seeds of hatred" and significant judgment towards others. Now that these have faded tremendously and she has also developed much more love for herself, she finds that she has much more love and compassion for others.
- Trent finds that he has "much more intimate and insightful and compassionate and enjoyable" relationships now.
- Jeremy has come to "cherish the value of human relationship to a far greater degree." He no longer feels he has to save people, and therefore finds his relationships much more wholesome and easeful—"I don't have to take other's pain to be loved."

LASTING HARMS

All participants expressed that their psychotic process has led to many more lasting benefits than lasting harms overall; however, all except for Theresa expressed that they still do experience some lasting harms.

- Sam feels he still has some unresolved grief and trauma related to his psychosis and the many years that he feels he lost in some ways.
- Byron says that he can connect with the general sense of being a "wounded healer." He also has some mild regrets for having missed out on higher education as a result of his "spiritual experiences [being] more compelling."
- Cheryl describes her psychosis as having been "traumatic," and she still feels some residue from that, primarily in the form of occasional anxiety.

- Trent finds that he still has some challenges with social stigma. In particular, he finds some challenge holding the tension between his high regard for authenticity and not wanting to be judged harshly by others for having been diagnosed with a "mental illness."
- Jeremy finds that he experiences one lasting harm that is, ironically, a negative aspect of the benefit of a deeper connection with himself and a sense of purpose. Since he is so strongly compelled to devote himself to improving the mental health care system, he feels the cutting off of alternative life choices that this kind of devotion entails.

Appendix B

Evidence for the Interplay Between Unity and Duality in the Field of Physics

Over the past hundred years, beginning with Einstein's theories of relativity and continuing with the development of quantum mechanics, the field of physics has undergone a truly profound paradigm shift. It has taken long strides away from the Descartian/Newtonian view of the universe as a fundamentally dualistic and materialistic machine, the components of which are entirely isolated entities knocking into each other like so many billiard balls; and it has moved towards a view of the universe in which the quality of profound interconnectedness (often referred to as *entanglement* within the field of quantum mechanics) has become recognized as one of the most fundamental qualities of the universe. Erwin Schrodinger, one of the original founders of quantum mechanics, said that fundamental interconnectedness (entanglement) is not "*one* but rather *the* characteristic trait of quantum mechanics, the one that enforces its entire departure from classical lines of thought"[1]. Following are some of the most important discoveries that have led to this paradigm shift in the field.

The *Big Bang* theory, by far the most well established theory of the creation of our* universe, postulates that our universe initially began its life as an infinitely small, infinitely dense point, known as a singularity that burst forth initially into a uniform sea of radiation, eventually coalescing and differentiating into the myriad manifestations that we experience today, including, of course, our very own beings[2]. In other words, the Big Bang theory suggests that all dualistic manifestation in our universe initially manifested from a single nondual source

* I use the expression "*our* universe" here rather than "*the* universe" because it has been speculated that there may be many other universes that have also manifested in a similar manner, in what is known as the *multiverse* theory[3].

(a singularity) and have been undergoing continuous change on all levels ever since. The Big Bang theory, then, provides significant evidence for the principles of fundamental interconnectedness and impermanence existing within the fabric of our universe, as well as for the principle of the paradoxical coexistence of unity and duality.

Einstein's well established equation, $E=mc^2$, has lent significant validity to the idea that matter and energy—the two fundamental building blocks of all that we have observed in the universe—are merely different manifestations of the same fundamental material of the universe, a concept referred to as *mass-energy equivalence*[4], a concept that is in accord with the Big Bang theory in indicating profound interconnectedness within the universe. Research within the field of quantum mechanics has taken this concept one step further and concluded that not only are matter and energy comprised of the same fundamental material, but that all such material actually consists of *both* matter *and* energy simultaneously (that light and subatomic particles are paradoxically both waves and particles simultaneously)[5].

Perhaps one of the most startling findings further validating the concept of fundamental interconnectedness have arisen from successfully replicated experiments that have determined that by affecting one particle within the universe, a different particle in an entirely different region of the universe is immediately affected[6]. This finding may at first glance appear to violate Einstein's *theory of relativity*, which posits that nothing can travel faster than the speed of light, but this is the case only if it is assumed that some signal is transmitted between the two particles. The evidence suggests, however, that the theory of relativity is not in fact violated; rather, there appears to be some fundamental interconnectivity between these particles that transcends the dualistic properties of time and space altogether, a principle known as *nonlocal interconnectivity*[7]. Again, we see significant evidence of a fundamental interconnectedness in the universe and the paradoxical coexistence of dualistic and unitive qualities.

While the Big Bang theory provides us with evidence that the dualistic forms within the universe may have arisen from a nondual source on the macrocosmic level, there is other evidence suggesting that a very similar dynamic is taking place continuously and much more rapidly on the subatomic level. Research within the field of quantum mechanics has provided evidence suggesting that within the fundamental fabric of the universe (within all matter/energy and even within the vacuum of apparently empty space), there is a constantly seething activity in which subatomic particles known as *virtual strings* or *virtual particles* literally manifest from apparent nothingness (a nondual source?) as pairs involved in a dialectical relationship (*particle/antiparticle pairs*) and then proceed to annihilate each other[8], apparently disappearing back into the same nondual source from which they came*. It is speculated that these pairs of particles, comprising literally everything within the known universe, continuously manifest and disappear many trillions of times a second. These findings provide us with evidence of profound impermanence and the interplay between duality and unity taking place continuously throughout the entire fabric of the universe.

A number of contemporary theoretical physicists[9] have taken a particularly strong interest in the concept of fundamental interconnectedness, a concept that has gained significant momentum since the birth of the field of quantum mechanics. After taking into account the accumulating evidence within the field of theoretical physics, they have suggested that the sum total of this evidence points to a new paradigm that is significantly different than the paradigm of strict duality that predates it. John Battista summed up this new paradigm well, referring to it as the *holistic paradigm*: "Consciousness cannot be separated from matter, . . . the observer cannot be removed from what is observed, . . . [and quantum-actions are] information events that involve the interaction of parts of one interconnected, conscious universe"[10]. It is clear, then, that the findings of modern physics and of intensive mindfulness meditation practice have been remarkably similar with regard to the fundamental nature of the universe. In particular, we can say that they both generally agree that the fabric of our experience and of our very beings consist of an interconnected, impermanent sea of activity stemming from the interplay between unity and duality.

---

* This concept has striking parallels to the graphical depictions of core human experience illustrated in Figures 19.2 and 21.1, suggesting the intriguing possibility that the existence of our very being may be but a macrocosm of the very same process occurring at the subatomic level.

References*

Introductory Quotes

1. Deegan, 1996, p. 91.

Introduction: The Case of Sam

1. Hopper, Harrison, Janca & Satorius, 2007.

Part One: Deconstructing the Myths of Madness

Chapter 1: First, Some Terminology

1. American Psychiatric Association [APA], 2010; American Psychological Association, 2008.
2. APA, 2000a, p. 153.
3. APA, 2000a, p. 154.
4. APA, 2003
5. APA, 2010.
6. British Psychological Society [BPS], 2000.
7. APA, 2000b, p. 297.
8. Ibid.
9. APA, 2000b.
10. Heery, 1989; Romme, Escher, Dillon, Corstens, & Morris, 2009; Watkins, 2008.
11. APA, 2000b.
12. Heery, 1989; Romme et al., 2009; Watkins, 2008.
13. Rubik, 2002, p. 707.

Chapter 2: Myth #1: Schizophrenia is a Brain Disease

1. APA, 2010; National Institute of Mental Health [NIMH], 2010a; Hopper et al., 2007.
2. Siebert, 1999.
3. APA, 2010; NIMH, 2010a.
4. NIMH, 2010; Satcher, 1999.
5. Hoffman, 1979.
6. Wooley & Shaw, 1954.
7. Stabenau, Creveling, & Daly, 1970.
8. Carlsson & Lindqvist, 1963.
9. Healy, 2008.

* More details about these references can be found in the *Bibliography*.

10. McKenna, 1994.
11. Abi-Dargham, 2005.
12. Bentall, 2003.
13. Abi-Dargham, 2005.
14. Bentall, 2004.
15. Lee and Seeman, 1980.
16. Burt, Creese, & Snyder, 1977; Kornhuber et al., 1989; Mackay, 1982.
17. As quoted in Whitaker, 2002, p. 198.
18. Hyman & Nestler, 1996, p. 161.
19. Read, 2004.
20. As quoted in Bentall, 2004, pp. 310-11.
21. Johnstone, Frith, Crow, & Owens, 1988.
22. Wright et al., 2000.
23. Gaser, Nenadic, Buchsbaum, Hazlett, and Buchsbaum, 2004, p. 154.
24. Read, 2004.
25. Woodruff & Lewis, 1996.
26. Bentall, 2004.
27. Ibid.
28. Read, 2004; Siebert, 1999.
29. Lewine, 1998, p. 499.
30. Siebert, 1999.
31. Maguire et al., 2000.
32. Bremner, Randall, Scott, & Bronen, 1995.
33. Andersen et al., 2008.
34. NIMH, 2010a.
35. Breggin, 2008a; Joseph, 2004; Read, 2004.
36. Joseph, 2004.
37. Joseph, 2004, p. 69.
38. Ibid.
39. Joseph, 2002.
40. Joseph, 2004
41. Heston, 1966; Rosenthal, 1971; Tienari et al., 1987.
42. Wender, Rosenthal, Kety, Schulsinger, & Welner, 1974.
43. Kety, Wender, Jacobsen, & Ingraham, 1994.
44. Heston, 1966; Kety et al., 1994; Rosenthal, 1971; Tienari et al., 1987; Wender et al., 1974.
45. Joseph, 2004.
46. Joseph, 2004, p. 73.
47. Kety et al., 1968.
48. Rosenthal, 1971.
49. Joseph, 2004.
50. Ibid., p. 76.
51. Crow, 2007, p. 13.
52. Williams et al., 1999, p. 1729.
53. Crow, 2007.
54. Delisi, 2000, p. 190.
55. Joseph, 2004, p. 78.
56. Tienari, 1991.
57. Read, Perry, Moskowitz, & Connolly, 2001.
58. Robins, 1974.

59. Beck & van der Kolk, 1987.
60. Livingston, 1987.
61. Honig, Romme, Ensink, Escher, Pennings, & de Vries, 1998).
62. Karon, 2003.
63. Read , van Os, Morrison, & Ross, 2005.
64. Arseneaul, Cannon, Witton & Murray, (2004); Henquet et al. (2005); Semple, McIntosh & Lawrie (2005); Van Os et al. (2002).
65. Koseki & Nabeshima (2010); Lecomte et al. (2010); Mikami et al. (2003); Vardy & Kay (1983).
66. Kuepper et al. (2011b).
67. Konings et al. (2012).
68. Kuepper et al. (2011a).
69. Dickins & Flynn, 2001.
70. Kagan, Reznick, & Snidman, 1988
71. Andersen et al., 2008; Bremner et al., 1995; Maguire et al., 2000.
72. Siegel, 2007, p. 24.
73. NIMH, 2010a, Para. 1.
74. NIMH, 2010b, Para. 1.
75. APA, 2010, Para. 1.
76. APA, 2010, Para. 10.
77. Ibid.
78. Satcher, 1999, Para. 1.
79. Bentall, 2003, 2004; Breeding, 2008; Breggin, 2008a; Mosher, 2008; Read, 2004; Siebert, 1999; Szasz, 2008; Whitaker, 2010.
80. Breggin, 2008a, p. 18.

Chapter 3: Myth #2: "Schizophrenia" is a Valid Construct

1. British Psychological Society [BPS], 2000, p. 17.
2. Ibid., p. 18.
3. Geekie and Read, 2009.
4. Ibid., p. 143.
5. Ibid., p. 138.
6. Ibid.
7. BPS, 2000.

Chapter 4: Myth #3: People Cannot Fully Recover from Schizophrenia

1. Hopper et al., 2007, p. 37.
2. Siebert, 1999.
3. Ibid.

Chapter 5: Myth #4: Mainstream Psychiatric Treatment Greatly Increases Beneficial Outcomes

1. Hopper et al., 2007.
2. Hegarty, Baldessarini, Tohen, & Waternaux, 1994.
3. Hagen, Nixon, & Peters, 2010.
4. Breggin, 2008a.
5. Hagen et al., 2010.
6. Carpenter, McGlashan, & Strauss, 1977; Harding, Zubin, & Strauss, 1987; Harrow & Jobe, 2007; Hopper et al., 2007; Rappaport, 1978.
7. Whitaker, 2010.
8. APA, 2003.
9. Read, 2004.
10. Ibid.
11. Ibid.
12. Bola & Mosher, 2003; Carpenter et al., 1977; Chouinard, Jones, & Annable, 1978; Gur,

Maany, Mozley, Swanson, & Bilker, 1998; Harding et al., 1987; Harrow & Jobe, 2007; Rappaport, 1978; Schooler, Goldberg, Boothe, & Cole, 1967.

13. Guttmacher, 1964.
14. Schooler et al., 1967, p. 991.
15. Bola & Mosher, 2003; Carpenter et al., 1977; Rappaport, 1978.
16. Carpenter et al., 1977, p. 19.
17. Rappaport, 1978.
18. Ibid, p. 106.
19. Carpenter et al., 1977; Rappaport, 1978.
20. Harding et al.,1987.
21. McGuire, 2000.
22. Harrow & Jobe, 2007.
23. Harrow, Jobe, & Faull, 2012.
24. Chouinard et al., 1978.
25. Whitaker, 2010.
26. Chakos et al., 1994; Gur et al., 1998; Madsen, Keiding, Karle, Esbjerg, & Hemmingsen, 1998.
27. Gur et al., 1998.
28. APA, 1992.
29. Breggin, 2008a.
30. Breggin, 2008b; Galynker & Nazarian, 1997; Shear, Frances, & Weiden, 1983; Wirshing, Van Putten, Rosenberg, & Marder, 1992.
31. Hagen et al., 2010; Keefe, Bollini, & Silva, 1999.
32. Breggin, 2008a.
33. Hagen et al., 2010, p. 50.
34. Ibid.
35. Healy et al., 2006.
36. Arana, 2000; Joukamaa et al., 2006; Waddington, Youssef, & Kinsella, 1998.
37. National Association of State Mental Health Program Directors, 2006, p. 5.
38. Davies et al., 2007; Knable, Heinz, Raedler, & Weinberger, 1997; Mattes, 1997; Rosebush & Mazurek, 1999.
39. Cullen, Kumra, Regan, Westerman, & Schulz, 2008.
40. Department of Health and Human Services, 1993, p. 4.
41. Knable et al., 1997; Rosebush & Mazurek, 1999.
42. Mattes, 1997.
43. Geddes, Freemantle, Harrison, & Bebbington, 2000, p. 1371.
44. Lieberman et al., 2005.
45. Davies et al., 2007.
46. Whitaker, 2007.
47. Mehta & Farina, 1997; Read, Haslam, Sayce, & Davies, 2006; Read, 2007.
48. Geekie & Read, 2009; Read et al., 2006.
49. Markowitz, 2005.
50. Ibid.
51. Breggin, 2008a.
52. APA, 2000b; Herman, 1997; Levine, 1997.
53. Hagen et al., 2010.
54. Herman, 1997; Levine, 1997.
55. Geekie & Read, 2009; Markowitz, 2005; Mehta & Farina, 1997.
56. House, 2001; Perry, 1999; Warner, 1994.
57. Hopper et al., 2007.
58. Ibid.
59. Ibid.

60. Ibid.
61. Ibid.
62. Ibid.
63. Bola & Mosher, 2003; Mosher, 1999; Mosher & Hendrix, 2004.
64. Bola & Mosher, 2003.
65. Whitaker, 2010.
66. Perry, 1999; Seikkula, Aaltonen, Alakare, Haarakangas, Keränen & Lehtinen, 2006; Whitaker, 2007.
67. Mosher, 1999, p. 142.
68. Seikkula et al., 2006.
69. Ibid., p. 214.

Chapter 6: Summarizing the Research on Schizophrenia and Recovery

1. IMS Health, 2010.

Part Two: Alternative Understandings of Psychosis

[Introduction]

1. Arieti, 1978, p. 20.
2. Perry, 1999, p. 147.
3. Tooth et al., 2003.
4. Arieti, 1978; House, 2001; Karon & VandenBos, 1996; Laing, 1967; May, 1977; Mindell, 2008; Mosher & Hendrix, 2004; Perry, 1999.
5. Laing, 1967, pp. 114-115.
6. In an interview in Mackler, 2008.

Chapter 9: Transpersonal Psychology—Spiritual Emergency vs. Pathological Psychosis

1. Taylor, 1999.
2. Lajoie & Shapiro, 1992.
3. Cortright, 1997, p. 156.
4. Ibid., p. 160.
5. Turner, Lukoff, Barnhouse, and Lu, 1995.
6. Nelson, 1990; Turner et al., 1995; Wilber, 2000.
7. Assagioli, 1989.
8. Grof and Grof, 1989.
9. Nelson, 1990.
10. Wilber, 2000.
11. Cortright, 1997, p. 170.
12. Jackson, 2001.
13. Chadwick, 2001; Clarke, 2008, 2010; Jackson, 2001; Jackson & Fulford, 2002; Margree, 2002; Peters, 2001.
14. Cortright, 1997, p. 156.
15. Ibid., p. 160.

Chapter 10: Psychosis as a Renewal Process (John Weir Perry)

1. As quoted in an interview conducted by O'Callaghan, 2001.
2. Perry, 1999, p. 147.
3. Perry, 1974, 1976, 1987, 1999
4. Perry, 1999, p. 21.
5. Ibid., p. 130.
6. Ibid.
7. Ibid., pp. 134-5.
8. Ibid., p. 135.
9. Ibid., p. 80.

10. Ibid., p. 106.
11. Ibid.

Chapter 11: The Process Paradigm (Arnold Mindell)

1. Mindell, 2008.
2. Ibid., p. 56.
3. Ibid., p. 60.
4. Ibid., p. 61.
5. Ibid., p. 125.

Chapter 12: Seeing Through the Veil of our Cognitive Constructs (Isabel Clarke)

1. Clarke, 2008, 2010.
2. Teasdale & Barnard, 1993.
3. Clarke, 2010, p. 108.
4. Chadwick, 1992.
5. Laing, 1967.
6. Clarke, 2010, p. 111.
7. Clarke, 2001, p. 142.
8. Ibid.

Chapter 13: The Creative Process Gone Awry (Mike Jackson)

1. Jackson, 2001.
2. Ibid., p. 185.
3. Batson & Ventis, 1982.
4. Wallas, 1926.
5. Batson & Ventis, 1982.

Chapter 14: The Life Fear / DeathFear Dialectic (Otto Rank)

1. Rank, 1936.
2. Ibid., p. 124.
3. Ibid.

Chapter 15: Overwhelmed by Death Anxiety (Irvin Yalom)

1. Yalom, 1980.
2. Ibid., p. 29.
3. As cited in Yalom, 1980, p. 42.
4. Yalom, 1980, p. 43.
5. Kierkegaard, 1844/1967.
6. Ibid., p. 112.
7. Ibid., p. 122.
8. Ibid., p. 125.
9. Ibid., p. 127.
10. Ibid., p. 141.
11. Ibid.
12. Ibid., p. 142.

Chapter 16: When Overwhelming Anxiety is Insoluble on any Other Level (Rollo May)

1. May, 1977.
2. Ibid.
3. Ibid., p. 326.
4. Kierkegaard, 1844/1967.
5. May, 1977, p. 205.
6. Ibid., p. 206.
7. Ibid., p. 205-6.
8. Ibid.
9. Ibid., p. 228.

10. Ibid.
11. Ibid., p. 326.
12. Ibid.

Chapter 17: Overwhelming Exposure to the True Nature of the World (Ernest Becker)

1. Becker, 1973.
2. Ibid., p. 26.
3. Becker, 1973.
4. Ibid., p. 52.
5. Ibid., p. 66.
6. Ibid., p. 4.
7. Ibid.
8. Ibid., p. 5.
9. Ibid., p. 209.
10. Ibid., p. 94.
11. Ibid., p. 219.
12. Ibid., p. 220.
13. Ibid. p. 219.

Chapter 18: Toward a Paradigm Shift in the Way We View Personal Paradigm Shifts

1. Clarke, 2010, p. 111.
2. Laing, 1967; Mindell, 2008; Mosher & Hendrix, 2004; Perry, 1999
3. Clarke, 2001; Farber, 1993; Laing, 1967; Mindell, 2008; Perry, 1999.

Part Three: Arriving at an Integrative and Comprehensive Model of Psychosis

Chapter 19: The Foundation of the Duality Unity Integrative (DUI) Model

1. Rank, 1936, p. 124.
2. May, 1977, p. 228.
3. Yalom, 1980, p. 142.
4. Schneider, 1999.

Chapter 20: Our Experience of Duality

1. Rank, 1936.
2. May, 1977.
3. Yalom, 1980.
4. Mahler, Pine, & Bergman, 1973.
5. Wallin, 2007.
6. Rathus, 2006, p. 218.
7. Ibid., p. 219.
8. Ibid.
9. May, 1977, p. 228.
10. Schore, 2002; Slade, 1999; Wallin, 2007.
11. May, 1977, p. 228.
12. Schore, 2002; Slade, 1999; Wallin, 2007.
13. Schore, 2002; Wallin, 2007.
14. Yalom, 1980.
15. Karen, 1994.
16. Galotti, 2008.

Chapter 21: Our Experience of the Interplay between Duality and Unity

1. Greene, 2003; Hawking, 2005
2. Becker, 1973, 1975.
3. Clarke, 2001, 2008, 2010.
4. Thalbourne & Delin, 1994.
5. Batson and Ventis, 1982; Jackson, 2001.

6. Chadwick, 1992, 2001; Clarke, 2001; Frith, 1979; Hemsley, 1998.

Chapter 22: A Description of the Full Spectrum of Our Feelings

1. Clarke, 2001, 2008, 2010; Frith, 1979; Hemsley, 1998.
2. Maslow, 1968.

Chapter 23: The Fundamental Roles of the Psyche

1. Becker, 1973.
2. May, 1977.
3. Yalom, 1980.
4. Heery, 1989; Romme et al., 2009; Watkins, 2008.
5. Yalom, 1980, p. 112.
6. Schneider, 1999, 2008.
7. Perls, Hefferline, & Goodman, 1951.
8. Buber, 1923/1996.

Chapter 24: The Case of Cheryl

1. Hopper et al., 2007.
2. Mosher, 1999.

Part Four: Making Sense of Madness, From Onset to Full Recovery

Chapter 27: The Onset and Deepening of Psychosis

1. Perry, 1999, p. 21.
2. May, 1977, p. 228.
3. Yalom, 1980.
4. Ibid., p. 112.
5. Batson & Ventis, 1982.
6. Jackson, 2001.
7. Kendell & Gourlay, 1970.
8. Robbins, Locke, & Reiger, 1991.

Chapter 28: The Anomalous Experiences

1. Yalom, 1980, p. 112.
2. Perry, 1999.
3. Schneider, 2008.

Chapter 29: Recovery

1. Grof, 2001.

Chapter 31: Lasting Benefits

1. Tedeschi & Calhoun, 1996.

Conclusion: Towards a New Paradigm

1. Knapp, 2008; Podvoll, 1990.
2. Batch, Hayes, & Gallop, 2012; Draper, Velligan, & Tai, 2010; Gottdiener, 2007; Morrison, 2007; Seikkula et al., 2006.
3. Gardiner, 2009.

Appendix B: Evidence for the Interplay between Unity and Duality in the Field of Physics

1. As quoted in Bitbol, 1996.
2. Hawking, 2005; Sawyer, 1999.
3. Green, 2003.
4. Einstein, 1922/2003.
5. Bohm & Hiley, 1993.
6. Bohm & Hiley, 1993; Goswami, 1993.
7. Ibid.
8. Greene, 2003; Hawking, 2005.
9. Bohm & Hiley, 1993; Goswami, 1993.
10. Battista, 1996, pp. 203-204.

Bibliography

Abi-Dargham, A. (2005). *The dopamine hypothesis of schizophrenia.* Retrieved from http://www.schizophreniaforum.org/for/curr/AbiDargham/default.asp

Amendola, L., & Tsujikawa, S. (2010). *Dark energy: Theory and observations.* Cambridge, UK: Cambridge University Press.

American Psychiatric Association [APA]. (1992). *Tardive dyskinesia: A task force report of the American Psychiatric Association.* Washington, DC: Author. Retrieved from http://ajp.psychiatryonline.org/cgi/content/abstract/137/10/1163

American Psychiatric Association [APA]. (2000a). *Desk reference to the diagnostic criteria from DSM-IV-TR.* Arlington, VA: Author.

American Psychiatric Association [APA]. (2000b). *Diagnostic and statistical manual IV. (text rev.).* Washington, DC: Author.

American Psychiatric Association [APA]. (2003). *American Psychiatric Association statement on diagnosis and treatment of mental disorders.* Retrieved from the American Psychiatric Association Healthy Minds, Healthy Lives website: http://www.psych.org/MainMenu/Newsroom/NewsReleases/2003NewsReleases/mentaldisorders0339.aspx

American Psychiatric Association [APA]. (2010). *Schizophrenia.* Retrieved from The American Psychiatric Association Healthy Minds, Healthy Lives website: http://www.healthyminds.org/Main-Topic/Schizophrenia.aspx

American Psychological Association. (2008). *Topic: Schizophrenia.* Retrieved from http://www.apa.org/topics/topicschiz.html

Andersen, S., Tomada, A., Vincow, E., Valente, E., Polcari, A., & Teicher, M. (2008). Preliminary evidence for the sensitive periods in the effect of childhood sexual abuse on regional brain development. *The Journal of Neuropsychiatry and Clinical Neurosciences, 20*(3), 292-301. doi:10.1176/appi.neuropsych.20.3.292

Arana, G. (2000). An overview of side effects caused by typical antipsychotics. *Journal of Clinical Psychiatry, 61*(Suppl. 8), 5-11. Retrieved from http://psychrights.org/research/Digest/NLPs/RWhitakerAffidavit/AranaOverviewofTypicalSideEffects.pdf

Arieti, S. (1978). *On schizophrenia, phobias, depression, psychotherapy, and the farther shores of psychiatry.* New York, NY: Brunner/Mazel.

Arseneault, L., Cannon, M., Witton, J., & Murray, R. M. (2004). Causal association between cannabis and psychosis: Examination of the evidence. *British Journal Of Psychiatry, 184*(2),

110-117. doi:10.1192/bjp.184.2.110

Assagioli, R. (1989). Self-realization and psychological disturbances. In C. Grof & S. Grof (Eds.), *Spiritual emergency* (pp. 27-48). New York, NY: Tarcher.

Baars, B. J. (1997). *In the theater of consciousness: The workspace of the mind.* Oxford, UK: Oxford University Press.

Bach, P., Hayes, S., & Gallop, R. (2012). Long-term effects of brief acceptance and commitment therapy for psychosis. *Behavior Modification, 36*(2), 165-181.

Bassman, R. (2007). *A fight to be: A psychologist's experience from both sides of the locked door.* New York, NY: Tantamount Press.

Batson, C. P., & Ventis, L. W. (1982). *The religious experience.* Oxford, UK: Oxford University Press.

Battista, J. R. (1996). Contemporary physics and transpersonal psychiatry. In B. W. Scotton, A. B. Chinen, & J. R. Battista (Eds.), *Textbook of transpersonal psychiatry and psychology* (pp. 195-206). New York, NY: Basic Books.

Beck, J., and van der Kolk, B. (1987). Reports of childhood incest and current behavior of chronically hospitalized psychotic women. The *American Journal of Psychiatry, 29*, 789-794.

Becker, E. (1973). *The denial of death.* New York, NY: Free Press Paperbacks.

Becker, E. (1975). *Escape from evil.* New York, NY: The Free Press.

Beers, C. W. (1981). *A mind that found itself.* Pittsburgh, PA: University of Pittsburgh Press.

Bentall, R. P. (2003). *Madness explained: Psychosis and human nature.* London: Penguin Books.

Bentall, R. P. (2004). Schizophrenia as construed by Szazs and the neo-Kraepelinians. In J. A. Schaler, (Ed.), *Szasz under fire* (pp. 301-320). Peru, IL: Carus Publishing.

Bentz, V. M., & Shapiro, J. J. (1998). *Mindful inquiry in social research.* Thousand Oaks, CA: Sage Publications.

Bitbol, M. (1996). *Schrödinger's philosophy of quantum mechanics.* Dordrecht, The Netherlands: Kluwer Academic Publishers.

Bleuler, M. (1968). A 23-year longitudinal study of 208 schizophrenics and impressions in regard to the nature of schizophrenia. In D. Rosenthal & S. Kety (Eds.), *The transmission of schizophrenia* (pp. 3-12). New York, NY: Pergamon Press.

Bodhi, B. (n.d.). *The trilogy of anicca, dukkha, and anatta.* Retrieved from Beyond the Net website: http://www.beyondthenet.net/dhamma/dh_main22.htm

Bohm, D., & Hiley, B. J. (1993). *The undivided universe.* New York, NY: Routledge.

Bola, J., & Mosher, L. (2003). Treatment of acute psychosis without neuroleptics: Two-year outcomes from the Soteria project. *Journal of Nervous and Mental Disease, 191*(4), 219-229. doi:10.1097/00005053-200304000-00002

Breeding, J. (2008). To see or not to see "schizophrenia" and the possibility of full "recovery." *Journal of Humanistic Psychology, 48(4)*, 489-504. Retrieved from Sage Journals Online: http://jhp.sagepub.com.ezproxy.humanisticpsychology.org:2048/cgi/reprint/48/4/489

Breggin, P. (2008a). *Brain-disabling treatments in psychiatry: Drugs, electroshock and the psychopharmaceutical complex.* New York, NY: Springer Publication Company.

Breggin, P. (2008b). *Medication madness: The role of psychiatric drugs in cases of violence, suicide, and crime.* New York, NY: St. Martin's Press.

Bremner, J., Randall, P., Scott, T. M., & Bronen, R. A. (1995). MRI-based measurement of hippocampal volume in patients with combat-related posttraumatic stress disorder. *The American Journal of Psychiatry*, 152(7), 973-981. Retrieved from http://www.ncbi.nlm.nih.gov/pubmed/7793467

Brett, C. (2002). Psychotic and mystical states of being: Connections and distinctions. *Philosophy, Psychiatry & Psychology : PPP,* 9(4), 321-341. doi:10.1353/ppp.2003.0053

British Psychological Society [BPS]. (2000). *Recent advances in understanding mental illness and psychotic experiences. A report by The British Psychological Society Division of Clinical Psychology.* Leicester, UK: Author.

Buber, M. (1996). *I and Thou* (W. Kaufman, Trans.). New York: Touchstone. (Original work published 1923)

Burt, D., Creese, I., & Snyder, S. (1977). Antischizophrenic drugs: Chronic treatment elevates dopamine receptor binding in brain. *Science (New York, N.Y.), 196*(4287), 326-328. doi:10.1126/science.847477

Calabrese, J. D., & Corrigan, P. W. (2005). Beyond dementia praecox: Findings from long-term follow-up studies of schizophrenia. In R. O. Ralph & P. W. Corrigan (Eds.), *Recovery in mental illness* (pp. 85-99). Washington, D.C.: American Psychological Association.

Carlsson, A., & Lindqvist, M. (1963). Effect of chlorpromazine or haloperidol on formation of 3methoxytyramine and normetanephrine in mouse brain. *Acta Pharmacologica Et Toxicologica, 20,* 140-144. Retrieved from http://www.ncbi.nlm.nih.gov/pubmed/14060771

Carpenter, W., McGlashan, T., & Strauss, J. (1977). The treatment of acute schizophrenia without drugs: An investigation of some current assumptions. *The American Journal of Psychiatry, 134*(1), 14-20. Retrieved fromhttp://psychrights.org/Research/Digest/Chronicity/treatacute.pdf

Chadwick, P. K. (1992). *Borderline: A psychological study of paranoia and delusional thinking.* London: Routledge.

Chadwick, P. K. (2001). Sanity to supersanity to insanity: A personal journey. In I. Clarke (Ed.), *Psychosis and spirituality: Exploring the new frontier* (pp. 75-89). London, UK: Whurr.

Chakos, M., Lieberman, J., Bilder, R., & Borenstein, M. (1994). Increase in caudate nuclei volumes of first-episode schizophrenic patients taking antipsychotic drugs. *The American Journal of Psychiatry, 151*(10), 1430-1436. Retrieved from http://psychrights.org/research/Digest/Chronicity/caudate.pdf

Chouinard, G., Jones, B., & Annable, L. (1978). Neuroleptic-induced supersensitivity psychosis. *The American Journal of Psychiatry, 135*(11), 1409-1410. Retrieved from http://psychrights.org/research/Digest/Chronicity/nids.pdf

Ciompi, L. (1980). Catamnestic long-term study on the course of life and aging of schizophrenics. *Schizophrenia Bulletin, 6*(4), 606-618. Retrieved from http://schizophreniabulletin.oxfordjournals.org/content/6/4/606.full.pdf+html

Claridge, G. (2001). Spiritual experience: Healthy psychoticism? In I. Clarke (Ed.), *Psychosis and spirituality: Exploring the new frontier* (pp. 90-106). London, UK: Whurr.

Clarke, I. (2001). Psychosis and spirituality: The discontinuity model. In I. Clarke (Ed.), *Psychosis and spirituality: Exploring the new frontier* (pp. 129-142). London, UK: Whurr.

Clarke, I. (2008). *Madness, mystery and the survival of God.* Winchester, UK: O Books.

Clarke, I. (2010). Psychosis and spirituality: The discontinuity model. In I. Clarke (Ed.), *Psychosis and spirituality: Consolidating the new paradigm (2nd ed.)* (pp. 101-114). West Sussex, UK: Wiley-Blackwell.

Clover. (1999). *Escape from psychiatry: The autobiography of Clover.* Ignacio, CO: Rainbow Pots and Press.

Cortright, B. (1997). *Psychotherapy and spirit: Theory and practice in transpersonal psychotherapy.* Albany: State University of New York Press.

Crow, T. J. (2007). How and why genetic linkage has not solved the problem of psychosis: Review and hypothesis. *American Journal of Psychiatry, 164*(1), 13-21. Retrieved from the American Journal of Psychiatry website: http://ajp.psychiatryonline.org/cgi/content/full/164/1/13

Cullen, K., Kumra, S., Regan, J., Westerman, M., & Schulz, S. (2008). Atypical antipsychotics for treatment of schizophrenia spectrum disorders. *Psychiatric Times, 25*(3), 61-66. Retrieved from http://www.psychiatrictimes.com/schizophrenia/content/article/10168/1147536

Davies, L., Lewis, S., Jones, P., Barnes, T., Gaughran, F., Hayhurst, K., & … Lloyd, H. (2007). Cost-effectiveness of first- v. second-generation antipsychotic drugs: Results from a randomised controlled trial in schizophrenia responding poorly to previous therapy. *British Journal of Psychiatry, 191*, 14-22. doi:10.1192/bjp.bp.106.028654

Deegan, P. (1996). Recovery as a journey of the heart. *Psychiatric Rehabilitation Journal, 19*(3), 91. Retrieved from http://www.bu.edu/cpr/repository/articles/pdf/deegan1996.pdf

DeLisi, L. (2000). Critical overview of current approaches to genetic mechanisms in schizophrenia research. *Brain Research. Brain Research Reviews, 31*(2-3), 187-192. doi:10.1016/S0165-0173(99)00036-3

Department of Health and Human Services. (1993, December). [FDA approval letter to Janssen Research Foundation]. Retrieved from http://psychrights.org/research/Digest/NLPs/TempleRisperdal%281993%29.pdf

DeSisto, M., Harding, C., McCormick, R., Ashikaga, T., & Brooks, G. (1995). The Maine and Vermont three-decade studies of serious mental illness. I. Matched comparison of cross-sectional outcome. *The British Journal of Psychiatry: The Journal of Mental Science, 167*(3), 331-338. Retrieved from http://www.bu.edu/resilience/examples/desisto-etal1995a.pdf

Dickins, W. T., & Flynn, J. R. (2001). Heritability estimates versus large environmental effects: The IQ paradox resolved. *Psychological Review 108*, 346-369. Retrieved from http://www.apa.org/journals/features/rev1082346.pdf

Dorman, D. (2003). *Dante's cure*. New York, NY: Other Press.

Draper, M. L., Velligan, D. I., & Tai, S. S. (2010). Cognitive behavioral therapy for schizophrenia: A review of recent literature and meta-analyses. *Minerva Psichiatrica, 51*(2), 85-94.

Einstein, A. (2003). *The meaning of relativity.* (E. Plimpton, Trans.). New York, NY: Routledge. (Original work published 1922)

Farber, S. (1993). *Madness, heresy, and the rumor of angels: The revolt against the mental health system*. Chicago, IL: Open Court Publishing.

The four immeasurables: Love, compassion, joy, and equanimity. (2006). Retrieved from http://www.buddhism.kalachakranet.org/immeasurables_love_compassion_equanimity_rejoicing.html

Frith, C. (1979). Consciousness, information processing and schizophrenia. *British Journal of Psychiatry, 134*, 225-235. doi:10.1192/bjp.134.3.225

Galotti, K. M. (2008). *Cognitive psychology: In and out of the laboratory (4th ed.)*. USA: Thomson Wadsworth.

Galynker, I., & Nazarian, D. (1997). Akathisia as violence. *The Journal of Clinical Psychiatry, 58*(1), 31-32. Retrieved from http://www.antidepressantsfacts.com/Akathisia%20as%20Violence.htm

GAP (Group for the Advancement of Psychiatry). (1976). *Mysticism: Spiritual quest or psychic disorder?* New York, NY: GAP Publications.

Gardiner, H. (2009, October 22). Drug Makers Are Advocacy Group's Biggest Donors. *The New York Times*, p. A23.

Gaser, C., Nenadic, I., Buchsbaum, B. R., Hazlett, E. A., & Buchsbaum, M. S. (2004). Ventricular enlargement in schizophrenia related to volume reduction of the thalamus, striatum, and superior temporal cortex. *American Journal of Psychiatry, 161*,154-156. Retrieved from http://ajp.psychiatryonline.org/cgi/reprint/161/1/154

Geddes, J., Freemantle, N., Harrison, P., & Bebbington, P. (2000). Atypical antipsychotics in the treatment of schizophrenia: Systematic overview and meta-regression analysis. *British Medical Journal, 321*(7273), 1371-1376. doi:10.1136/bmj.321.7273.1371

Geekie, J., & Read, J. (2009). *Making sense of madness: Contesting the meaning of schizophrenia*. New York, NY: Routledge.

Goswami, A. (1993). *The self-aware universe: How consciousness creates the material world*. New

York, NY: Jeremy P. Tarcher/Putnam.

Gottdiener, W. H. (2007). Psychodynamic psychotherapy for schizophrenia: Empirical support. In J. Read, L. R. Mosher, & R. P. Bentall (Eds.), *Models of madness: Psychological, social and biological approaches to schizophrenia* (pp. 307-318). New York, NY: Routledge.

Greenberg, J. (1964). *I never promised you a rose garden.* Chicago: Signet.

Greene, B. (2003). *The elegant universe: Superstrings, hidden dimensions, and the quest for the ultimate theory.* New York, NY: W. W. Norton & Company.

Grof, C., & Grof, S. (1989). *Spiritual emergency.* New York, NY: Tarcher.

Grof, S. (1986). *Beyond the brain: Birth, death, and transcendence in psychotherapy.* Albany, NY: State University of New York Press.

Grof, S. (2000). *Psychology of the future: Lessons from modern consciousness research.* Albany, NY: State University of New York Press.

Grof, S. (2001). *LSD psychotherapy.* Sarasota, FL. Multidisciplinary Association for Psychotherapy Studies.

Gur, R., Maany, V., Mozley, D., Swanson, C., & Bilker, W. (1998). Subcortical MRI volumes in neuroleptic-naive and treated patients with schizophrenia. *The American Journal of Psychiatry, 155*(12), 1711-1717. Retrieved from http://www.madinamerica.com/madinamerica.com/Schizophrenia_files/subcortical.pdf

Guttmacher, M. (1964). Phenothiazine treatment in acute schizophrenia; Effectiveness: The National Institute of Mental Health Psychopharmacology Service Center Collaborative Study Group. *Archives of General Psychiatry, 10,* 246-261. Retrieved from http://archpsyc.ama-assn.org/cgi/reprint/10/3/246

Hagen, B. F., Nixon, G., & Peters, T. (2010). The greater of two evils? How people with transformative psychotic experiences view psychotropic medications. *Ethical Human Psychology and Psychiatry: An International Journal of Critical Inquiry, 12*(1), 44-59.

Harding, C., Zubin, J., & Strauss, J. (1987). Chronicity in schizophrenia: Fact, partial fact, or artifact? *Hospital & Community Psychiatry, 38*(5), 477-486. Retrieved from http://psychservices.psychiatryonline.org/cgi/reprint/38/5/477

Harrow, M., & Jobe, T. (2007). Factors involved in outcome and recovery in schizophrenia patients not on antipsychotic medications: A 15-year multifollow-up study. *Journal of Nervous and Mental Disease, 195*(5), 406-414. Retrieved from http://www.madinamerica.com/madinamerica.com/Schizophrenia_files/OutcomeFactors.pdf

Harrow, M., Jobe, T. H., & Faull, R. N. (2012). Do all schizophrenia patients need antipsychotic treatment continuously throughout their lifetime? A 20-year longitudinal study. *Psychological Medicine, First View Articles,* 1-11.doi: 10.1017/S0033291712000220

Hart, W. (1987). *The art of living: Vipassana meditation as taught by S. N. Goenka.* New York, NY: HarperCollins.

Hawking, S. (2005). *A briefer history of time.* New York, NY: Random House.

Healy, D. (2008). *Psychiatric drugs explained* (5th ed.). London: Elsevier.

Healy, D., Harris, M., Tranter, R., Gutting, P., Austin, R., Jones-Edwards, G, & Roberts, A. (2006). Lifetime suicide rates in treated schizophrenia: 1875-1924 and 1994-1998 cohorts compared. *British Journal of Psychiatry, 188*(3), 223-228. doi:10.1192/bjp.188.3.223

Heery, M. W. (1989). Inner voice experiences: An exploratory study of thirty cases. *The Journal of Transpersonal Psychology, 21*(1), 73-82.

Hegarty, J., Baldessarini, R., Tohen, M., & Waternaux, C. (1994). One hundred years of schizophrenia: A meta-analysis of the outcome literature. *The American Journal of Psychiatry, 151*(10), 1409-1416. Retrieved from http://psychrights.org/research/Digest/Chronicity/100years.pdf

Hemsley, D. (1998). The disruption of the 'sense of self' in schizophrenia: Potential links with disturbances of information processing. *British Journal of Medical Psychology, 71*(2),

115-124. Retrieved from PsycINFO database. (Accession No. 1998-02915-001)

Henquet, C., Krabbendam, L., Spauwen, J., Kaplan, C., Lieb, R., Wittchen, H., & Van Os, J. (2005). Prospective cohort study of cannabis use, predisposition for psychosis, and psychotic symptoms in young people. *BMJ: British Medical Journal (International Edition), 330*(7481), 11-14.

Herman, J. (1997). *Trauma and recovery: The aftermath of violence—from domestic abuse to political terror.* New York, NY: Basic Books.

Heston, L. (1966). Psychiatric disorders in foster home reared children of schizophrenic mothers. *The British Journal of Psychiatry: The Journal of Mental Science, 112*(489), 819-825. doi:10.1192/bjp.112.489.819

Hoffman, A. (1979). *LSD: My problem child.* Sarasota, FL: MAPS.

Honig, A., Romme, M., Ensink, B., Escher, S., Pennings, M., & de Vries, M. (1998). Auditory hallucinations: A comparison between patients and nonpatients. *The Journal of Nervous and Mental Disease, 186,* 646-651.

Hopper, K., Harrison, G., Janca, A., & Sartorius, N. (2007). *Recovery from schizophrenia: An international perspective: A report from the WHO Collaborative Project, The International Study of schizophrenia.* New York, NY: Oxford University Press

House, R. (2001). Psychopathology, psychosis and the kundalini: Postmodern perspectives on unusual subjective experience. In I. Clarke (Ed.), *Psychosis and spirituality: Exploring the new frontier* (pp. 75-89). London: Whurr Publishers.

Huber, G., Gross, G., & Schuttler, R. (1975). A long-term follow-up study of schizophrenia: Psychiatric course of illness and prognosis. *Acta Psychiatrica Scandinavica,* 52(1), 49-57. doi:10.1111/j.1600-0447.1975.tb00022.x

Hyman, S., & Nestler, E. (1996). Initiation and adaptation: A paradigm for understanding psychotropic drug action. *The American Journal of Psychiatry, 153*(2), 151-162. Retrieved from http://www.ncbi.nlm.nih.gov/pubmed/8561194

IMS Health. (2010). *IMS Health Reports U.S. Prescription Sales Grew 5.1 Percent in 2009, to $300.3 Billion.* Retrieved from IMS Health website: http://www.imshealth.com/portal/site/imshealth/menuitem.a46c6d4df3db4b3d88f611019418c22a/?vgnextoid=d690a27e9d5b7210VgnVCM100000ed152ca2RCRD

Jackson, M. (2001). Psychotic and spiritual experience: A case study comparison. In I. Clarke (Ed.), *Psychosis and spirituality: Exploring the new frontier* (pp. 165-190). London: Whurr Publishers.

Jackson, M. C., & Fulford, K. W. M. (2002). Psychosis good and bad: Values-based practice and the distinction between pathological and nonpathological forms of psychotic experience. *Philosophy, Psychiatry & Psychology : PPP, 9*(4), 321-341. doi:10.1353/ppp.2003.0059

Johnstone, E. C., Frith, C. D., Crow, T. J., & Owens, D. C. (1988). The Northwick Park "functional" psychosis study: Diagnosis and treatment response. *The Lancet,(8603),* 119. Retrieved from Research Library Core. (Document ID: 8739104)

Joseph, J. (2002). Twin studies in psychiatry and psychology: Science or pseudoscience? *Psychiatric Quarterly, 73*(1), 71-82.doi:10.1023/A:1012896802713

Joseph, J. (2004). Schizophrenia and heredity: Why the emperor has no genes. In J. Read, L. R. Mosher, & R. P. Bentall (Eds.), *Models of madness: Psychological, social and biological approaches to schizophrenia* (pp. 67-83). New York, NY: Routledge.

Joukamaa, M., Heliövaara, M., Knekt, P., Aromaa, A., Raitasalo, R., & Lehtinen, V. (2006). Schizophrenia, neuroleptic medication and mortality. *British Journal of Psychiatry, 188*(2), 122-127. doi:10.1192/bjp.188.2.122

Kagan, J., Reznick, S., & Snidman, N. (1988). Biological bases of childhood shyness. *Science 240*(4849), 167-71. Retrieved from Research Library Core. (Document ID: 1789076)

Karen, R. K. (1994). *Becoming attached: First relationships and how they shape our capacity to love.* Oxford, UK: Oxford University Press.

Karon, B. P. (2003). The tragedy of schizophrenia without psychotherapy. *Journal of the American Academy of Psychoanalysis, 31*(1), 89-119. doi:10.1521/jaap.31.1.89.21931

Karon, B. P., & VandenBos, G. (1996). *Psychotherapy of schizophrenia: The treatment of choice.* Lanham, MD: Rowman & Littlefield Publishing, Inc.

Keefe, R., Bollini, A., & Silva, S. (1999). Do novel antipsychotics improve cognition? A report of a meta-analysis. *Psychiatric Annals, 29*(11), 623-629. Retrieved from http://psychrights.org/research/Digest/NLPs/RWhitakerAffidavit/KeefeDoNovelAntipsychotics ImproveCognition.pdf

Kendell, R. E., & Gourlay, J. A. (1970). The clinical distinction between the affective psychoses and schizophrenia, *British Journal of Psychiatry, 117*, 261-266.

Kety, S., Wender, P., Jacobsen, B., & Ingraham, L. (1994). Mental illness in the biological and adoptive relatives of schizophrenic adoptees: Replication of the Copenhagen study in the rest of Denmark. *Archives of General Psychiatry, 51*(6), 442-455. Retrieved from http://ajp.psychiatryonline.org/cgi/reprint/140/6/720

Kierkegaard, S. (1967). *The concept of dread* (W. Lowrie, Ed. & Trans.). Princeton, NJ: Princeton University Press. (Original work published 1844)

Knable, M., Heinz, A., Raedler, T., & Weinberger, D. (1997). Extrapyramidal side effects with risperidone and haloperidol at comparable D2 receptor occupancy levels. *Psychiatry Research: Neuroimaging, 75*(2), 91-101. doi:10.1016/S0925-4927(97)00023-1

Knapp, C. (2008). Windhorse therapy: Creating environments that rouse the energy of health and sanity. In F. J. Kaklauskas, S. Nimanheminda, L. Hoffman, and S. J. MacAndrew (Eds.), *Brilliant sanity: Buddhist approaches to psychotherapy* (pp. 275-297). Colorado Springs, CO: University of the Rockies Press.

Konings, M. M., Stefanis, N. N., Kuepper, R. R., de Graaf, R. R., Have, M., van Os, J. J., & ... Henquet, C. C. (2012). Replication in two independent population-based samples that childhood maltreatment and cannabis use synergistically impact on psychosis risk. *Psychological Medicine, 42*(1), 149-159. doi:10.1017/S0033291711000973

Kornhuber, J., Riederer, P., Reynolds, G., Beckmann, H., Jellinger, K., & Gabriel, E. (1989). 3H-spiperone binding sites in post-mortem brains from schizophrenic patients: Relationship to neuroleptic drug treatment, abnormal movements, and positive symptoms. *Journal of Neural Transmission, 75*(1), 1-10. doi:10.1007/BF01250639

Koseki, T., & Nabeshima, T. (2010). [Phencyclidine abuse, dependence, intoxication, and psychosis]. *Nihon Rinsho. Japanese Journal Of Clinical Medicine, 68*(8), 1511-1515.

Kuepper, R. R., van Os, J. J., Lieb, R. R., Wittchen, H. U., & Henquet, C. C. (2011a). Do cannabis and urbanicity co-participate in causing psychosis? Evidence from a 10-year follow-up cohort study. *Psychological Medicine, 41*(10), 2121-2129. doi:10.1017/S0033291711000511

Kuepper, R. R., Van Os, J. J., Lieb, R. R., Wittchen, H. U., Höfler, M., & Henquet, C. (2011b). Continued cannabis use and risk of incidence and persistence of psychotic symptoms: 10 year follow-up cohort study. *BMJ: British Medical Journal (Overseas & Retired Doctors Edition), 342*(7796), 537. doi:10.1136/bm|.d738

Laing, R.D. (1967). *The politics of experience.* New York: Pantheon Books.

Lajoie, D. H., & Shapiro, S. I. (1992). Definitions of transpersonal psychology: The first twenty-three years. *Journal of Transpersonal Psychology, 24*(1), 79-98. Retrieved from Journal of Transpersonal Psychology website: http://jhp.sagepub.com

Lecomte, T., Mueser, K. T., MacEwan, W. G., Laferrière-Simard, M., Thornton, A. E., Buchanan, T., & ... Honer, W. G. (2010). Profiles of individuals seeking psychiatric help for psychotic symptoms linked to methamphetamine abuse—Baseline results from the MAPS (methamphetamine and psychosis study). *Mental Health And Substance Use: Dual Diagnosis, 3*(3), 168-181. doi:10.1080/17523281.2010.504645

Lee, T., & Seeman, P. (1980). Elevation of brain neuroleptic/dopamine receptors in schizophrenia. *The American Journal of Psychiatry, 137*(2), 191-197. Retrieved from http://ajp.psychiatryonline.org/cgi/reprint/137/2/191

Levine, P. (1997). *Waking the tiger: Healing trauma*. Berkeley, CA: North Atlantic Books.

Lewine, R. (1998). Epilogue. In M. F. Lenzenweger & R. H. Dworkin (Eds.), *Origin and development of schizophrenia* (pp. 493-503). Washington, DC: American Psychological Association.

Lieberman, J., Stroup, T., McEvoy, J., Swartz, M., Rosenheck, R., Perkins, D., & ... Hsiao, J. (2005). Effectiveness of antipsychotic drugs in patients with chronic schizophrenia. *The New England Journal of Medicine, 353*(12), 1209-1223. doi:10.1056/NEJMoa051688

Livingston, R. (1987). Sexually and physically abused children. *The Journal of the American Academy of Child and Adolescent Psychiatry*, 26: 413-415.

Lukoff, D. (1985). The diagnosis of mystical experiences with psychotic features. *Journal of transpersonal Psychology 17*(2), 155-181. Retrieved from The Journal of Transpersonal Psychology web site: http://jhp.sagepub.com

Mackay, A. (1982). Increased brain dopamine and dopamine receptors in schizophrenia. *Archives of General Psychiatry, 39*(9), 991-997. Retrieved from http://archpsyc.ama-assn.org/cgi/reprint/39/9/991

Mackler, D. (Producer). (2008). *Take these broken wings: Recovery from Schizophrenia without medication* [DVD]. Available from www.iraresoul.com

Madsen, A., Keiding, N., Karle, A., Esbjerg, S., & Hemmingsen, R. (1998). Neuroleptics in progressive structural brain abnormalities in psychiatric illness. *Lancet, 352*(9130), 784-785. doi:10.1016/S0140-6736(05)60678-2

Maguire, E., Gadian, D., Johnsrude, I., Good, C., Ashburner, J., Frackowiak, R., & Frith, C. (2000). Navigation-related structural change in the hippocampi of taxi drivers. *Proceedings of the National Academy of Sciences of the United States of America, 97*(8), 4398-4403. doi:10.1073/pnas.070039597

Mahler, M. S., Pine, F., & Bergman, A. (1973). *The Psychological Birth of the Human Infant*, New York: Basic Books.

Margree, V. (2002). Normal and abnormal: Georges Canguilhem and the question of mental pathology. *Philosophy, Psychiatry & Psychology : PPP, 9*(4), 299-312. doi:10.1353/ppp.2003.0056

Markowitz, F. E. (2005) Sociological models of recovery. In R. O. Ralph & P. W. Corrigan (Eds.), *Recovery in mental illness* (pp. 85-99). Washington, D.C.: American Psychological Association.

Marneros, A., Deister, A., Rohde, A., & Steinmeyer, E. (1989). Long-term outcome of schizoaffective and schizophrenic disorders: A comparative study: I. Definitions, methods, psychopathological and social outcome. *European Archives of Psychiatry & Neurological Sciences, 238*(3), 118-125. doi:10.1007/BF00450998

Maslow, A. H. (1968). *Toward a psychology of being* (2nd ed.). New York: D. Van Nostrand.

Mattes, J. (1997). Risperidone: How good is the evidence for efficacy? *Schizophrenia Bulletin, 23*(1), 155-161. Retrieved from http://schizophreniabulletin.oxfordjournals.org/content/23/1/155.full.pdf

May, R. (1977). *The meaning of anxiety.* New York: W. W. Norton & Company.

McGlashan, T. (1984a). The Chestnut Lodge follow-up study. I. Follow-up methodology and study sample. *Archives of General Psychiatry, 41*(6), 573-585. Retrieved from http://archpsyc.ama-assn.org/cgi/reprint/41/6/573

McGlashan, T. (1984b). The Chestnut Lodge follow-up study. II. Long-term outcome of schizophrenia and the affective disorders. *Archives of General Psychiatry, 41*(6), 586-601. Retrieved from http://archpsyc.ama-assn.org/cgi/reprint/41/6/586

McGuire, P. (2000). New hope for people with schizophrenia. *American Psychological Association's Monitor on Psychology, 31* (2), 24-26.

McKenna, P. J. (1994). *Schizophrenia and related syndromes.* Oxford, UK: Oxford University Press.

Mehta, S., & Farina, A. (1997). Is being 'sick' really better? Effect of disease view of mental disorder on stigma. *Journal of Social and Clinical Psychology, 16*(4), 405-419.

Mikami, T., Naruse, N., Fukura, Y., Ohkubo, H., Ohkubo, T., Matsuura, M., & ... Kojima, T. (2003). Determining vulnerability to schizophrenia in methamphetamine psychosis using exploratory eye movements. *Psychiatry And Clinical Neurosciences, 57*(4), 433-440. doi:10.1046/j.1440-1819.2003.01143.x

Mindell. A. (2008). *City shadows: Psychological interventions in psychiatry.* New York, NY: Routledge.

Modrow, J. (2003). *How to become a schizophrenic: The case against biological psychiatry.* Lincoln, NE: Writers Club Press.

Moghaddam, F. M., Taylor, D.M., & Wright, S. C. (1993). *Social psychology in cross-cultural perspective.* New York, NY: W. H. Freeman and Company.

Moore, T., Zammit, S., Lignford-Hughes, A., Barnes, T., Jones, P. B., Burke, M., & Lewis, G. (2007). Cannabis use and risk of psychotic or affective mental health outcomes: a systematic review. *Lancet, 370*(9584), 319-328.

Morrison, A. P. (2007). Cognitive therapy for people with psychosis. (2007). In J. Read, L. R. Mosher, & R. P. Bentall (Eds.), *Models of madness: Psychological, social and biological approaches to schizophrenia* (pp. 291-306). New York, NY: Routledge.

Mosher, L. R. (1999). Soteria and other alternatives to acute psychiatric hospitalization: A personal and professional review. *The Journal of Nervous and Mental Disease, 187*, 142-149.

Mosher. L. R., & Hendrix, V. (with Fort, D. C.) (2004). *Soteria: Through madness to deliverance.* USA: Authors.

National Association of State Mental Health Program Directors [NASMHPD]. (2006). *Morbidity and Mortality in People with Serious Mental Illness.* Retrieved from http://www.nasmhpd.org/general_files/publications/med_directors_pubs/Technical%20Report%20on%20Morbidity%20and%20Mortaility%20-%20Final%2011-06.pdf

National Institute of Mental Health [NIMH]. (2010a). *Schizophrenia.* Retrieved from http://www.nimh.nih.gov/health/publications/schizophrenia/complete-publication.shtml

National Institute of Mental Health [NIMH]. (2010b). *How is schizophrenia treated.* Retrieved from http://www.nimh.nih.gov/health/publications/schizophrenia/how-is-schizophrenia-treated.shtml

Nelson, J. (1990). *Healing the split.* Albany, New York: State University of New York Press.

Nixon, G., Hagen, B. F., & Peters, T. (2009). Psychosis and transformation: A phenomenological inquiry. *International Journal of Mental Health and Addiction.* doi: 10.1007/s11469-009-9231-3

Nixon, G., Hagen, B. F., & Peters, T. (2010). Recovery from psychosis: A phenomenological inquiry. *International Journal of Mental Health and Addiction.* doi: 10.1007/s11469-010-9271-8

O'Callaghan, M. (2001). [Interview of J. W. Perry]. *Mental breakdown as healing process.* Retrieved from http://www.global-vision.org/interview/perry.html

Ogawa, K., Miya, M., Watarai, A., & Nakazawa, M. (1987). A long-term follow-up study of schizophrenia in Japan—with special reference to the course of social adjustment. *British Journal of Psychiatry, 151*,758-765. doi: 10.1192/bjp.151.6.758

Perls, F., Hefferline, R., & Goodman, P. (1951). *Gestalt therapy: Excitement and growth in the human personality.* Gouldsboro, ME: The Gestalt Journal Press.

Perry, J. W. (1974). *The far side of madness.* Englewood Cliffs, NJ: Prentice-Hall.

Perry, J. W. (1976). *Roots of renewal in myth and madness.* San Francisco: Jossey-Bass.

Perry, J. W. (1987). *The self in psychotic process.* Dallas, TX: Spring Publications.

Perry, J. W. (1999). *Trials of the visionary mind.* State University of New York Press.

Peters, E. (2001). Are delusions on a continuum? The case of religious and delusional beliefs. In I. Clarke (Ed.), *Psychosis and spirituality: Exploring the new frontier* (pp. 191-207). London, UK: Whurr.

Podvoll, E. (1990). *Recovering sanity: A compassionate approach to understanding and treating psychosis.* Boston: Shambhala Publications, Inc.

Rank, O. (1936). *Will therapy*. New York, NY: W. W. Norton & Company.

Rappaport, M. (1978). Are there schizophrenics for whom drugs may be unnecessary or contraindicated? *International Pharmacopsychiatry, 13*(2), 100-111. Retrieved May 26, 2010, from PsycInfo database.

Rathus, S. A. (2006). *Childhood and adolescence: Voyages in development.* Belmont, Canada: Thompson Wadsworth.

Read, J., Perry, B. D., Moskowitz, A., & Connolly, J. (2001). The contribution of early traumatic events to schizophrenia in some patients: A traumagenic neurodevelopmental model. *Psychiatry, 64*(4), 319-345.

Read, J. (2004). Biological psychiatry's lost cause. In J. Read, L. R. Mosher, & R. P. Bentall, (Eds.), *Models of madness: Psychological, social and biological approaches to schizophrenia* (pp. 57-65). New York: Routledge.

Read, J., van Os., J., Morrison, A. P., & Ross, C. A. (2005). Deviant children grown up: A sociological and psychiatric study of sociopathic personality. *Acta Psychiatric Scandinavia, 112,* 330-350. DOI:10.1111/j.1600-0447.2005.00634.x

Read, J., Haslam, N., Sayce, L., Davies, E. (2006). Prejudice and schizophrenia: A review of the "Mental illness is an Illness like any other" approach.' *Acta Psychiatrica Scandinavica, 114,* 303-318.

Read, J. (2007). Why promoting biological ideology increases prejudice against people labeled "schizophrenic." *Australian Psychologist, 42,* 118-128.

Robbins, L. N., Locke, B. Z., & Reiger, D. A. (1991). An overview of psychiatric disorders in America, in L. N. Robins & B. Z. Locke (eds.), *Psychiatric Disorders in America* (pp. 328-366). New York: Free Press.

Robins, L. (1974). *Deviant children grown up: A sociological and psychiatric study of sociopathic personality*. Malabar, FL: R. E. Krieger Pub. Co.

Romme, M., Escher, S., Dillon, J., Corstens, D., & Morris, M. (2009). *Living with voices: 50 stories of recovery.* Herfordshire, UK: PCCS Books.

Rosebush, P., & Mazurek, M. (1999). Neurologic side effects in neuroleptic-naive patients treated with haloperidol or risperidone. *Neurology, 52*(4), 782-785. Retrieved from http://www.neurology.org/content/54/7/1542.3.full.pdf+html

Rosenthal, D. (1971). The adopted-away offspring of schizophrenics. *The American Journal of Psychiatry, 128*(3), 307-311. Retrieved from http://ajp.psychiatryonline.org/cgi/reprint/128/3/307

Rubik, B. (2002). The biofield hypothesis: Its biophysical basis and role in medicine. *The Journal of Alternative and Complementary Medicine, 8*(6), pp. 703-717.

Satcher, D. (1999). *Etiology of schizophrenia.* Retrieved from http://www.surgeongeneral.gov/library/mentalhealth/chapter4/sec4_1.html

Sawyer, K. (1999, October). Unveiling the universe. *National Geographic, 196*(4), 8. Retrieved from Academic Search Premier database. (Accession no. 2325654)

Schneider, K. J. (1999). *The paradoxical self: Toward an understanding of our contradictory nature*. Amherst, NY: Humanity Books.

Schneider, K. J. (2008). Theory of the Existential-Integrative (EI) approach. In K. J. Schneider (Ed.), *Existential-Integrative Psychotherapy* (pp. 35-48). New York, NY: Taylor & Francis Group.

Schooler, N., Goldberg, S., Boothe, H., & Cole, J. (1967). One year after discharge: Community adjustment of schizophrenic patients. *The American Journal of Psychiatry, 123*(8), 986-995. Retrieved from http://www.madinamerica.com/madinamerica.com/Schizophrenia_files/Schooler%283%29.PDF

Schore, A. N. (2002). Advances in neuropsychoanalysis, attachment theory, and trauma research: Implications for self psychology. *Psychoanalytic Inquiry, 22*, 433-484.

Sechehaye, M. (1951). *Autobiography of a schizophrenic girl.* New York, NY: Meridian.

Seikkula, J., Aaltonen, J., Alakare, B., Haarakangas, K., Keränen, J., & Lehtinen, K. (2006). Five-year experience of first-episode nonaffective psychosis in open-dialogue approach: Treatment principles, follow-up outcomes, and two case studies. *Psychotherapy Research, 16*(2), 214-228. doi:10.1080/10503300500268490.

Shear, M., Frances, A., & Weiden, P. (1983). Suicide associated with akathisia and depot fluphenazine treatment. *Journal of Clinical Psychopharmacology, 3*(4), 235-236. doi:10.1097/00004714-198308000-00006

Siebert, A. (1999). *Brain disease hypothesis for schizophrenia disconfirmed by all evidence.* Retrieved from http://psychrights.org/states/Alaska/CaseOne/180Day/Exhibits/Wnotbraindisease.pdf

Siegel, D. J. (2007). *The mindful brain.* New York, NY: W. W. Norton & Company.

Slade, A. (1999). Attachment theory and research: Implications for the theory and practice of individual psychotherapy with adults. In J. Cassidy & P. R. Shaver (Eds.), *Handbook of attachment: Theory, research, and clinical applications* (pp. 575-594). New York: Guilford press.

Stabenau, J., Creveling, C., & Daly, J. (1970). The 'pink spot,' 3, 4-dimethoxyphenylethylamine, common tea, and schizophrenia. *The American Journal of Psychiatry, 127*(5), 611-616. Retrieved from http://ajp.psychiatryonline.org/cgi/reprint/127/5/611

Szasz, T. (2008). *Psychiatry: The science of lies.* Syracuse, NY: Syracuse University Press.

Taylor, E. (1999). *Shadow culture: Psychology and Spirituality in America.* Washington, D. C.: Counterpoint.

Teasdale, J.D., & Barnard, P. J. (1993) *Affect, cognition and change: Remodeling depressive thought.* Hove, U.K.: Lawrence Erlbaum Associates.

Tedeschi, R. G., & Calhoun, L. G. (1996). The posttraumatic growth inventory: Measuring the positive legacy of trauma. *Journal of Traumatic Stress, 9*(3), 455-471.

Thalbourne, M. A., & Delin, P. S. (1994). A common thread underlying belief in the paranormal, creative personality, mystical experience and psychopathology. *Journal of Parapsychology, 58*, 3-38.

Thornhill, H., Clare, L., & May, R. (2004). Escape, enlightenment and endurance: Narratives of recovery from psychosis. *Anthropology & Medicine, 11*(2), pp. 181-199. doi:10.1080/13648470410001678677

Tienari, P., Lahti, I., Sorri, A., Naarala, M., Moring, J., Wahlberg, K., & Wynne, L. (1987). Genetic and Psychosocial Factors in Schizophrenia: The Finnish Adoptive Family Study. *Schizophrenia Bulletin, 13*(3), 477-484. doi:10.1016/0022-3956(87)90091-4

Tienari, P. (1991). Interaction between genetic vulnerability and family environment. *Acta Psychiatrica Scandinavica 84*, 460–465.

Tobert, N. (2001). *The polarities of consciousness.* In I. Clarke (Ed.), *Psychosis and spirituality: Exploring the new frontier* (pp. 129-142). London, UK: Whurr.

Tooth, B., Kalyanasundaram, V., Glover, H., & Momenzadah, S. (2003). Factors consumers identify as important to recovery from schizophrenia. *Australasian Psychiatry, 11*, S70-S77.

doi:10.1046/j.1440-1665.11.s1.1.x

Tsuang, M., & Winokur, G. (1975). The Iowa 500: Field work in a 35-year follow-up of depression, mania, and schizophrenia. *The Canadian Psychiatric Association Journal / La Revue de l'Association des psychiatres du Canada, 20*(5), 359-365. Retrieved from http://psycnet.apa.org/?fa=main.doiLanding&uid=1976-07323-001

Turner, R. P., Lukoff, D., Barnhouse, R. T. & Lu, F. G. (1995). Religious or spiritual problem. *Journal of Nervous and Mental Disease, 183*(7), 435-443.

van Os, J., Bak, M., Hanssen, M., Bijl, R., de Graaf, R., & Verdoux, H. (2002). Cannabis use and psychosis: a longitudinal population-based study. *American Journal Of Epidemiology, 156*(4), 319-327.

Vardy, M., & Kay, S. (1983). LSD psychosis or LSD-induced schizophrenia? A multimethod inquiry. *Archives Of General Psychiatry, 40*(8), 877-883.

Volkmar, A., & Gottwalz, E. (2007). (2007). In J. Read, L. R. Mosher, & R. P. Bentall (Eds.), *Models of madness: Psychological, social and biological approaches to schizophrenia* (pp. 335-347). New York, NY: Routledge.

Waddington, J., Youssef, H., & Kinsella, A. (1998). Mortality in schizophrenia: Antipsychotic polypharmacy and absence of adjunctive anticholinergics over the course of a 10-year prospective study. *British Journal of Psychiatry, 173*, 325-329. doi:10.1192/bjp.173.4.325

Wallas, G. (1926). *The art of thought*. New York: Harcourt.

Watkins, J. (2008). *Hearing voices: A common human experience*. South Yarra, UK: Michelle Anderson Publishing.

Wallin, D. J. (2007). *Attachment in psychotherapy*. New York: The Guilford Press.

Warner, R. (1994). *Recovery from schizophrenia: Psychiatry and political economy (2nd ed.)*. London: Routledge.

Wender, P., Rosenthal, D., Kety, S., Schulsinger, F., & Welner, J. (1974). Crossfostering: A research strategy for clarifying the role of genetic and experiential factors in the etiology of schizophrenia. *Archives of General Psychiatry, 30*(1), 121-128. Retrieved from http://archpsyc.ama-assn.org/cgi/reprint/30/1/121

Whitaker, R. (2002). *Mad in America*. New York: Basic Books.

Whitaker, R. (2007). *Affidavit of Robert Whitaker*. Retrieved from http://psychrights.org/litigation/WhitakerAffidavit.pdf

Whitaker, R. (2010). *Anatomy of an epidemic: Magic bullets, psychiatric drugs, and the astonishing rise of mental illness in America*. New York, NY: Crown Publishers.

Wiersma, D., Nienhuis, F., Slooff, C., & Giel, R. (1998). Natural course of schizophrenic disorders: A 15-year follow-up of a Dutch incidence cohort. *Schizophrenia Bulletin, 24*(1), 75-85. Retrieved from http://schizophreniabulletin.oxfordjournals.org/cgi/reprint/24/1/75.pdf

Wilber, K. (2000). *Integral Psychology*. Boston, MA: Shambhalla Publications, Inc.

Williams, N., Rees, M., Holmans, P., Norton, N., Cardno, A., Jones, L., & ... Owen, M. (1999). A two-stage genome scan for schizophrenia susceptibility genes in 196 affected sibling pairs. *Human Molecular Genetics, 8*(9), 1729-1739. doi:10.1093/hmg/8.9.1729

Williams, P. (2011). *A multiple-case study exploring personal paradigm shifts throughout the psychotic process from onset to full recovery*. (Doctoral dissertation, Saybrook Graduate School and Research Center, 2011). Retrieved from http://gradworks.umi.com/34/54/3454336.html

Wirshing, W., Van Putten, T., Rosenberg, J., & Marder, S. (1992). Fluoxetine, akathisia, and suicidality: Is there a causal connection? *Archives of General Psychiatry, 49*(7), 580-581. Retrieved from http://archpsyc.ama-assn.org/cgi/reprint/49/7/580

Woodruff, P. W. R., & Lewis, S. (1996). Structural brain imaging in schizophrenia. In S. Lewis & N. Higgins (Eds.), *Brain imaging in psychiatry*. Oxford, UK: Blackwell.

Wooley, D. E., & Shaw, E. (1954). A biochemical and pharmacological suggestion about certain mental disorders. *Proceedings of the National Academy of Sciences USA*, 40, 228-231. Retrieved from http://www.ncbi.nlm.nih.gov/pmc/articles/ PMC527980/pdf/pnas00731-0006.pdf

Wright, I., Rabe-Hesketh, S., Woodruff, P., David, A., Murray, R., & Bullmore, E. (2000). Meta-analysis of regional brain volumes in schizophrenia. *The American Journal of Psychiatry*, *157*(1), 16-25. Retrieved from http://ajp.psychiatryonline.org/cgi/reprint/157/1/16.pdf

Yalom, I. D. (1980). *Existential psychotherapy*. USA: Basic Books.

Glossary*

affect-images. Sensory perceptions and emotions, often of an anomalous nature, that are intricately bound together.

agranulocytosis. The potentially fatal loss of white blood cells, a side effect of antipsychotic drug use.

akathisia. The condition in which one feels overwhelming agitation and restlessness on the inside while feeling trapped in a body that is heavily sedated and unresponsive, a common side effect of antipsychotic drug use.

ambivalent attachment. An *insecure attachment* style (see *attachment theory*) characterized by excessive concern with being abandoned and/or isolated (also sometimes referred to as *anxious attachment*). In adults, this is sometimes referred to as *preoccupied attachment*. Within the context of the *self/other dialectic*, this corresponds to excessive *isolation anxiety* and a *window of tolerance* that is skewed towards the *other pole*. (See also *attachment theory*)

anomalous belief. See *anomalous experience.*

anomalous experience. A subjective experience (typically either a belief or a sensory perception such as a sound, a vision, a taste, a smell, or a tactile sensation) that is considered invalid within the framework of *consensus reality* (according to the particular individual's society or group). These may or may not cause distress and/or limitation (see *psychotic experience*).

anomalous perception. See *anomalous experience.*

antipsychotics. Originally referred to as *major tranquilizers*, these powerful

* The terms that are italicized within these definitions represent other terms that are listed within this glossary.

drugs suppress various neural networks within the brain, leading to a general dulling of one's energy and experience, and carrying with them the potential for severe and even fatal side effects. (see also *typical antipsychotics* and *atypical antipsychotics*)

anxious attachment. See *ambivalent attachment.*

attachment theory. A psychological theory that emphasizes the patterns of relational bonding between individuals, referred to as *attachment styles.* It is generally agreed within the field that there are only a few predominant attachment styles: *secure attachment,* and three types of *insecure attachment* styles—*ambivalent attachment, avoidant attachment,* and *disorganized attachment.*

atypical antipsychotics. The second generation of antipsychotics, these were developed with the hope of carrying less side effects than the first generation of antipsychotics, the *typical antipsychotics.* Unfortunately, while the side effect profiles of these are somewhat different, they are arguably not any safer, and potentially even more dangerous. Like the typicals, they target the D2 receptors of the dopamine pathways, but they also target other neural pathways, including the serotonin and glutamate pathways. (see also *antipsychotics* and *typical antipsychotics*)

auditory hallucinations. *Anomalous perceptions* that are heard, typically in the form of voices.

avoidant attachment. An *insecure attachment* style (see *attachment theory*) characterized by excessive concern with being overwhelmed by too much connection or intimacy with others. In adults, this is sometimes referred to as *dismissive attachment.* Within the context of the *self/other dialectic,* this corresponds to excessive *engulfment anxiety* and a *window of tolerance* that is skewed towards the *self pole.* (See also *attachment theory*)

axis of rapprochement, the. This represents the spectrum of possible subjective experiences that we can have between the two poles of the *self/other dialectic,* ranging from extreme *isolation anxiety* near the *self pole* to extreme *engulfment anxiety* near the *other pole.* At any given moment, our current situation (or at least our perception of it) determines where our subjective experience lies along this axis.

bipolar disorder. A formal diagnosis within the DSM for the condition in which one's mood swings relatively dramatically between mania and depression. There are two major types listed in the DSM, with the main difference between them being the strength of the manic episode(s). Stronger manic episodes frequently include psychotic experiences.

brief psychotic disorder. A formal diagnosis within the DSM for a psychotic condition in which "signs of the disturbance" persist for at least one day but no more than one month.

cognitive constructs. The particular cognitive framework that each of us develops continuously throughout our lives, which allows us to interpret the sensory data that we take in from the world. In other words, our cognitive constructs are our thoughts, interpretations, understandings, categories, etc., that aid us in distinguishing the various phenomena that we encounter in the world.

compassion. Feeling connected with the suffering of others along with an altruistic desire to provide relief from their suffering. (See also the *four immeasurables*)

consensus reality. The set of beliefs and experiences considered to be valid according to an individual's society or group.

delusion. An anomalous belief. See *anomalous experiences.*

delusional disorder. A formal diagnosis within the DSM for the persistence of "nonbizarre delusions" persisting for at least a month and which are not accompanied by other psychotic experiences or significant limitations.

dementia praecox. Meaning literally "senility of the young," this is a term coined by Emil Kreapelin in the late 1800's to refer to his hypothesis that psychotic experiences are the manifestations of a diseased brain. This term, along with the theory behind it, eventually evolved into today's diagnostic label, *schizophrenia,* even though this theory remains unsubstantiated.

dialectic. Meaning literally "the tension between two conflicting or interacting forces, elements, or ideas," this is an important concept in most existentially-oriented theories.

dialectical monism. A term used in the West to refer to the long-standing Eastern understanding (found particularly within Hinduism, Taoism, Buddhism, and their various derivatives) that all dualistic manifestations of the world arise from a fundamental *unity.*

dialectical tension. The tension we experience that is an inherent aspect of our dualistic existence (see *duality*). There are several different ways we can think about this tension, all of which are synonymous: as the tension between *self* and *other;* as the tension between our longing for connection and our longing for autonomy; and as the tension between our fear of isolation and our fear of being engulfed by *other.* Within the *self/other dialectic,* this is represented by the tension between our two diametrically opposed existential anxieties—the *engulfment anxiety* associated with the *other pole,* and the *isolation anxiety* associated with the *self pole.* A relative balance in the dialectical tension is necessary for the survival of the self. A degree of tension at either extreme (either too high or too low) can represent a severe existential threat to the self and potentially result in *psychosis*. We subjectively feel the compulsion to maintain this balance by our desire to continuously keep our experience within a tolerable range between the two *existential anxieties* (i.e., within our *window of tolerance*). (See the *DUI* summary at the end of Chapter 23).

dismissive attachment. See *avoidant attachment.*

disorganized attachment. An *insecure attachment* style (see *attachment theory*) characterized by an apparent mix of the other two *insecure attachment* styles—*avoidant attachment* and *ambivalent attachment.* An individual exhibiting this style will often experience excessive concern with being overwhelmed by both too much connection/intimacy and also with being abandoned/isolated. Within the context of the *self/other dialectic*, this most likely corresponds to both excessive *engulfment anxiety* and excessive *isolation anxiety*, and therefore with a relatively narrow *window of tolerance.* (See also *attachment theory*)

dualistic feelings. In the context of the *DUI model*, these are the myriad feelings directly associated with the fears and desires of the *self/other dialectic*—the *engulfment anxiety*, the *isolation anxiety*, and the tension between them. (See *unitive feelings* and the *self/other dialectic*)

duality. The apparent experience of a "self" separate from "other", and a "this" separate from "that." This can also be seen as the subject/object split. While this is how most of us experience the world at the apparent level, it is fundamentally an illusion that any one manifestation of the world is entirely separate from any other.

Duality/Unity Integrative (DUI) Model, the. This is the overarching theoretical framework that includes the *self/other dialectic* existing dynamically within the fundamental qualities of *impermanence* and *interconnectedness* (i.e., that we experience ourselves as a dualistic "self" separate from "others" in spite of the fact that the world is fundamentally impermanent and interconnected). This model highlights the two core and closely related existential dilemmas with which we all struggle as a result of finding ourselves existing within this particular configuration: (1) the need to find some tenable balance between autonomy and connection; and (2) the need to maintain the existence of a dualistic self in world that is fundamentally nondual. (See the DUI model summary at the end of Chapter 23)

DUI model, the. See the *Duality/Unity Integrative model.*

dukkha. A Pali word used to refer to the suffering that is inherent within our existence. We do not have a word in the English language that captures this concept well, with "suffering" perhaps being the closest but still missing the mark in many ways. Within the context of the *DUI model*, dukkha is essentially synonymous with *dialectical tension*. (See also the *three marks of existence*)

ego strength. The degree of stability and well-foundedness of one's sense of self.

engulfment anxiety. One of the two existential anxieties of the *self/other dialectic*, this refers to our fear of being "engulfed" by "other," of losing our sense of self and autonomy as a result of too much connection. This fear corresponds with a desire for more autonomy.

entanglement. A modern physics term used to refer to the property of profound interconnectedness within the universe. (See also *interconnectedness*)

equanimity. The capacity to maintain a balanced state of mind in the midst of challenging circumstances, in contrast to *resilience* (see *resilience*).

existential. Referring to the most fundamental layer of our subjective experience—that which is most directly associated with existence of the self at this most fundamental level.

existential anxiety. The anxiety or suffering inherent within our existence—inherent because it is directly associated with the process by which we maintain our existence (see *dialectical tension*).

existential dilemma/crisis. See *existential anxiety* and the *Duality/Unity Integrative model.*

folie à deux (shared psychotic disorder). A formal diagnosis within the DSM for the condition in which more than one person share a common anomalous belief system.

four immeasurables, the. Drawn from the Buddhist literature, this term refers to the four qualities of subjective experience most directly associated with the direct experience and acceptance of the fundamental nature of the world (the *three marks of existence*). These four qualities are *unconditional love, compassion, sympathetic joy,* and *equanimity.*

full recovery. The condition of having achieved a *homeodynamic* balance in which the overall distress and limitation (and not necessarily the anomaly) associated with one's subjective experiences is the same or less than that which preceded the psychosis.

groundlessness. The subjective experience of profound instability with regard to one's sense of self and the world in general. This is often associated with *transliminal experiences.*

hallucination. An anomalous perception. See *anomalous experiences.*

hero systems. A term coined by existential thinker Ernest Becker referring to the various strategies we each develop in order to deny our mortality. These systems allow us to feel that we are a part of something larger and less mortal than our limited bodily self. We have an almost endless choice of hero systems, such as religions, nations, families, organizations, particular value systems, etc.

heroic striving. The powerful compulsion to act as a hero, which typically manifests as the desire to fight against evil and/or for the benefit of others.

homeodynamic. The condition of having returned to a dynamic balance appropriate to a newly integrated experience. This is in contrast to the term *homeostatic*, which implies the return to the condition that existed prior to experiencing a stressor. Homeodynamic is an important concept in recovery, since it suggests that recovery often does not entail the return to one's former condition but rather entails a transformation into an altogether new way of being as a result of having integrated such powerful experiences. (See also *full recovery*)

identification. The tendency for an individual to personally identify with the powerful archetypal *affect-images* that often manifest within psychosis, perhaps literally believing oneself to be, for example, the Virgin Mother or the Second Coming of Christ. This is in contrast to *projection* (see *projection*).

ignorance. The degree to which *wisdom* is lacking within one's experience (see *wisdom*).

immortality projects. See *hero systems.*

impermanence. See the *three marks of existence.*

insecure attachment. Attachment styles (see *attachment theory*) that are associated with excessive anxiety with regard to intimate relationships. It's generally agreed that there are three different types of insecure attachment styles—*ambivalent attachment, avoidant attachment,* and *disorganized attachment.* In the context of the *self/other dialectic*, these are associated with a relatively narrow and/or skewed *window of tolerance.*

interconnectedness. The concept that all manifestations of the world are fundamentally interconnected—merely being just different manifestations of a common whole. The subjective experience of this is common within *transliminal experiences*. (See also the *three marks of existence*)

isolation anxiety. One of the two existential anxieties within the *self/other dialectic*, this refers to our fear of being isolated and/or abandoned. This fear corresponds with a desire for more connection and intimacy.

medical model, the. A theoretical system based on the idea that various distressing psychological conditions are caused by discrete "mental illnesses." Advocates of this model hope that we will someday confirm that these so called "mental illnesses" are caused by various discrete physiological diseases of the brain. (See Chapter 1 for a more complete description)

mindfulness meditation. A western term used to refer to *Vipassana*, the particular practice of meditation discovered by the Buddha. The essential components of mindfulness meditation are (1) to maintain one's concentration fixed on some aspect of present experience (which is often one's natural breath or other physical sensation), and (2) to maintain *equanimity* with regard to this experience.

mystical experience. See *unitive experience.*

neuroleptic malignant syndrome (NMS). A life threatening neurological disorder caused by the use of antipsychotic drugs. It can develop extremely rapidly, going from the first sign of symptoms to death within just a few days in rare cases. Survivors often suffer from irreversible brain damage.

neuroreceptor. The part of a neuron that receives *neurotransmitters*. (See also *neurotransmitter*)

neurotransmitter. The chemical messenger that is passed from one neuron to another, allowing communication. Antipsychotics and other psychoactive drugs work by directly interfering with neurotransmission (typically either by blocking *neuroreceptors* from being able to receive neurotransmitters, or by preventing the reuptake of neurotransmitters once they have been emitted).

nonduality. A term used to refer to the idea that the world, and all manifestations within it, are profoundly interconnected and ephemeral in spite of our general experience of ourselves and the world as dual (as a "self" that exists separately from "others", and one object existing separately from "other" objects).

observing ego, the. This refers to the capacity to maintain a relatively detached awareness of one's experience. In the context of *psychosis*, this is an important component in determining how far out of alignment one's own construction of the world is from that of *consensus reality*.

organismic wisdom. The innate wisdom within all living organisms and living systems that "knows" what is necessary to maintain survival and aspire continuously towards health and growth. (See also *psyche*)

***other* pole, the.** One of the two poles of the *self/other dialectic*, the *other* pole is associated with our experience of "other" and the corresponding *engulfment anxiety*.

paranoia. The experience that one is being watched over and/or persecuted by malevolent entities.

peak experience. Coined by transpersonal psychologist Abraham Maslow, this term generally refers to a *transliminal experience* that is not overwhelming and so results in primarily positive *unitive experiences* and possibly even lasting growth.

personal paradigm. One's experience and understanding of the world and of oneself, it's closely related to the term *cognitive constructs* (see *cognitive constructs*).

phenomenological. Meaning literally "pertaining to one's direct subjective experience." This is a very important concept for the study of psychosis and other kinds of highly subjective experiences.

phenomenology. The study of direct subjective experience. This is in contrast to positivistic inquiry (or positivism), in which all importance is given to so called objective observation (that which is directly observable to the researcher), and virtually none is given to one's subjective experience.

posttraumatic growth. This refers to the strengths and benefits that result from having worked through a traumatic experience. We see clear evidence of significant posttraumatic growth for those who have recovered from psychosis.

preoccupied attachment. See *ambivalent attachment.*

projection. The tendency for an individual to project onto others and/or the world in general the powerful archetypal *affect-images* that often manifest within psychosis, especially those of a particularly benevolent or malevolent nature. This is in contrast to *identification* (see *identification*).

psyche, the. That aspect of a living organism or living system that contains self agency and continuously strives for the survival and growth of the organism. It is closely related to *organismic wisdom*, in that we can say that organismic wisdom is the inherent wisdom that guides the psyche (see *organismic wisdom*). It may also be useful to think of organismic wisdom and the psyche as two sides of the same coin, with organismic wisdom representing the guidance/direction and the psyche representing the energy/power.

psychosis. An ongoing condition in which *psychotic experiences* predominate. This condition also typically entails significant instability with regards to one's psychotic experiences.

psychotic experience. An *anomalous experience* that causes significant distress and/or limitation. (See also *anomalous experience*)

rapprochement. Our perpetual dilemma of trying to find some tolerable middle ground between autonomy and connection, or to put it in the context of *the self/other dialectic*, between our *isolation anxiety* and our *engulfment anxiety*.

reaction formation. A term used in psychoanalytic theory that refers to the attempt to master unacceptable feelings or impulses by exaggerating one's behavior or beliefs in the opposite direction.

recovery. The condition of experiencing a general diminishment of the distressing and/or limiting aspect(s) of one's anomalous experiences.

resilience. The capacity to regain the balance of one's mind after having lost it, in contrast to *equanimity* (see *equanimity*).

samsara. A Pali and Sanskrit term meaning literally "continuous flowing," it refers to the ongoing cycles of existence in its many forms—birth/death, creation/destruction, integration/disintegration, etc.

schizophrenia. A formal diagnosis within the DSM for a psychotic condition in which "signs of the disturbance" persist for at least six months, which includes at least one month of active symptoms unless "formally treated." There are numerous different types listed in the DSM, including catatonic, paranoid, disorganized, undifferentiated, and residual.

schizophreniform disorder. A formal diagnosis within the DSM for a psychotic condition in which "signs of the disturbance" persist for at least one month but not more than six months.

secure attachment. The attachment style (see *attachment theory*) considered to be the healthiest, characterized by minimal anxiety both within intimate relationship and while spending time alone. In the context of *the self/other dialectic*, this is associated with a relatively wide and centered *window of tolerance*.

self/other dialectic, the. The configuration of our psyche with regard to our experience of *duality* at the most fundamental (*existential*) level. (See the DUI summary at the end of Chapter 23)

***self* pole, the.** One of the two poles of the *self/other dialectic*, the *self* pole is associated with our experience of "self" and the corresponding *isolation anxiety*.

spiritual emergence. A term from the field of transpersonal psychology, this refers to the condition in which altered states of consciousness and other *anomalous experiences* may be experienced, but individuals are able to cope with them in a way that does not significantly interfere with other aspects of their lives. This is in contrast to *spiritual emergency* (see below).

spiritual emergency. In spiritual emergency, one finds oneself more or less overwhelmed by altered states and *anomalous experiences*, which is in contrast to *spiritual emergence*, in which one is generally not overwhelmed by such experiences. While a number of people in the field suggest that spiritual emergency and *psychosis* are entirely distinct phenomena, the case study research suggests that they likely *are* a common phenomenon, but that in spiritual emergency, an *observing ego* is more present, whereas in *psychosis*, one is more prone to becoming completely detached from *consensus reality*. The case study research shows that even this is not so black and white, since many people experience a fluctuation in the strength of one's *observing ego* to a greater or lesser degree (see Chapter 9).

spiritual experiences. See *unitive experience.*

sympathetic joy. Taking delight in the joy of others and the joy inherent in life. (See also the *four immeasurables*)

tardive dyskinesia. A disorder of the voluntary nervous system caused by permanent brain damage as a result of antipsychotic drug use. The resulting symptoms are uncontrollable movements of the tongue and other parts of the body, which result in difficulties with speaking, eating, walking, and even sitting still.

three marks of existence, the. An expression coming from the Buddhist literature, this refers to the three most fundamental qualities of our subjective experience of the world—*impermanence* (anicca), *interconnectedness* (anatta), and the suffering inherent in attempting to maintain the existence of a *dualistic* self in a world that is fundamentally *nondual* (*dukkha*).

transliminal experience. A relatively strong and/or sudden *unitive experience.* Depending upon our ability to integrate this experience, we may experience genuine growth as we integrate the *unity* principle into our *personal paradigm*, or we may experience this as an overwhelming threat to our sense of self, resulting in a counter-reaction of terror and possibly even *psychosis*. (see Figures 22.1 and 22.2).

typical antipsychotics. The first generation of antipsychotics, these primarily block the D2 receptors in the dopamine pathways of the brain. (See also *antipsychotics* and *atypical antipsychotics*)

unconditional love. A sense of selfless love and deep connection with all other beings. (See also the *four immeasurables*)

unitive experience. Any subjective experience associated with our direct experience of unity. These generally include unitive feelings and other types of expansive, liberating, and/or euphoric experiences. However, if this kind of experience occurs too strongly and/or suddenly to be properly integrated (i.e., an overwhelming *transliminal experience*), there is the potential for this kind of experience to be extremely threatening to our sense of self, resulting in a counter-reaction of terror and possibly even psychosis. (See Figures 22.1 and 22.2).

unitive feelings. In the context of the *DUI model*, these are the feelings that are naturally associated with the direct experience of the fundamental quality of unity that exists beneath our dualistic experience. (See also the *four immeasurables* and also the DUI summary at the end of Chapter 23)

unity. The fundamental quality of profound *interconnectedness* that exists beneath the apparent experience of *duality* (see *duality*).

Vipassana. A Pali term meaning literally "clear seeing," it refers to the type of meditation that the Buddha discovered, generally known in the West as *mindfulness meditation.*

window of tolerance. Within the *self/other dialectic,* this represents the region along the *axis of rapprochement* on which one's experience is tolerable—neither the *engulfment anxiety* or the *isolation anxiety* is overwhelming. Depending upon one's history, conditioning, and perhaps other factors, an individual may develop a *window of tolerance* that is relatively narrow, relatively wide, or skewed in one direction or the other.

wisdom. In the context of the *DUI model,* and borrowing from the Buddhist tradition, wisdom refers to the successful integration into our *personal paradigm* (to greater or lesser degrees) of the fundamental qualities of the world—profound impermanence and interconnectedness, along with a corresponding degree of equanimity with regard to the existential anxiety that is a necessary part of our existence.

Resources

Peer Support Organizations

Community Consortium ["Building inclusive communities for people with psychiatric disabilities"]: www.community-consortium.org

Family Outreach and Response Program [support and education for families and friends]: www.familymentalhealthrecovery.org

Freedom Center ["A support and activism community run by and for people labeled with severe 'mental disorders.' We call for compassion, human rights, self-determination, and holistic options."]: www.freedom-center.org

The Hearing Voices Network USA ["A partnership between individuals who hear voices or have other extreme or unusual experiences, professionals and allies in the community, all of whom are working together to change the assumptions made about these phenomenon and create supports, learning and healing opportunities for people across the country."]: www.hearingvoicesusa.org

The Icarus Project ["We are a network of people living with and/or affected by experiences that are often diagnosed and labeled as psychiatric conditions. We believe these experiences are mad gifts needing cultivation and care, rather than diseases or disorders."]: www.theicarusproject.net

INTAR [International Network Toward Alternatives and Recovery—"gathers prominent survivors, professionals, family members, and advocates from around the world to work together for new clinical and social practices towards emotional distress and what is often labeled as psychosis."]: www.intar.org)

InterVoice ["A network [that] focuses on solutions that improve the life of people who hear voices, for those who are distressed by the experience."]: http://www.intervoiceonline.org

MindFreedom International ["A nonprofit organization that unites 100 sponsor and affiliate grassroots groups with thousands of individual members to win human rights and alternatives for people labeled with psychiatric disabilities."]: www.mindfreedom.org

National Association of Peer Specialists ["We strive to make peer specialists an important component in mental health treatment. We also advocate for better working conditions, compensation and the adoption of recovery practices."]: www.naops.org

National Coalition for Mental Health Recovery ["Ensuring that consumer/survivors have a major voice in the development and implementation of health care, mental health, and social policies at the state and national levels, empowering people to recover and lead a full life in the community."]: www.ncmhr.org

The National Empowerment Center ["NEC is a consumer/survivor/expatient-run organization [that carries] a message of recovery, empowerment, hope and healing to people with lived experience with mental health issues, trauma, and and/or extreme states."]: www.power2u.org

Non-Medical Model Practitioners/Therapists/Clinics/Helplines

Associated Psychological Health Services [an alternative mental health clinic located in Wisconsin]: www.abcmedsfree.com

Common Ground [computer application that facilitates communication with mental health care workers]: www.patdeegan.com

INTAR's directory of crisis centers: intar.org/category/resources/crisis-respites

MindFreedom International's directory of mental health practitioners: www.mindfreedom.org/directory

National Empowerment Center's directory of peer-run crisis centers: www.power2u.org/peer-run-crisis-services

Safe Harbor's directory of alternative mental health practitioners: www.alternativementalhealth.com/directory/search.asp

Sequoia Psychotherapy Center [an alternative mental health clinic located in central California]: www.medsfree.com

Soteria Network [a group devoted to setting up alternative residential facilities]: www.soterianetwork.org.uk

Spiritual Crisis Network [support for those undergoing spiritual crisis]: http://www.spiritualcrisisnetwork.org.uk/

Warmlines [directory of peer-run listening lines]: www.warmline.org

Windhorse Associates [home-based holistic support model]:
- In central Colorado: www.windhorsecommunityservices.com
- In central California and Massachusetts: www.windhorseimh.org

Books and Film

Theory and/or History of Madness

Bentall, R. P. (2003). *Madness explained: Psychosis and human nature*. London: Penguin Books.

Boyle, M. (1992; expanded 2nd ed., 2003). *Schizophrenia—A scientific delusion?* London: Routledge.

Chadwick, P. (1997). *Schizophrenia: The positive perspective*. London: Routledge.

Chadwick, P. K. (1992). *Borderline: A psychological study of paranoia and delusional thinking*. London: Routledge.

Clarke, I. (2008). *Madness, mystery and the survival of God*. Winchester, UK: O Books.

Clarke, I. (Ed.) (2010). *Psychosis and spirituality: Consolidating the new paradigm (2nd ed.)*. West Sussex, UK: Wiley-Blackwell.

Farber, S. (1993). *Madness, heresy, and the rumor of angels: The revolt against the mental health system*. Chicago, IL: Open Court Publishing.

Farber, S. (2012). *The spiritual gift of madness: The failure of psychiatry and the rise of the mad pride movement*. Rochester, VT: Inner Traditions.

Geekie, J., & Read, J. (2009). *Making sense of madness: Contesting the meaning of schizophrenia*. New York, NY: Routledge.

Goodbread, Joseph (2008). *Living on the edge: The mythical, spiritual, and philosophical roots of social Marginality (Health and human development)*. Hauppauge, NY: Nova Science Publishers, Inc.

Grof, C., & Grof, S. (Eds.) (1989). *Spiritual emergency*. New York, NY: Tarcher.

Hornstein, Gail A. (2009). *Agnes's jacket: A psychologist's search for the meanings of madness*. New York: Rodale Books.

Icarus Project, The. (Ed.) (n.d.). *Navigating the space between brilliance and madness: A roadmap and reader to bipolar worlds* (available from www.theicarusproject.net)

ISPS. [Psychosis book series—a very comprehensive and scholarly series of books on psychosis]. (Available at www.routledgementalhealth.com/isps/)

Laing, R.D. (1967). *The politics of experience*. New York: Pantheon Books.

Laing, R. D. (1969). *The divided self*. London: Penguin Group.

Levine, B. (2001). *Commonsense rebellion: Debunking psychiatry, confronting society: An A to Z guide to rehumanizing our lives.* New York: Continuum.

Mindell. A. (2008). *City shadows: Psychological interventions in psychiatry.* New York, NY: Routledge.

Perry, J. W. (1999). *Trials of the visionary mind.* State University of New York Press.

Podvoll, E. (1990). *Recovering sanity: A compassionate approach to understanding and treating psychosis.* Boston: Shambhala Publications, Inc.

Read, J., Mosher, L. R., & Bentall, R. P. (Eds.) (2004). *Models of madness: Psychological, social and biological approaches to schizophrenia.* New York: Routledge.

Watters, Ethan. (2010). *Crazy like us: The globalization of the American psyche.* New York: Free Press.

Whitaker, R. (2002). *Mad in America.* New York: Basic Books.

Biographies and Autobiographies

Bassman, R. (2007). *A fight to be: A psychologist's experience from both sides of the locked door.* New York, NY: Tantamount Press.

Beers, C. W. (1981). *A mind that found itself.* Pittsburgh, PA: Univ. of Pittsburgh Press.

Clover. (1999). *Escape from psychiatry: The autobiography of Clover.* Ignacio, CO: Rainbow Pots and Press.

[DVD] Cotton, T. (Director) *There is a fault in reality?* [An exploration of the stories of three people who struggled with psychosis]. Available from www.tigerlilyfilms.com

Dorman, D. (2003). *Dante's cure.* New York, NY: Other Press.

Greenberg, J. (1964). *I never promised you a rose garden.* Chicago: Signet.

Modrow, J. (2003). *How to become a schizophrenic: The case against biological psychiatry.* Lincoln, NE: Writers Club Press.

Penney, D., & Stastny, P. *The lives they left behind: Suitcases from a state hospital attic.* New York: Bellevue Literary Press.

[DVD] Rosenthal, K. (Producer). (2010). *Crooked Beauty* [a poetic documentary of Ashley McNamara's transformative journey from psych ward patient to pioneering mental health advocacy]. Available from www.crookedbeauty.com

Sechehaye, M. (1951). *Autobiography of a schizophrenic girl.* New York, NY: Meridian.

Alternative Treatment Models

Bloom, S. (1997). *Creating sanctuary: Towards the evolution of safe communities.* London: Routledge.

Chadwick, P., Birchwood, M. J. & Trower, P. (1999). *Cognitive Therapy for Delusions, Voices and Paranoia (Wiley Series in Clinical Psychology).* New York: John Wiley and Sons.

Coleman, R., Smith, M. and Good, J. (2003). *Psychiatric first aid in psychosis: A handbook for nurses, carers and people distressed by psychotic experience (2nd ed.).* Lewis, Scotland: P&P Press (available from www.workingtorecovery.co.uk).

Greek, M. (2012). *Schizophrenia: A blueprint for recovery.* Athens, OH: Milt Greek.

Karon, B. P., & VandenBos, G. (1996). *Psychotherapy of schizophrenia: The treatment of choice.* Lanham, MD: Rowman & Littlefield Publishing, Inc.

Larkin, W. and Morrison, A. (2005). *Trauma and psychosis: New directions for theory and therapy.* Routledge: London.

[DVD] Mackler, D. (Producer). (2008). *Take these broken wings: Recovery from Schizophrenia without medication.* (Available from www.iraresoul.com)

[DVD] Mackler, D. (Producer). (2011). *Healing homes: An alternative, Swedish model for healing psychosis.* (Available from www.iraresoul.com)

[DVD] Mackler, D. (Producer). (2011). *Open Dialogue: An alternative, Finnish approach to healing psychosis.* (Available from www.iraresoul.com)

Mackler, D., & Morrissey, M. (2010). *A Way Out of Madness: Dealing with Your Family After You've Been Diagnosed with a Psychiatric Disorder.* Bloomington, IN: AuthorHouse.

Mosher. L. R., & Hendrix, V. (with Fort, D. C.) (2004). *Soteria: Through madness to deliverance.* USA: Authors.

[DVD] Undercurrents (Producer). *Evolving Minds—Psychosis and spirituality.* (Available from www.undercurrents.org/minds.html)

Stastny, P. & Lehmann, P. (2007). *Alternatives beyond psychiatry* (available from www.peter-lehmann-publishing.com).

Support Specifically for Hearing Voices

Coleman, R. & Smith, M. (2006) *Working with voices—Victim to victor (2nd ed).* Lewis, Scotland: P&P Press (available from www.workingtorecovery.co.uk).

Corstens, D., May, R. and Longden, E. (2007). *Talking with voices: The voice dialoguing manual* (available from www.intervoiceonline.org).

Deegan, Patricia. (n.d.) *Coping with voices: Self-help strategies for people who hear voices that are distressing* (available from National Empowerment Center, www.power2u.org).

Downs, Julie (ed.). (2001). *Coping with voices and visions and starting and supporting hearing voices groups.* Manchester, UK: Hearing Voices Network (available from www.hearing-voices.org).

Escher, Sandra and Marius Romme. (2010). *Children hearing voices: What you need to know and what you can do* (available from www.pccs-books.co.uk).

Romme, M. and S. Escher. (2000). *Making sense of voices: A guide for mental health professionals working with voice-hearers (includes the Maastricht Interview).* London: MIND Publications.

Romme, Marius, Sandra Escher, Jacqui Dillon, Dirk Corstens, and Mervyn Morris (eds.). (2009). *Living with Voices: 50 Stories of Recovery* (available from www.pccs-books.co.uk).

Smith, Daniel. (2007). *Muses, madmen, and prophets: Rethinking the history, science, and meaning of auditory hallucination.* New York: Penguin Press.

Watkins, John. (1998). *Hearing Voices: A Common Human Experience.* Melbourne, Australia: Hill of Content.

Psychiatric Drugs

Bentall, R. (2009). *Doctoring the mind.* New York: New York University Press

Breggin, P. (2008). *Brain-disabling treatments in psychiatry: Drugs, electroshock and the psychopharmaceutical complex.* New York, NY: Springer Publication Company.

Breggin, P. (2008). *Medication madness: The role of psychiatric drugs in cases of violence, suicide, and crime.* New York, NY: St. Martin's Press.

Healy, D. (2012). *Pharmageddon.* Berkeley and Los Angeles: University of California Press.

The Icarus Project and Freedom Center (n.d.). *Harm reduction guide to coming off psychiatric drugs* (available from www.theicarusproject.net)

Lehmann, P. (Ed.). (2002). *Coming off psychiatric drugs* (available from www.peter-lehmann-publishing.com).

Whitaker, R. (2010). *Anatomy of an epidemic: Magic bullets, psychiatric drugs, and the astonishing rise of mental illness in America.* New York: Crown Books.

Nonduality

Bohm, D., & Hiley, B. J. (1993). *The undivided universe.* New York, NY: Routledge.

Capra, Fritjof. *The Tao of physics: An exploration of the parallels between modern physics and eastern mysticism (5th ed., updated).* Boston: Shambhalla Publications.

[DVD] Chasse, B., & Vicente, M. (Directors). (2004). *What the bleep do we know!?* [documentary]. USA: 20th Century Fox.

Goswami, A. (1993). *The self-aware universe: How consciousness creates the material world.* New York, NY: Jeremy P. Tarcher/Putnam.

Mindell, A. (2000). *Quantum Mind: The edge between physics and psychology.* Portland, Oregon: Lao Tse Press.

Prendergast, J. J., Fenner, P., & Krystal, S. (Eds.). (2003). *The sacred mirror: Nondual wisdom and psychotherapy.* St. Paul, Minnesota: Paragon House.

Prendergast, J. J., & Bradford, K. G. (Eds.) (2007). *Listening from the heart of silence: Nondual wisdom and psychotherapy, Volume 2.* St. Paul, Minnesota: Paragon House.

[DVD] Shadyac, T. (Director). (2009). *I am.* USA: Gaiam.

Attachment Theory

Karen, R. K. (1994). *Becoming attached: First relationships and how they shape our capacity to love.* Oxford, UK: Oxford University Press.

Schore, A. (1994). *Affect regulation and the origin of the self: The neurobiology of emotional development.* Hillsdale, NJ: Lawrence Erlbaum Associates, Inc.

Siegel, D.J., & Hartzell, M. (2003). *Parenting from the inside out: How a deeper self-understanding can help you raise children who thrive.* New York: Penguin Putnam.

Siegel, D.J. (2012). *The developing mind, second edition: How relationships and the brain interact to shape who we are.* New York: Guilford Press

Solomon, M., & Siegel, D.J. (Eds.). (2003). *Healing trauma: Attachment, mind, body and brain.* New York: W.W. Norton & Company.

Wallin, D. J. (2007). *Attachment in psychotherapy.* New York: The Guilford Press.

Western Existential Theory

Becker, E. (1973). *The denial of death.* New York, NY: Free Press Paperbacks.

May, R. (1977). *The meaning of anxiety.* New York: W. W. Norton & Company.

Schneider, K. J. (1999). *The paradoxical self: Toward an understanding of our contradictory nature.* Amherst, NY: Humanity Books.

Schneider, K. J. (2009). *Awakening to awe: Personal stories of profound transformation.* United Kingdom: Jason Aronson.

Yalom, I. D. (1980). *Existential psychotherapy.* USA: Basic Books.

Mindfulness and Buddhist-based Psychology

Boyce, B. & Shambhala Sun (2011). *The mindfulness revolution: Leading psychologists, scientists, artists, and meditation teachers on the power of mindfulness in daily life.* New York: Random House Publications, Inc.

Germer, C. K., Siegel, R. D., & Fulton, P. R. (Eds.). (2005). *Mindfulness and psychotherapy.* New York: The Guilford Press.

Goldstein, J (1994). *Insight meditation: The practice of Freedom.* Boston & London: Shambhalla

Gunaratana, B. H. (2002) . *Mindfulness in plain English.* Somerville, MA: Wisdom Publications.

Hart, W. (1987). *The art of living: Vipassana meditation as taught by S. N. Goenka.* New York, NY: HarperCollins.

Kabat-Zinn, J. (1990). *Full catastrophe living: Using the wisdom of your body and mind to face stress, pain, and illness.* New York: Bantam Dell.

Kabat-Zinn, J. (2012). *Mindfulness for beginners: Reclaiming the present moment--and your life.* Boulder, CO: Sounds True, Inc.

Kaklauskas, F. J., Nimanheminda, S., Hoffman, S, & MacAndrew, S. J. (Eds.). (2008). *Brilliant sanity: Buddhist approaches to psychotherapy.* Colorado Springs, CO: University of the Rockies Press.

Kornfield, J. (2008). The wise heart: A guide to the universal teachings of Buddhist psychology. New York: Bantam Books.

Nisker, W. (1998). *Buddha's Nature: A practical guide to discovering your place in the cosmos.* New York: Bantam Books.

Siegel, D.J. (2007). *The mindful brain: Reflection and attunement in the cultivation of well-being.* New York: W.W. Norton & Company.

Siegel, D. J. (2010). *The mindful therapist: A clinician's guide to mindsight and neural integration.* New York: W.W. Norton & Company.

Welwood, J. (2002). *Toward a psychology of awakening: Buddhism, psychotherapy, and the path of personal and spiritual transformation.* Boston & London: Shambhala.

RADIO

Madness Radio archive [a treasure trove of interviews with many leaders in the field]: www.theicarusproject.net/Madness_Radio

HUMAN RIGHTS FOR MENTAL HEALTH CONSUMERS

Disability Rights International [an organization "dedicated to promoting the human rights and full participation in society of people with disabilities worldwide."]: www.disabilityrightsintl.org

NARPA [National Association for Rights Protection and Advocacy—an organization advocating for the human rights of mental health consumers]: www.narpa.org

PsychRights: Law Project for Psychiatric Rights [a public interest law firm dedicated to protecting the rights of those with psychiatric diagnoses]: www.psychrights.org

World Network of Users and Survivors of Psychiatry [an international organization that advocates for the human rights of mental health consumers]: www.wnusp.rafus.dk

Blogs

Beyond Meds: Alternatives to Psychiatry [numerous articles and other resources devoted to the topic of psychiatric drugs and alternatives to psychiatry]: www.beyondmeds.com

Mad in America: Science, Psychiatry, and Community [a hub of numerous blogs from many of the leaders in the recovery and alternatives to psychiatry movement]: www.madinamerica.com/writers

Recovery from Schizophrenia [Ron Unger's blog with numerous recovery-oriented articles and resources]: www.recoveryfromschizophrenia.org

Successful Schizophrenia [numerous resources related to recovery and the challenging the medical model paradigm]: www.successfulschizophrenia.org

Additional Resources For Therapists and Clinicians

ICTP (Institute for Cognitive Therapy for Psyhosis): www.sites.google.com/site/ictpsychosis

ISEPP (International Society for Ethical Psychology and Psychiatry) ["Our main purpose is to examine and disseminate information concerning the impact of psychiatric theory and practice upon personal freedom, liberty, and a moral/spiritual conception of humanity"]: www.psychintegrity.org

ISPS (The International Society for Psychological and Social Approaches to Psychosis) [An organization devoted to the promotion of psychotherapy, research, and education related to psychotic disorders, with an emphasis on psychological and social interventions]:

- In the U.S.: www.isps-us.org
- International: www.isps.org

MindFreedom International's directory [to be listed as an alternative mental health care provider]: www.mindfreedom.org/directory

Index

A

B

C

D

M

N

O

Q

R

S

T

U

V

W

Y

About the Author

In the midst of a successful career as a hang gliding instructor and competition pilot (winning a World Champion title and multiple National Champion titles), Paris Williams suddenly found himself plunged into a profound struggle with experiences that would have likely resulted in the diagnosis of a psychotic disorder.

Fortunately, he managed to avoid becoming entangled within the psychiatric system, and he instead embarked upon a journey of healing and self discovery, attempting to resolve his own personal crisis while aspiring to support others going through similar crises. He has since spent over a decade deeply exploring both Eastern and Western understandings of mind and consciousness, studying intensive meditation from a number of different meditation masters around the world, earning a Ph.D. in Clinical Psychology, working in numerous settings supporting people struggling with challenging and extreme experiences, and conducting a series of pioneering research studies at Saybrook University on recovery from schizophrenia and other psychotic disorders.

Paris currently lives in the San Francisco Bay Area with his wife, Toni, working as a psychologist and still occasionally taking to the sky.

Lightning Source UK Ltd.
Milton Keynes UK
UKOW05f2122151215

264791UK00004B/239/P